Birkhäuser

Static & Dynamic Game Theory: Foundations & Applications

More information about this series at http://www.springer.com/series/10200

Leon A. Petrosyan • Vladimir V. Mazalov •
Nikolay A. Zenkevich
Editors

Frontiers of Dynamic Games

Game Theory and Management,
St. Petersburg, 2017

Editors
Leon A. Petrosyan
St. Petersburg State University
St. Petersburg, Russia

Vladimir V. Mazalov
Institute of Applied Mathematical Research
Karelian Research Center of RAS
Petrozavodsk, Russia

Nikolay A. Zenkevich
Graduate School of Management
St. Petersburg State University
St. Petersburg, Russia

ISSN 2363-8516 ISSN 2363-8524 (electronic)
Static & Dynamic Game Theory: Foundations & Applications
ISBN 978-3-030-06563-8 ISBN 978-3-319-92988-0 (eBook)
https://doi.org/10.1007/978-3-319-92988-0

Printed on acid-free paper

This book is published under the imprint Birkhäuser, www.birkhauser-science.com by the registered company Springer International Publishing AG part of Springer Nature.
The registered company address is: Gewerbestrasse 11, 6330 Cham, Switzerland

Preface

Game theory is an area of applied mathematics that models the interaction between agents (called players) to find the optimal behavior that each player has to adopt to maximize his or her reward when such prize depends not only on the individual choices of a player (or a group of players) but also on the decisions of all agents involved in the system.

Nowadays, game theory is an extremely important tool for economic theory and has contributed to a better understanding of human behavior in the process of decision-making in situations of conflict. In its beginnings, game theory was a tool to understand the behavior of economic systems, but currently it is used in many fields, such as biology, sociology, political science, military strategy, psychology, philosophy and computer science. In all these areas, game theory is perhaps the most sophisticated and fertile paradigm that applied mathematics can offer to analyze the process of making a decision under real-world conditions.

The conflicts between rational beings that distrust each other, or between competitors that interact and influence each other, constitute the object of study of game theory. Such studies are based on rigorous mathematical analyses; nevertheless, they arise naturally from the observation of a conflict from a rational point of view. For the theorists in our field, a "game" is a conflictive situation in which competing interests of individuals or institutions prevail, and in that context, each party influences the decisions that the others will make; thus, the result of the conflict is determined by the decisions taken by all the actors. In the so-called canonical form, a game takes place when an individual pursues an objective when other individuals concurrently pursue other (overlapping or conflicting) objectives, and in the same time these objectives cannot be reached by individual actions of one decision maker. The problem is to determine each player's optimal decision (with respect to some predetermined criterion), how such decisions interact among each other, and what are the properties of the outcome brought about by these choices.

The contents of this volume are primarily based on selected talks presented at the 11th International Conference "Game Theory and Management" 2017 (GTM2017) held in Saint Petersburg State University, in Saint Petersburg, Russia, from 28 to 30 June 2017. Each chapter in this volume has passed a rigorous reviewing process, as is the case for the journals on applied mathematics. It is worth mentioning that the predecessors of this conference (GTM2007-GTM2016) were held in Saint Petersburg State University and were supported by the International Society of Dynamic Games—Russian Chapter. The conference unites the game theorists of two schools: the classical school founded by J. V. Neumann and O. Morgenstern, and the school of differential games first introduced by R. Isaacs. GTM has succeeded to achieve this goal along the years, and this can be seen by taking a look at the list of our plenary speakers: R. Aumann, T. Bashar, G. J. Olsder, J. Nash, R. Selten, F. Kidland, R. Myerson, D. W. K. Yeung, G. Zaccour, E. Maskin, S. Jorgensen, D. Schmeidler, A. Tarasyev, H. Moulin, D. Novikov, A. Haurie, G. Owen, A. Newman, P. Bernhard, J. Weibull, B. Monien, S. Zamir, S. Aseev, S. Hart, M. Breton, R. Laraki, and others (among whom the authors of this preface have the honor to appear).

The present volume proves that GTM offers an interactive program on a wide range of the latest developments in game theory and management. It includes recent advances in topics with high future potential and existing developments in classical fields.

I wish to thank all of the associate editors and reviewers for their outstanding contribution. Without them, this book would have not been possible.

St. Petersburg, Russia — Leon A. Petrosyan
Petrozavodsk, Russia — Vladimir V. Mazalov
St. Petersburg, Russia — Nikolay A. Zenkevich
March, 2018

Contents

Contributors

Maria Garmash National Research University Higher School of Economics, St. Petersburg, Russia

George Geronikolaou Democritus University of Thrace, Komotini, Greece

Ekaterina V. Gromova Saint Petersburg State University, Saint Petersburg, Russia

Elena Gubar Saint Petersburg State University, Saint Petersburg, Russia

Simon Hoof Paderborn University, Department of Economics, Paderborn, Germany

Anatolii Kleimenov Krasovskii Institute of Mathematics and Mechanics UrB RAS, Yekaterinburg, Russia

Nicolas Klein Université de Montréal and CIREQ, Département de Sciences Économiques, Montréal, QC, Canada

Ekaterina A. Kolpakova Krasovskii Institute of Mathematics and Mechanics UrB RAS, Yekaterinburg, Russia

Alexei Korolev National Research University Higher School of Economics, St. Petersburg, Russia

Nikolay Krasovskii Krasovskii Institute of Mathematics and Mechanics UrB RAS, Yekaterinburg, Russia

Evgeniy Krupennikov Krasovskii Institute of Mathematics and Mechanics UrB RAS, Yekaterinburg, Russia

Ural Federal University, Yekaterinburg, Russia

Suriya Kumacheva Saint Petersburg State University, Saint Petersburg, Russia

Dmitrii Lozovanu Institute of Mathematics and Computer Science of Moldova Academy of Sciences, Chisinau, Moldova

Vladimir Matveenko National Research University Higher School of Economics, St. Petersburg, Russia

Maria Nastych National Research University Higher School of Economics, St. Petersburg, Russia

Konstantinos G. Papadopoulos Aristotle University of Thessaloniki, Thessaloniki, Greece

Mathieu Parenti European Center for Advanced Research in Economics and Statistics (ECARES), Brussels, Belgium

Ovanes Petrosian Saint Petersburg State University, Saint Petersburg, Russia

Leon A. Petrosyan Saint Petersburg State University, Saint Petersburg, Russia

Artem Sedakov Saint Petersburg State University, Saint Petersburg, Russia

Erfang Shan School of Management, Shanghai University, Shanghai, People's Republic of China

Alexander Sidorov Novosibirsk State University, Novosibirsk, Russia

Sobolev Institute for Mathematics, Novosibirsk, Russia

Krzysztof Szajowski Wrocław University of Science and Technology, Wrocław, Poland

Alexander Tarasyev Krasovskii Institute of Mathematics and Mechanics UrB RAS, Yekaterinburg, Russia

Ural Federal University, Yekaterinburg, Russia

Jacques-Francois Thisse CORE-UCLouvain, Louvain-la-Neuve, Belgium

NRU-Higher School of Economics, Moscow, Russia

CEPR, Washington, D.C., USA

Galina Tomilina Saint Petersburg State University, Saint Petersburg, Russia

Dmitrii Volf Saint Petersburg State University, Saint Petersburg, Russia

Guang Zhang School of Management, Shanghai University, Shanghai, People's Republic of China

Ekaterina Zhitkova Saint Petersburg State University, Saint Petersburg, Russia

Chapter 1
Countervailing Power with Large and Small Retailers

George Geronikolaou and Konstantinos G. Papadopoulos

Abstract When concentration in the retail market increases, retailers gain more market power towards the suppliers and they hence can achieve better wholesale prices. In the 1950s, Galbraith introduced the concept of countervailing power claiming that lower wholesale prices will pass on to consumer as lower retail prices. Consequently higher concentration may turn out to be beneficial for consumers. In this model where a monopolistic supplier sells an intermediate good to M large retailers who are Cournot competitors and a competitive fringe consisting of N retailers, we show that higher concentration does not decrease retail prices and results solely to a reallocation of profits between the supplier and large retailers, thus invalidating Galbraith's conjecture. The same result carries on when the exogenously given level of bargaining power of large retailers increases.

1.1 Introduction

In 1952, John Kenneth Galbraith in his book "American Capitalism: The Concept of Countervailing Power" [8], introduced the concept of countervailing power as an inherent power in market economies, which works to the benefit of consumers in oligopolistic markets, i.e. in markets with small number of sellers who have the ability to manipulate prices.

In every market there are buyers and sellers. According to Galbraith, if one side of the market (e.g. the seller) enjoys gains of monopoly power then the other side (the buyer) will defend against the monopolization by developing its own monopoly power. As a result, the two forces will cancel each other to the benefit of the

G. Geronikolaou
Democritus University of Thrace, Komotini, Greece
e-mail: ggeronik@ierd.duth.gr

K. G. Papadopoulos (✉)
Aristotle University of Thessaloniki, Thessaloniki, Greece
e-mail: kpap@econ.auth.gr

L. A. Petrosyan et al. (eds.), *Frontiers of Dynamic Games*,
Static & Dynamic Game Theory: Foundations & Applications,
https://doi.org/10.1007/978-3-319-92988-0_1

consumer. Big supermarkets like Carrefour in Europe or Wall-Mart in the U.S. are common examples in the business literature. Because of their size they manage to offset the power of suppliers, buying cheaper and offering their products at lower prices to the consumer.

The concept of countervailing power served as an additional argument in favor of the free market economy, able to perform a role akin to the "invisible hand" of Adam Smith, despite fierce criticism addressed by the economists in Galbraith's time. However, until the 1990s there was no theoretical mathematical model to confirm or invalidate the beneficial role of countervailing power for consumers.

Over the last 20 years, the consequences of the countervailing power returned to the research scene, mainly because of the development of Industrial Organization. There has been a growing research interest both from a theoretical point of view,[1] and from the U.S. and E.U. competition authorities (reports Federal Trade Commission [6, 7], UK Competition Commission [3, 4], The European Commission [11]) on the consequences of horizontal mergers or acquisitions among retailers on consumer prices. It is known that in horizontal mergers retailers increase the countervailing power towards their suppliers-producers. This increase is reflected in practice by their ability to achieve better contractual terms towards suppliers, for example in terms of various discounts, better wholesale prices, better franchising terms. In the literature, the impact of countervailing power to consumers and the level of social welfare remains open, mostly because theoretical results are model specific.

Generally, competition authorities take a sceptic stance against mergers because they usually lead to higher concentration and more market power for firms to the detriment of consumers. In the context of vertical industrial relations higher concentration at the retail level has a double effect. On the one hand it increases the market power of retailers towards the consumers as sellers of final goods (oligopolistic power), on the other hand it increases their market power as buyers of the intermediate good (oligopsonistic/buyer power) against the supplier. The main research question is whether the reduction in costs for retailers will translate to lower prices for consumers or higher profits for the downstream firms. The answer is not obvious because it depends on the market structure (number and size of firms, production technologies), the degree of competition on the market of final goods (i.e. among retailers), the market of intermediate goods (among suppliers) and equally importantly the vertical contractual relations among firms and the type of contracts they use in their transactions (linear and non-linear contracts).

In this paper, we construct a model which consists of a monopolist of an intermediate good who sells it to a small number of large retailers M, who are Cournot competitors, as well as a large number of small retailers N, who are price takers. Consumers are represented by a demand function for the homogenous final good.

[1] See for instance [9] and [10] and the references therein.

The model is constructed as a three stage game. In the first stage, the supplier chooses unilaterally a non-linear contract with the competitive firms and in the second stage it negotiates simultaneously with the large retailers about their respective contracts. We will assume that the outcome of negotiations is given by the generalized Nash bargaining program. We also assume that large retailers are symmetric, that is, they have the same technology. In the third stage large retail firms will strategically choose their quantities as Cournot players, taking as given the supply function of a competitive fringe.

As a solution to the above game we will use the concept of subgame perfect Nash equilibrium. Once we calculate the equilibrium we will do comparative statics with respect to the number of large retailers (that serves here as a proxy for countervailing power) and their degree of bargaining power so that we can clarify the effects of countervailing power on consumer prices and welfare. Our model combines the dominant firm-competitive fringe model with the Cournot model.

Our work shows that when concentration, as measured by the number of large retailers, increases, consumer prices stay constant. This result is interesting for two reasons. First, because it is contrary to the standard Cournot model result where prices go down when the number of sellers increases and second, because it provides a theoretical argument against Galbraith's conjecture about the beneficial role of countervailing power. In fact we show that, in our context, countervailing power, represented either as the level of bargaining power or concentration, cannot benefit consumers, even in the presence of competition in the retail level, i.e. the price taking competitive fringe. Competition is a prerequisite in the models of Von Ungern-Sternberg [12], Dobson and Waterson [5] and Chen [1] for countervailing power to function for the benefit of consumers.

1.2 The Model

A single supplier denoted by s produces and sells an intermediate product to $M + N$ retailers, M symmetric large retailers and a competitive fringe consisting of N symmetric retailers.[2] The number of firms is exogenous but we assume that $M < N$ and there is no possibility of entry in the market. Retailers transform the intermediate good to a final one on a 1–1 basis, suffering some retail cost. We normalize the supplier's cost to zero, without loss of generality.

Let m denote a large retailer and n a fringe firm. A large retailer has constant marginal retailing cost $MC_m = c_m$. The retail cost function of a fringe retailer is

[2] We use capital letters M and N to denote the set, the last element of the set or the cardinality of the set, depending on the context. We use small letters m and n to denote a typical element or an index of the set M and N respectively. So we adopt the convention $M = \{1, \ldots, m, \ldots, M\}$ and $N = \{1, \ldots, n, \ldots, N\}$.

$C(q_n) = kq_n^2/2$ and so the fringe firm faces an increasing marginal cost, $MC(q_n)$ with $MC'(q_n) > 0$ and $MC(0) = 0$, where q_n is the quantity of output produced by the fringe retailer. The overall marginal cost of a unit, including the input wholesale price w, is $w_m + c_m$ for the large retailer and $w_n + MC(q_n)$ for the fringe retailer. $AC(q_n)$ denotes the average retail cost function of a fringe retailer.

Consumers are represented by the inverse demand function $p(Q) = a - bQ$, for $a, b > 0$. We denote the total quantity purchased by large retailers by $Q_M = \sum_{m=1}^{M} q_m$, where q_m is the quantity bought by a single retailer m. The total quantity purchased by the small retailers is $Q_N = \sum_{n=1}^{N} q_n$, where q_n is the quantity bought by a single retailer n. Hence the total quantity is denoted by $Q = Q_M + Q_N$.

The timing of the game is the following:

At $t = 1$, the supplier makes a take-it-or-leave-it offer to each one of the fringe retailers simultaneously, a pair (F_n, w_n) consisting of a fee F_n which is independent of the quantity purchased and a wholesale price w_n for one unit of the intermediate good. The contract is binding once signed.

At $t = 2$, the supplier and the large retailers bargain simultaneously over a two-part tariff (F_m, w_m).

At $t = 3$, large retailers play a Cournot game among themselves, that is they choose how much to sell taking as given the supply function of the fringe retailers. Given their quantity choice, total quantity Q is sold at price p. Given p, each fringe retailer chooses how much input quantity q_n to buy at w_n and then sell at the final good price p.

1.3 Equilibrium

The concept of equilibrium that we use is that of subgame perfect Nash equilibrium. We proceed by backward induction.

At $t = 3$, each fringe retailer chooses how much to sell to consumers given the retail price p and the (F_n, w_n) contract that is already signed with the supplier. The fringe retailer's problem is

$$\max_{q_n} \pi_n = [p - AC_n(q_n) - w_n]q_n - F_n \tag{1.1}$$

Average cost is $kq_n/2$. Solving the first order condition $d\pi_n/dq_n = 0$, we obtain $p - w_n = MC(q_n)$ with $d^2\pi_n/dq_n = -k < 0$ and so the supply function of a fringe firm and the total fringe supply are respectively

$$q_n^* = \frac{p - w_n}{k}, \tag{1.2}$$

$$Q_N = N\frac{p - w_n}{k} = N\frac{a - b(Q_M + Q_N) - w_n}{k}. \tag{1.3}$$

Solving for Q_N we obtain

$$Q_N = N\frac{a - bQ_M - w_n}{k + bN}. \quad (1.4)$$

Let Q_{-m} be the quantity chosen by all M retailers expect m. Then a large Cournot retailer chooses quantity q_m so that

$$\max_{q_m} \pi_m = [p(q_m + Q_{-m} + Q_N(q_m)) - c_m - w_m]q_m - F_m, \quad (1.5)$$

taking as given the choices of the rest of the large retailers Q_{-m} and the total quantity supplied by the competitive fringe from (1.4). The reaction function of a large retailer is (see Appendix)

$$q_m(Q_{-m}) = \frac{ak - (c_m + w_m)(k + bN) - bkQ_{-m} + bNw_n}{2bk}$$

and since the large retailers are symmetric, $Q_{-m} = (M - 1)q_m$ we obtain

$$q_m^*(w_m, w_n) = \frac{ak - (c_m + w_m)(k + bN) + bNw_n}{bk(1 + M)} \quad (1.6)$$

which makes the total quantity $Q_M^* = Mq_m^*$. Given (1.3), we obtain the consumer price as a function of the wholesale prices w_m and w_n.

$$p(w_m, w_n) = \frac{ak + M(k + bN)(c_m + w_m) + bNw_n}{(1 + M)(k + bN)} \quad (1.7)$$

At $t = 2$ the supplier bargains simultaneously with the set of large retailers over the (F_m, w_m) contract. We assume that the bargaining outcome is represented by the maximization of the following generalized Nash bargaining program where $\gamma_m \in (0, 1)$ is the degree of bargaining power of a large retailer m and $\sigma = 1 - \sum_{m=1}^{M} \gamma_m$ that of the supplier, so that $\sigma + \sum_{m=1}^{M} \gamma_m = 1$.

$$\max_{(F_m, w_m)_{m=1}^{M}} [\pi_s(F_m, w_m) - \bar{\pi}_s]^{\left(1 - \sum_{m-1}^{M} \gamma_m\right)} \prod_{m=1}^{M} [\pi_m(F_m, w_m) - \bar{\pi}_m]^{\gamma_m} \quad (1.8)$$

where the profit of the supplier is

$$\pi_s(F_m, w_m) = M(F_m + w_m q_m) + N(F_n + w_n q_n), \quad (1.9)$$

and the profit function of a large retailer is

$$\pi_m(F_m, w_m) = [p - c_m - w_m]q_m - F_m, \quad (1.10)$$

while $\bar{\pi}_s$ and $\bar{\pi}_m$ denote the players' disagreement payoffs. Since a large retailer is unable to produce the final good without the provision of the essential input by the supplier, its disagreement payment is zero, $\bar{\pi}_m = 0$, whilst the outside option of the supplier in case negotiations break down is the profit that can be obtained by supplying only to the fringe retailers, $\bar{\pi}_s = N(F_n + w_n q_c) > 0$, where $q_c = (a - w_n)/(k + bN)$ is the quantity sold at the market clearing retail price $p_c = (ak + bNw_n)/(k + bN)$. The solution to (1.8) is given by (see Appendix)

$$F_m = \frac{[(ak - c_m(k + bN)]^2(1 - M^2\gamma)}{4bkM^2(k + bN)}, \tag{1.11}$$

$$w_m(w_n) = \frac{(M - 1)[ak - c_m(k + bN)] + 2bMNw_n}{2M(k + bN)}. \tag{1.12}$$

At the first stage of the game, $t = 1$, the supplier decides about the take-it-or-leave-it offer for the fringe retailers, taking as given $p(w_m, w_n)$, $Q_N(w_m, w_n)$, $Q_M(w_m, w_n)$, F_m, $w_m(w_n)$ from the next stages. The supplier's problem is

$$\max_{w_n} \pi_s(w_n) = M[F_m + w_m(w_n)q_m(w_n)] + N[F_n + w_n q_n(w_n)], \tag{1.13}$$

$$s.t. \qquad F_n = [p(w_n) - AC_f(q_n(w_n)) - w_n]q_n(w_n).$$

or equivalently

$$\max_{w_n} \pi_s(w_n) = M[F_m + w_m(w_n)q_m(w_n)] + N[p(w_n) - AC_f(q_n(w_n))]q_n(w_n),$$

which gives the following optimal contract

$$w_n^* = \frac{1}{2}\left(a - \frac{c_m(k + bN)}{k + 2bN}\right),$$

$$F_n^* = \frac{1}{2k}\left(\frac{c_m(k + bN)}{k + 2bN}\right)^2.$$

Given the optimal contract (F_n^*, w_n^*) at $t = 1$ we may now calculate the following values at the subgame perfect Nash equilibrium

$$w_m^* = \frac{1}{4}\left[c_m\left(-3 + \frac{2}{M} + \frac{k}{k + 2bN}\right) + 2a\left(1 - \frac{k}{M(k + bN)}\right)\right],$$

$$p^* = \frac{1}{4}\left(2a + c_m + \frac{c_m k}{k + 2bN}\right),$$

$$q_n^* = \frac{c_m(k + bN)}{k(k + 2bN)},$$

$$q_m^* = \frac{ak - c_m(k + bN)}{2bkM},$$

$$Q_N^* = \frac{Nc_m(k+bN)}{k(k+2bN)},$$

$$Q_M^* = \frac{ak - c_m(k+bN)}{2bk},$$

$$Q_M^* + Q_N^* = \frac{k(a-c_m) + bN(2a-c_m)}{2b(k+2bN)}.$$

Equilibrium profits, consumer surplus CS, total profits PS, and total surplus TS are

$$\pi_n^* = 0,$$

$$\pi_m^* = \gamma \frac{[ak - c_m(k+bN)]^2}{4bk(k+bN)},$$

$$\pi_s^* = \frac{1}{4bk}\left(a^2k - 2ac_mk + \frac{c_m^2(k+bN)(k+3bN)}{k+2bN} - \frac{M[ak - c_m(k+bN)]^2\gamma}{k+bN}\right),$$

$$CS^* = \frac{[c_m(k+bN) - a(k+2bN)]^2}{8b(k+2bN)^2},$$

$$PS^* = \frac{1}{4bk}\left(a^2k - 2ac_mk + \frac{c_m^2(k+bN)(k+3bN)}{k+2bN}\right),$$

$$TS^* = \frac{3a^2}{8b} - \frac{ac_m(3k+5bN)}{4b(k+2bN)} + \frac{c_m^2(k+bN)(3k^2+11bkN+12b^2N^2)}{8bk(k+2bN)^2}$$

1.4 The Effects of Concentration and Bargaining Power on Retail Prices

In this model, countervailing power is represented by the degree of bargaining power γ of each large retailer, as in [1] or [2] and alternatively, by the degree of downstream concentration, which is given by the number of symmetric large retailers M, as in [12]. We also consider the effects of the fringe size N on equilibrium values. We obtain the following propositions.

Proposition 1.1 *When the number of large retailers decreases, each large retailer obtains a lower wholesale price from the supplier and sells a higher quantity of the final good, while the fringe quantity, total quantity, consumer price, industry profits and total welfare remain constant.*

Proof In order to guarantee positive quantities at equilibrium, we have to assume that $N < k(a-c_m)/bc_m$ because $q_m^* = [ak - c_m(k+bN)]/2bkM$. Calculating derivatives at equilibrium we have:

$$\frac{\partial w_m^*}{\partial M} = \frac{ak - c_m(k+bN)}{2M^2(k+bN)} > 0,$$

$$\frac{\partial q_m^*}{\partial M} = \frac{-ak + c_m(k+bN)}{2kbM^2} < 0,$$

$$\frac{\partial q_n^*}{\partial M} = \frac{\partial Q_M^*}{\partial M} = \frac{\partial Q_N^*}{\partial M} = 0,$$

$$\frac{\partial p^*}{\partial M} = \frac{\partial PS^*}{\partial M} = \frac{\partial TS^*}{\partial M} = 0.$$

When the number of large retailers decreases, they obtain more bargaining power towards the supplier, so that they can achieve a lower wholesale price. Nevertheless, a lower wholesale price does not lead to a lower retail price at equilibrium. There are two conflicting effects. On the one hand, a lower wholesale price leads to a higher individual production, while on the other hand, fewer large retailers face relaxed competition and want to produce less. The first effect dominates so that large retailers increase individual production up to the level where total production, and hence retail price, remain constant. Moreover, the profit of a large retailer will not change, despite of the fact that its per unit profit increases due to the lower wholesale price. Any higher profit obtained is completely captured by a higher fixed fee charged by the supplier. However, when large retailers become fewer, the supplier's profit increases, because the supplier collects fewer, yet higher fees. These findings are summarized in the following proposition.

Proposition 1.2 *When the number of large retailers changes, the profits of the supplier and the consolidated profits of the large retailers move to opposite directions and their change is of the same magnitude.*

Proof $\partial(M\pi_m^*)/\partial M = \gamma[(ak - c_m(k+bN)]^2/4bk(k+bN) > 0$ and $\partial\pi_s^*/\partial M = -\gamma[(ak - c_m(k+bN)]^2/4bk(k+bN) < 0$ so $(\partial(M\pi_m^*)/\partial M)(\partial\pi_s^*/\partial M) < 0$ and $\partial(M\pi_m^*)/\partial M + \partial\pi_s^*/\partial M = 0$.

At equilibrium, the size of the pie is invariant to the number of large retailers. The effect of a change of the number of retailers results solely in a redistribution of profits.

Next, we examine the effect of a change of the exogenously given level of bargaining power of the large retailers. The following proposition summarizes the neutrality of countervailing power:

Proposition 1.3 *When the degree of bargaining power γ of large retailers increases, they pay a lower fee F_m and their profits increase. They do not obtain any lower wholesale price and the retail price does not change at equilibrium.*

Proof $-\partial\pi_m^*/\partial\gamma = \partial F_m^*/\partial\gamma = -[(ak - c_m(k+bN)]^2/4bk(k+bN) < 0, \partial w_m^*/\partial\gamma = \partial p^*/\partial\gamma = 0.$

The proposition suggests that a higher level of bargaining power will lead to a lower fee for the large retailer, not a lower wholesale price. Consequently, the retail price will not change either. This is due to the fact that the level of wholesale price is set so as to maximize multilateral profits, hence a change in bargaining power can only lead to a reallocation of profits.

1.5 Conclusion

In this work we examine a particular retail market structure consisting of a set of large retailers with power over wholesale and retail prices and a set of small retailers who are price takers. We use the number of large retailers as a measure of the degree of concentration in the market. When concentration in the market increases, equilibrium wholesale prices become lower, nevertheless we show that equilibrium consumer prices and welfare remain constant. Since higher concentration is tantamount to higher countervailing power, we prove, contrary to Galbraith's argument, that countervailing power is not effective in this model.

Alternatively, we use the degree of bargaining power as a proxy for countervailing power. Keeping the number of larger retailers constant, when large retailers obtain greater bargaining power exogenously, they achieve a lower fee, not a lower wholesale price. This is due to the fact that wholesale prices maximize the multilateral profits of the supplier and the large retailers in the negotiation process. Consequently, even if there were a positive pass-through rate from wholesale prices to consumer price, the consumer price cannot fall because the wholesale prices remain constant. Again, countervailing power is neutral.

Acknowledgements Konstantinos Papadopoulos gratefully acknowledges Research Grant no 87937 from the Aristotle University of Thessaloniki Research Committee.

Appendix

Derivation of the Reaction Function of the Large Retailer $q_m(Q_{-m})$

Using (1.4), the profit function of a Cournot retailer as defined in (1.5) can be written as

$$\begin{aligned}\pi_m &= [a - b(q_m + Q_{-m} + Q_N(q_m)) - c_m - w_m]q_m - F_m \\ &= \left[a - b\left(q_m + Q_{-m} + \frac{N[a - b(Q_{-m} + q_m) - w_n]}{k + bN}\right) - c_m - w_m\right] q_m - F_m\end{aligned}$$

which we differentiate with respect to q_m to obtain

$$q_m(Q_{-m}) = \frac{ak - (c_m + w_m)(k + bN) - bkQ_{-m} + bNw_n}{2bk}.$$

Derivation of Bargaining Outcome (1.11) and (1.12)

Let $\tilde{\pi}_s = Mw_m q_m + N(F_n + w_n q_n)$ so that from (1.9) we can write $\pi_s(F_m, w_m) = \tilde{\pi}_s + MF_m$. Let $\tilde{\pi}_m = [p - c_m - w_m]q_m$ so that from (1.10) we can write $\pi_m(F_m, w_m) = \tilde{\pi}_m - F_m$. Cournot retailers are symmetric so $\sum_{m-1}^{M} \gamma_m = M\gamma_m$ and given that $\bar{\pi}_m = 0$, (1.8) reduces to

$$\max_{(F_m, w_m)} \left[\tilde{\pi}_s + MF_m - \bar{\pi}_s\right]^{(1-M\gamma_m)} \left[\tilde{\pi}_m - F_m\right]^{M\gamma_m}. \tag{1.14}$$

The first order condition with respect to F_m is

$$\begin{aligned} 0 = {} & M(1 - M\gamma_m)\left[\tilde{\pi}_s + MF_m - \bar{\pi}_s\right]^{-M\gamma_m} \left[\tilde{\pi}_m - F_m\right]^{M\gamma_m} \\ & - M\gamma_m \left[\tilde{\pi}_s + MF_m - \bar{\pi}_s\right]^{(1-M\gamma_m)} \left[\tilde{\pi}_m - F_m\right]^{(M\gamma_m - 1)} \\ \Rightarrow {} & (1 - M\gamma_m)/\gamma_m = (\tilde{\pi}_s + MF_m - \bar{\pi}_s)/(\tilde{\pi}_m - F_m) \end{aligned}$$

or

$$F_m = \tilde{\pi}_m - \gamma_m (M\tilde{\pi}_m + \tilde{\pi}_s - \bar{\pi}_s). \tag{1.15}$$

If we substitute $\tilde{\pi}_m = [p - c_m - w_m]q_m$, $\tilde{\pi}_s = Mw_m q_m + N(F_n + w_n q_n)$ and $\bar{\pi}_s = N(F_n + w_n q_c)$ in (1.15) where $q_c = (a - w_n)/(k + bN)$ is the quantity sold at the market clearing retail price $p_c = (ak + bNw_n)/(k + bN)$ we end up with (1.11).

In order to find w_m that solves (1.14), we introduce (1.15) in the objective function in (1.14) and we rearrange terms so that

$$\begin{aligned} & \left[\tilde{\pi}_s + MF_m - \bar{\pi}_s\right]^{(1-M\gamma_m)} \left[\tilde{\pi}_m - F_m\right]^{M\gamma_m} \\ & = [(1 - M\gamma_m)^{(1-M\gamma_m)} \gamma_m^{M\gamma_m}] (M\tilde{\pi}_m + \tilde{\pi}_s - \bar{\pi}_s). \end{aligned}$$

Consequently, the maximization problem can be written as

$$\max_{w_m} M\tilde{\pi}_m + \tilde{\pi}_s - \bar{\pi}_s \tag{1.16}$$

because $(1-M\gamma_m)^{(1-M\gamma_m)}\gamma_m^{M\gamma_m}$ is a constant. Notice also that $\bar{\pi}_s$ does not depend on w_m, so, in fact, w_m maximizes the multilateral profits of the supplier with the M Cournot retailers (efficiency of Nash bargaining solution). So

$$\begin{aligned}M\tilde{\pi}_m + \tilde{\pi}_s - \bar{\pi}_s &= M(p - c_m - w_m)q_m + Mw_mq_m + N(F_n + w_nq_n)\\ &\quad -N(F_n + w_nq_c)\\ &= M[p - c_m]q_m + Nw_n(q_n - q_c)\\ &= M\left[a - b(Mq_m(w_m) + Nq_n(w_m) - c_m\right]q_m(w_m)\\ &\quad +Nw_n(q_n - q_c)\\ &= NF_n + \frac{1}{bk(1+M)^2(k+bN)}Z\end{aligned}$$

where $Z = a^2k^2M + M(k+bN)^2(c_m+w_m)(c_m-Mw_m) + bMN(k+bN)[c_m(M-1)+2Mw_m]w_n - bN[k(1+M)^2+bM^2N]w_n^2 + ak[M(k+bN)((M-1)w_m - 2c_m) + b(1+3M)Nw_n]$. The maximization of (1.16) with respect to w_m will give (1.12).

References

1. Chen, Z.: Dominant retailers and the countervailing-power hypothesis. RAND J. Econ. **34**, 612–25 (2003)
2. Christou, C., Papadopoulos, K.G.: The countervailing power hypothesis in the dominant firm-competitive fringe model. Econ. Lett. **126**, 110–113 (2015)
3. Competition Commission: Supermarkets: a report on the supply of groceries from multiple stores in the United Kingdom. Cm 4842 (2000)
4. Competition Commission: The supply of grocery in the UK market investigation: provisional findings report (2007)
5. Dobson, P.W., Waterson, M.: Countervailing power and consumer prices. Econ. J. **107**(441), 418–30 (1997)
6. Federal Trade Commission: Report on the FTC workshop on slotting allowances and other marketing practices in the grocery industry (2001)
7. Federal Trade Commission: Slotting allowances in the retail grocery industry: selected case studies in five product categories (2003)
8. Galbraith, J.K.: American Capitalism: The Concept of Countervailing Power. Houghton Mifflin, Boston (1952)
9. Gaudin, G.: Vertical bargaining and retail competition: what drives countervailing power? Econ. J. (2017). https://doi.org/10.1111/ecoj.12506
10. Inderst, R., Mazzarotto, N.: Buyer power in distribution. In: Collins W.D. (ed.) ABA Antitrust Section Handbook Issues in Competition Law and Policy. ABA Book Publishing, Chicago (2008)
11. The European Commission: Buyer Power and Its Impact on Competition in the Food Retail Distribution Sector of the European Union, prepared by Dobson and Consulting, DGIV Study Contract No. IV/98/ETD/078 (1999)
12. Von Ungern-Sternberg, T.: Countervailing power revisited. Int. J. Ind. Organ. **12**, 507–519 (1996)

Chapter 2
Dynamic Voluntary Provision of Public Goods: The Recursive Nash Bargaining Solution

Simon Hoof

Abstract Grim trigger strategies can support any set of control paths as a cooperative equilibrium, if they yield at least the value of the noncooperative Nash equilibrium. We introduce the recursive Nash bargaining solution as an equilibrium selection device and study its properties by means of an analytically tractable n-person differential game. The idea is that the agents bargain over a tuple of stationary Markovian strategies, before the game has started. It is shown that under symmetry the bargaining solution yields efficient controls.

2.1 Introduction

Most noncooperative differential games lack Pareto efficiency. That is, all agents can increase their individual payoffs if they agree to coordinate controls. However, in order to attain the socially optimal outcome at least two conditions must be fulfilled: (1) the agents form the grand coalition to derive the efficient controls and (2) payoffs must be transferable and distributed in such a way that every agent benefits from cooperation.[1]

Here we study a mechanism which implements the Pareto efficient outcome as a bargaining solution. The crucial difference to the classic cooperative approach is that agents do not mutually agree to maximize overall payoffs and distribute them appropriately, but bargain over the controls. In order to support the resulting controls as an equilibrium we fix grim trigger strategies. If an agent defects on the agreement, all agents switch to their noncooperative Nash equilibrium strategies [7].

---

[1]See Yeung and Petrosyan [9] for a recent treatment on subgame consistent cooperation in differential games.

S. Hoof (✉)
Paderborn University, Department of Economics, Paderborn, Germany
e-mail: simon.hoof@upb.de

L. A. Petrosyan et al. (eds.), *Frontiers of Dynamic Games*,
Static & Dynamic Game Theory: Foundations & Applications,
https://doi.org/10.1007/978-3-319-92988-0_2

Sorger [6] proposed the recursive Nash [4] bargaining solution for difference games. We introduce a continuous time analogon and apply it to the differential game of public good provision of Fershtman and Nitzan [2]. Considering the noncooperative equilibrium they showed that the public good is underprovided with respect to the efficient solution. The result, however, crucially depends on the linearity of the Markovian strategies. This simplification makes the game analytically tractable and yields a unique steady state.[2]

This note contributes to the literature on cooperative agreements in noncooperative differential games. It is well known that grim trigger strategies can support a set of control paths as equilibria, if they payoff dominate the noncooperative Nash equilibrium. The Nash bargaining solution can then be used as an equilibrium selection device. Since bargaining problems are defined in the payoff space we need to construct a value under agreement. In games with transferable payoffs one can simply fix the efficient value of the grand coalition and define an imputation. Here, however, we do not assume that the grand coalition forms and jointly maximizes payoffs. But we can define the agreement value in terms of a stationary Hamilton-Jacobi-Bellman equation (HJBe), if the agents stick to the agreement strategies over the entire time interval. The agreement strategies are then determined by the Nash bargaining solution.

The remainder of the paper is organized as follows: Sect. 2.2 presents the problem, Sect. 2.3 the solution concept and Sect. 2.4 concludes.

2.2 Problem Statement

The model is essentially the most rudimentary version of Fershtman and Nitzan [2].[3] Let $x(t) \in X := [0, \frac{1}{2}]$ denote the stock of a pure public good at time $t \in R_+$. We could think of x being the total contribution to some joint project carried out by n agents. Each agent $i \in N := \{1, 2, \dots, n\}$ can partially control the evolution of the state according to the state equation

$$\dot{x}(t) = f(x(t), u(t)) = \sum_{i \in N} u_i(t) - \delta x(t) \tag{2.1}$$

$$x_0 := x(0) \in X \tag{2.2}$$

where $u(t) := (u_i(t))_{i \in N} \in \times_{i \in N} U_i =: U \subset R^n$ denotes the investment (control) vector and $\delta \in (0, 1]$ is the deprecation rate. In the context of the joint project, $u_i(t)$ then denotes the contribution rate of any agent $i \in N$. We consider quadratic payoffs

[2]Wirl [8] showed that within the set of nonlinear Markovian strategies the Nash equilibrium is nonunique and that the efficient steady state is potentially reachable.

[3]See also Dockner et al. [1, Ch. 9.5] for a textbook treatment.

of the form

$$F_i(x(t), u_i(t)) = x(t)(1 - x(t)) - \frac{1}{2}u_i(t)^2 \tag{2.3}$$

such that the game is linear quadratic and thus possesses a closed form solution [1, Ch. 7.1]. Note that the instantaneous payoff function is monotonously increasing in the state $\frac{\partial F_i(x,u_i)}{\partial x} > 0$ for all $x \in X$. The state is thus a pure public good and each agent benefits by its provision. With costly investment, however, there exists a trade-off between increasing the stock and minimizing costs. This trade-off defines a public good game. Each agent wants the others to invest, such that one can free ride on the effort of the other agents. This behavior results in an inefficiently low overall investment level. The objective functional for each agent $i \in N$ is then given by the stream of discounted payoffs

$$J_i(u(s), t) := \int_t^{\infty} e^{-r(s-t)} F_i(x(s), u_i(s))ds \tag{2.4}$$

where $r > 0$ denotes the time preference rate.

2.3 Solution Concepts

In what follows we consider a stationary setup and hence save the time argument t frequently. First we will derive the efficient collusive solution of joint payoff maximization. The efficient value is an upper bound on the agreement value. We then derive the noncooperative Nash equilibrium which serves as the disagreement value for the bargaining solution. The noncooperative equilibrium value is a lower bound on the agreement value. Any cooperative agreement lies in the set of strategies which support payoffs between the noncooperative Nash and efficient value. The noncooperative equilibrium strategies also serve as threats for deviations from the agreed upon bargaining solution.

2.3.1 Collusive Solution

Assume all agents agree to cooperate and jointly maximize overall payoffs. The value function for the efficient solution then reads

$$C(x(t)) := \max_{u(s) \in U} \sum_{i \in N} J_i(u(s), t). \tag{2.5}$$

The optimal controls must satisfy the stationary HJBe

$$rC(x) = \max_{u \in U} \left\{ \sum_{i \in N} F_i(x, u_i) + C'(x) f(x, u) \right\}. \tag{2.6}$$

The maximizers of the right hand side of (2.6) are $u_i = C'(x)$ for all $i \in N$. Substituting the maximizers into the HJBe yields

$$rC(x) = nx(1 - x) + \frac{n}{2} C'(x)^2 - C'(x)\delta x. \tag{2.7}$$

Theorem 2.1 *If we consider symmetric stationary linear strategies of the form* $\hat{u}_i = \alpha x + \beta$ *for all* $i \in N$ *where* α *and* β *are constants, then there exists a* unique *quadratic solution to* (2.7)

$$C(x) = \frac{\alpha}{2} x^2 + \beta x + \gamma \tag{2.8}$$

with

$$\alpha := \frac{1}{2n} \left(r + 2\delta - \sqrt{(r + 2\delta)^2 + 8n^2} \right), \tag{2.9}$$

$$\beta := \frac{n}{\delta - n\alpha + r}, \tag{2.10}$$

$$\gamma := \frac{n}{2r} \beta^2. \tag{2.11}$$

Proof Substitute the guess (2.8) and thus $C'(x) = \alpha x + \beta$ into (2.7)

$$r \left(\frac{\alpha}{2} x^2 + \beta x + \gamma \right) = nx(1 - x) + \frac{n}{2} (\alpha x + \beta)^2 - (\alpha x + \beta)\delta x. \tag{2.12}$$

This optimality condition must hold at any $x \in X$. Evaluate (2.12) at $x = 0$, which yields γ

$$r\gamma = \frac{n}{2} \beta^2 \iff \gamma = \frac{n}{2r} \beta^2. \tag{2.13}$$

Taking the derivative of (2.12) gives

$$r(\alpha x + \beta) = n(1 - 2x) + \alpha n(\alpha x + \beta) - \delta(2\alpha x + \beta). \tag{2.14}$$

Again, at $x = 0$ we have

$$r\beta = n + \alpha n\beta - \delta\beta \iff \beta = \frac{n}{\delta - \alpha n + r}. \tag{2.15}$$

Resubstituting β in (2.14) and solving for α yields

$$\alpha = \frac{1}{2n}\left(r + 2\delta \pm \sqrt{(r + 2\delta)^2 + 8n^2}\right). \tag{2.16}$$

Note that the state dynamics become $\dot{x}(t) = (n\alpha - \delta)x(t) + n\beta$. There exists a unique and globally asymptotically stable steady state at $x = -n\beta/(n\alpha - \delta)$ if $n\alpha - \delta < 0$ holds, which is ensured for the negative root of (2.16).

2.3.2 *Noncooperative Equilibrium*

The collusive solution implies two restrictive assumptions. The grand coalition must form and payoffs must be transferable in order to split the total payoff.[4] Let us assume that the collusive solution is not feasible. If this is the case we consider a noncooperative differential game and each agent maximizes his individual payoffs. The noncooperative Markovian strategies are denoted by $\phi_i : X \to U_i$ and satisfy[5]

$$\phi_i(x(s)) \in \arg \max_{u_i(s) \in U_i} J_i(u_i(s), \phi_{-i}(x(s)), t). \tag{2.17}$$

where $\phi_{-i} := (\phi_j)_{j \in N \setminus \{i\}}$. A noncooperative Nash equilibrium is then defined as follows.

Definition 2.1 The strategy tuple $\phi(x(s)) := (\phi_i(x(s)))_{i \in N} \in U$ is a noncooperative Nash equilibrium if the following holds

$$J_i(\phi(x(s)), t) \geq J_i(u_i(s), \phi_{-i}(x(s)), t) \quad \forall u_i(s) \in U_i, \ \forall i \in N. \tag{2.18}$$

Denote by

$$D_i(x(t)) := J_i(\phi(x(s)), t) \tag{2.19}$$

the noncooperative disagreement value.

---

[4]The latter assumption is not too prohibitive. If payoffs were not transferable the individual cooperative value is simply given by $C_i(x_t) = J_i(\hat{u}(s), t)$ where $\hat{u}(s)$ are the Pareto efficient controls. It turns out that in the symmetric setup $C_i(x_t) = \frac{C(x_t)}{n}$ which would also be the result under an equal sharing rule with transferable payoffs.

[5]See e.g. Dockner et al. [1, Ch. 4] for the theory on noncooperative differential games.

Theorem 2.2 *If we consider symmetric stationary linear strategies of the form* $\phi_i(x) = \omega x + \lambda$ *for all* $i \in N$ *where* ω *and* λ *are constants, then there exists a unique quadratic solution to* (2.19)

$$D_i(x) = \frac{\omega}{2}x^2 + \lambda x + \mu \tag{2.20}$$

with

$$\omega := \frac{1}{2(2n-1)}\left(r + 2\delta - \sqrt{(r+2\delta)^2 + 8(2n-1)}\right), \tag{2.21}$$

$$\lambda := \frac{1}{\delta - (2n-1)\omega + r}, \tag{2.22}$$

$$\mu := \frac{2n-1}{2r}\lambda^2. \tag{2.23}$$

Proof The proof follows the same steps as Theorem 2.1.

The noncooperative equilibrium, however, is generally not efficient. It can be shown eventually that the collusive solution yields a cooperation dividend such that the value under cooperation always exceeds the noncooperative value, i.e., $C(x) > \sum_{i \in N} D_i(x)\ \forall x \in X$. The investment levels and thus the provision of the public good are inefficiently low. This result is standard in public good games and due to free riding. It is rational to assume that the agents do not want to stick to the fully noncooperative equilibrium, but increase overall efficiency by exploiting the cooperation dividend.

2.3.3 Bargaining Solution

It was shown by Tolwinski et al. [7][6] that any control path $\tilde{u}_i^t := (\tilde{u}_i(s))_{s \geq t}$, $i \in N$ can be supported as an equilibrium if the control profiles are agreeable and defection from the agreement is punished.[7] Let $\sigma_i : X \rightarrow U_i$ denote a Markovian strategy that generates $\tilde{u}_i$. Suppose the agents agree on some strategy profile $\sigma(x) := (\sigma_i(x))_{i \in N}$ at $\underline{t} < 0$ before the game has started. If the agents agree from t onwards, the agreement value is defined as

$$A_i(x(t)) = J_i(\sigma(x(s)), t). \tag{2.24}$$

[6]See also Dockner et al. [1, Ch. 6].

[7]Agreeability is a stronger notion than time consistency. In the former the agreement payoff dominates the noncooperative play for any state while in the latter only along the cooperative path. Time consistency was introduced by Petrosjan [5] (originally 1977) and agreeability by Kaitala and Pohjola [3]. See also Zaccour [10] for a tutorial on cooperative differential games.

Definition 2.2 A strategy tuple $\sigma(x)$ is agreeable at $\underline{t}$ if

$$A_i(x(t)) \geq D_i(x(t)) \quad \forall t, \ \forall x(t), \ \forall i \tag{2.25}$$

such that every agent benefits the agreement in comparison to the noncooperative equilibrium.

If this inequality was not about to hold there exists an agent who rather switches to the noncooperative equilibrium, because it payoff dominates the agreement. The condition, also refereed to dynamic individual rationality, is necessary but not sufficient for dynamic stability of an agreement. An agent might deviate from the agreement if he benefits from it.

Now we construct the history dependent non-Markovian grim trigger strategies $\tau_i : [0, \infty) \to U_i$ that support $\sigma_i(x)$ as an equilibrium. Given some agreement strategy profile $\sigma(x)$ the agents can solve the differential equation (2.1) for the agreement trajectory of the state

$$x^a(t) := x_0 + \int_0^t f(x(s), \sigma(x(s)))ds. \tag{2.26}$$

Suppose the agents perfectly observe the state and can recall the history of the state $(x(s))_{s\in[0,t]}$. If they observe that an agent deviates in t, they can impose punishment with delay $t + \epsilon$. Now the grim strategies read

$$\tau(s) = \begin{cases} \sigma(x(s)) & \text{for } s \in [t, t+\epsilon] \quad \text{if } x(l) = x^a(l) \ \forall l \in [0, t], \\ \phi(x(s)) & \text{for } s \in [t+\epsilon, \infty) \quad \text{if } x(t) \neq x^a(t). \end{cases} \tag{2.27}$$

That is, if the agents observe that another player deviated at t from the agreement they implement their noncooperative equilibrium strategies from $t + \epsilon$ onwards. Let $d \in N$ denote a potential defector who deviates from $\sigma(x)$ at t. In the interval $s \in [t, t+\epsilon]$ he maximizes his payoff against the agreement strategies of the opponents. From $t+\epsilon$ onwards he receives the discounted disagreement payoff. Let $V_d(x(t); \epsilon)$ denote the value of the defector defined as

$$\begin{aligned} V_d(x(t); \epsilon) := & \max_{(u_d(s))_{s\in[t,t+\epsilon]}} \int_t^{t+\epsilon} e^{-r(s-t)} F_d(x(s), u_d(s))ds \\ & + e^{-r\epsilon} D_d(x(t+\epsilon)) \\ \text{s.t.} \quad & \dot{x}(s) = f(x(s), u_d(s), \sigma_{-d}(x(s))) \quad (s \in [t, t+\epsilon]). \end{aligned} \tag{2.28}$$

The threat is effective if

$$A_i(x(t)) \geq V_i(x(t); \epsilon) \quad \forall x(t), \ \forall i \tag{2.29}$$

holds and every agent benefits the agreement over defecting on the agreement. Now we can always fix an $\epsilon \in (0, \overline{\epsilon}]$ such that (2.29) holds. Suppose punishment can be implemented instantly $\epsilon = 0$. Equation (2.29) then becomes

$$A_i(x(t)) \geq V_i(x(t); 0) = D_i(x(t)) \tag{2.30}$$

which is true by the definition of individual rational agreements. Let $\overline{\epsilon}$ denote a threshold such that (2.29) holds with equality

$$A_i(x(t)) = V_i(x(t); \overline{\epsilon}). \tag{2.31}$$

Then the threat is effective for all $\epsilon \in (0, \overline{\epsilon}]$. The threat is also credible, because after defection occurs all agents switch to their noncooperative equilibrium strategies and thus have no unilateral incentive to deviate from the punishment by the definition of an equilibrium. The grim trigger strategies and a sufficiently small punishment delay guarantee that the agents stick to the initial agreement over the entire time horizon.

Differentiating (2.24) w.r.t. time yields a representation of the agreement value in terms of the stationary HJBe

$$A_i'(x(t))\dot{x}(t) = -F_i(x(t), \sigma_i(x(t))) + \int_t^\infty re^{-r(s-t)} F_i(x(s), \sigma_i(x(s)))ds \tag{2.32}$$

$$\iff \quad rA_i(x) = F_i(x, \sigma_i(x)) + A_i'(x) f(x, \sigma(x)) \tag{2.33}$$

This gives us a stationary definition for the agreement value. Next we want to determine a particular strategy profile $\sigma(x)$ by the Nash bargaining solution. Fix the excess demand function as follows

$$E_i(x, \sigma(x)) := \frac{1}{r}[F_i(x, \sigma_i(x)) + A_i'(x) f(x, \sigma(x))] - D_i(x). \tag{2.34}$$

That is, each agent claims an amount which exceeds his disagreement value. Since each agent will only agree on some bargaining strategy if it gives him at least his disagreement value, we must restrict the control set. The set of individual rational strategies is then defined as

$$\Omega(x) := \{\sigma(x) \in U \mid E_i(x, \sigma(x)) \geq 0 \quad \forall i \in N\}. \tag{2.35}$$

Note that these are all stationary representations. That is, the actual time instance t is not important, but state $x(t)$. Since the relation holds for all $t \in R$, we saved the time argument. We are now in the position to state our main result and show how to solve for the bargaining strategy $\sigma(x)$.

Theorem 2.3 *For the fully symmetric case the agreement strategies that solve the Nash bargaining product*

$$\sigma^N(x) \in \arg \max_{\sigma(x)\in\Omega(x)} \prod_{i\in N} E_i(x, \sigma(x)) \tag{2.36}$$

yield the Pareto optimal controls.

Proof The first order conditions for $j \in N$ of (2.36) is given by

$$\begin{aligned}
& 0 = \frac{\partial \prod_{i\in N} E_i(x, \sigma(x))}{\partial \sigma_j(x)} \\
\iff \quad & 0 = \frac{1}{r} \sum_{i\in N} \left[\frac{\partial E_i(x, \sigma(x))}{\partial \sigma_j(x)} \prod_{k\in N\backslash\{i\}} E_k(x, \sigma(x)) \right] \\
\iff \quad & 0 = \frac{\partial E_j(x, \sigma(x))}{\partial \sigma_j(x)} \prod_{k\in N\backslash\{j\}} E_k(x, \sigma(x)) \\
& \quad + \sum_{i\in N\backslash\{j\}} \left[\frac{\partial E_i(x, \sigma(x))}{\partial \sigma_j(x)} \prod_{k\in N\backslash\{i\}} E_k(x, \sigma(x)) \right] \\
\iff \quad & 0 = (-\sigma_j(x) + A_j'(x)) \prod_{k\in N\backslash\{j\}} E_k(x, \sigma(x)) \\
& \quad + \sum_{i\in N\backslash\{j\}} \left[A_i'(x) \prod_{k\in N\backslash\{i\}} E_k(x, \sigma(x)) \right].
\end{aligned} \tag{2.37}$$

Under symmetry, we must have $E_i(\cdot) =: \overline{E}(\cdot)$, $A_i'(\cdot) =: \overline{A}'(\cdot)$ and $\sigma_i(\cdot) =: \overline{\sigma}(\cdot)$ for all $i \in N$. The first order condition then becomes

$$(-\overline{\sigma}(x) + n\overline{A}'(x))\overline{E}(x, \overline{\sigma}(x))^{n-1} = 0 \quad \iff \quad \overline{\sigma}(x) = n\overline{A}'(x). \tag{2.38}$$

Since $\overline{E}(\cdot) = 0 \Leftrightarrow \overline{A}(\cdot) = \overline{D}(\cdot)$ implies that all agents stick to the disagreement strategy we can neglect this case here. Now substitute the maximizer $\overline{\sigma}(x) = n\overline{A}'(x)$ into (2.33) which gives

$$\begin{aligned}
r\overline{A}(x) &= x(1-x) - \frac{1}{2}\overline{\sigma}(x)^2 + \frac{\overline{\sigma}(x)}{n}(n\overline{\sigma}(x) - \delta x) \\
&= x(1-x) + \frac{1}{2}\overline{\sigma}(x)^2 - \frac{\delta}{n}\overline{\sigma}(x)x.
\end{aligned} \tag{2.39}$$

Take the derivative with respect to x

$$r\overline{A}'(x) \overset{(2.38)}{=} \frac{r}{n}\overline{\sigma}(x) = 1 - 2x + \overline{\sigma}(x)\overline{\sigma}'(x) - \frac{\delta}{n}(\overline{\sigma}'(x)x + \overline{\sigma}(x)). \tag{2.40}$$

We claimed that the agreement strategies satisfy the efficient solution and are thus given by $\overline{\sigma}(x) = \alpha x + \beta$ with $\overline{\sigma}'(x) = \alpha$. Equation (2.40) becomes

$$\frac{r}{n}(\alpha x + \beta) = 1 - 2x + (\alpha x + \beta)\alpha - \frac{\delta}{n}(2\alpha x + \beta). \tag{2.41}$$

This relation must hold at any $x \in X$. At $x = 0$, the equation simplifies to

$$\frac{r}{n}\beta = 1 + \beta\alpha - \frac{\delta}{n}\beta \quad \Longleftrightarrow \quad \beta = \frac{n}{\delta - n\alpha + r} = (2.10). \tag{2.42}$$

Now substitute β into (2.41) and solve for α, which then is identical with (2.9). Since the controls and thus dynamics are identical under the collusive and bargaining solution, the values must be identical as well.

2.4 Conclusion

We studied the recursive Nash bargaining solution for symmetric differential games. It was shown by an analytically tractable example that the bargaining solution yields the Pareto efficient outcome of full cooperation. In an accompanying paper the author also wants to investigate asymmetric games and compare different solution concepts (e.g. Kalai-Smorodinsky and Egalitarian solution). Especially for the case of asymmetric discounting the recursive bargaining solution can be useful, because then efficient controls are not derivable in the standard way by joint payoff maximization.

Acknowledgements I thank Mark Schopf and participants of the *Doktorandenworkshop der Fakultät für Wirtschaftswissenschaften* for valuable comments. This work was partially supported by the German Research Foundation (DFG) within the Collaborative Research Center "On-The-Fly Computing" (SFB 901).

References

1. Dockner, E.J., Jørgensen, S., Long, N.V., Sorger, G.: Differential Games in Economics and Management Science. Cambridge University Press, Cambridge (2000)
2. Fershtman, C., Nitzan, S.: Dynamic voluntary provision of public goods. Eur. Econ. Rev. **35**, 1057–1067 (1991)

3. Kaitala, V., Pohjola, M.: Economic development and agreeable redistribution in capitalism: efficient game equilibria in a two-class neoclassical growth model. Int. Econ. Rev. **31**(2), 421–438 (1990)
4. Nash, J.F., Jr.: The bargaining problem. Econometrica **18**(2), 155–162 (1950)
5. Petrosjan, L.A.: Agreeable solutions in differential games. Int. J. Math. Game Theory Algebra **2–3**, 165–177 (1997)
6. Sorger, G.: Recursive Nash bargaining over a productive asset. J. Econ. Dyn. Control **30**(12), 2637–2659 (2006)
7. Tolwinski, B, Haurie, A., Leitmann, G.: Cooperative equilibria in differential games. J. Math. Anal. Appl. **119**(1–2), 182–202 (1986)
8. Wirl, F.: Dynamic voluntary provision of public goods: extension to nonlinear strategies. Eur. J. Polit. Econ. **12**, 555–560 (1996)
9. Yeung, D.W.K., Petrosyan, L.A.: Subgame Consistent Cooperation. Springer, Singapore (2016)
10. Zaccour, G.: Time consistency in cooperative differential games: a tutorial. Inf. Syst. Oper. Res. (INFOR) **46**(1), 81–92 (2008)

Chapter 3
Altruistic, Aggressive and Paradoxical Types of Behavior in a Differential Two-Person Game

Anatolii Kleimenov

Abstract A non-antagonistic positional (feedback) differential two-person game is considered in which each of the two players, in addition to the usual normal (nor) type of behavior oriented toward maximizing own functional, can use other types of behavior. In particular, it is altruistic (alt), aggressive (agg) and paradoxical (par) types. It is assumed that in the course of the game players can switch their behavior from one type to another. In this game, each player simultaneously with the choice of positional strategy selects the indicator function defined over the whole time interval of the game and taking values in the set $\{nor, alt, agg, par\}$. Player's indicator function shows the dynamics for changing the type of behavior that this player adheres to. The concept of BT-solution for such game is introduced. The use by players of types of behaviors other than normal can lead to outcomes more preferable for them than in a game with only normal behavior. An example of a game with the dynamics of simple motion on a plane and phase constraints illustrates the procedure for constructing BT-solutions.

3.1 Introduction

In the present paper we consider a non-antagonistic positional (feedback) differential two-person game (see, for example, [1, 3, 12]), for which emphasis is placed on the case where each of the two players, in addition to the normal (nor), type of behavior oriented on maximizing their own functional, can use other types of behavior introduced in [5, 9], such as altruistic (alt), aggressive (agg) and paradoxical (par) types. It is assumed that during the game, players can switch their behavior from one type to another. The idea of using the players to switch their behavior from one type to another in the course of the game was applied to the game

---

A. Kleimenov (✉)
Krasovskii Institute of Mathematics and Mechanics UrB RAS, Yekaterinburg, Russia
e-mail: kleimenov@imm.uran.ru

L. A. Petrosyan et al. (eds.), *Frontiers of Dynamic Games*,
Static & Dynamic Game Theory: Foundations & Applications,
https://doi.org/10.1007/978-3-319-92988-0_3

with cooperative dynamics in [9] and for the repeated bimatrix 2×2 game in [6], which allowed to obtain new solutions in these games.

It is also assumed that in the game each player chooses the indicator function determined over the whole time interval of the game and takes values in the set $\{nor, alt, agg, par\}$, simultaneously with the choice of the positional strategy. Player's indicator function shows the dynamics for changing the type of behavior that this player adheres to. Rules for the formation of controls are introduced for each pair of behaviors of players.

The formalization of positional strategies in the game is based on the formalization and results of the general theory of positional (feedback) differential games [4, 10, 11]. The concept of the BT-solution is introduced.

The idea to switch players' behavior from one type to another in the course of the game is somewhat similar to the idea of using trigger strategies [2]. This is indicated by the existence of punishment strategies in the structure of decision strategies. However, there are significant differences. In this paper, we also use more complex switching, namely, from one type of behavior to another, changing the nature of the problem of optimization—from non-antagonistic games to zero-sum games or team problem of control and vice versa.

An example of a game with dynamics of simple motion on a plane and a phase constraint in two variants is proposed. In the first variant we assume that the first and second players can exhibit altruism towards their partner for some time periods. In the second variant, in addition to the assumption of altruism of the players, we also assume that each player can act aggressively against other player for some periods of time, and a case of mutual aggression is allowed. In both variants sets of BT-solutions are described. This paper is a continuation of [6–8].

3.2 Some Results from the Theory of Non-antagonistic Positional Differential Games (NPDG) of Two Persons

The contents of this section can be found in [4]. In what follows, we use the abbreviated notation NPDG to denote non-antagonistic positional (feedback) differential game.

Let the dynamics of the game be described by the equation

$$\dot{x} = f(t, x, u, v), \quad t \in [t_0, \vartheta], \quad x(t_0) = x_0, \tag{3.1}$$

where $x \in R^n$, $u \in P \in comp(R^p)$, $v \in Q \in comp(R^q)$; ϑ is the given moment of the end of the game.

Player 1 (P1) and Player 2 (P2) choose controls u and v, respectively.

Let G be a compact set in $R^1 \times R^n$ whose projection on the time axis is equal to the given interval $[t_0, \vartheta]$. We assume, that all the trajectories of system (3.1), beginning at an arbitrary position $(t_*, x_*) \in G$, remain within G for all $t \in [t_*, \vartheta]$.

It is assumed that the function $f : G \times P \times Q \mapsto R^n$ is continuous over the set of arguments, satisfies the Lipschitz condition with respect to x, satisfies the condition of sublinear growth with respect to x and satisfies the saddle point condition in the small game [10, 11]

$$\max_{u \in P} \min_{v \in Q} s^T f(t, x, u, v) = \min_{v \in Q} \max_{u \in P} s^T f(t, x, u, v) \tag{3.2}$$

for all $(t, x) \in G$ and $s \in R^n$.

Both players have information about the current position of the game $(t, x(t))$. The formalization of positional strategies and the motions generated by them is analogous to the formalization introduced in [10, 11], with the exception of technical details [4].

Strategy of Player 1 is identified with the pair $U = \{u(t, x, \varepsilon), \beta_1(\varepsilon)\}$, where $u(\cdot)$ is an arbitrary function of position (t, x) and a positive precision parameter $\varepsilon > 0$ and taking values in the set P. The function $\beta_1 : (0, \infty) \longmapsto (0, \infty)$ is a continuous monotonic function satisfying the condition $\beta_1(\varepsilon) \to 0$ if $\varepsilon \to 0$. For a fixed ε the value $\beta_1(\varepsilon)$ is the upper bound step of subdivision the segment $[t_0, \vartheta]$, which Player 1 applies when forming step-by-step motions. Similarly, the strategy of Player 2 is defined as $V = \{v(t, x, \varepsilon), \beta_2(\varepsilon)\}$.

Motions of two types: approximated (step-by-step) ones and ideal (limit) ones are considered as motions generated by a pair of strategies of players. Approximated motion $x[\cdot, t_0, x_0, U, \varepsilon_1, \Delta_1, V, \varepsilon_2, \Delta_2]$ is introduced for fixed values of players' precision parameters ε_1 and ε_2 and for fixed subdivisions $\Delta_1 = \{t_i^{(1)}\}$ and $\Delta_2 = \{t_j^{(2)}\}$ of the interval $[t_0, \vartheta]$ chosen by P1 and P2, respectively, under the conditions $\delta(\Delta_i) \leq \beta_i(\varepsilon_i)$, $i = 1, 2$. Here $\delta(\Delta_i) = \max_k (t_{k+1}^{(i)} - t_k^{(i)})$. A limit motion generated by the pair of strategies (U, V) from the initial position (t_0, x_0) is a continuous function $x[t] = x[t, t_0, x_0, U, V]$ for which there exists a sequence of approximated motions

$$\{x[t, t_0^k, x_0^k, U, \varepsilon_1^k, \Delta_1^k, V, \varepsilon_2^k, \Delta_2^k]\}$$

uniformly converging to $x[t]$ on $[t_0, \vartheta]$ as $k \to \infty$, $\varepsilon_1^k \to 0$, $\varepsilon_2^k \to 0$, $t_0^k \to t_0$, $x_0^k \to x_0$, $\delta(\Delta_i^k) \leq \beta_i(\varepsilon_i^k)$.

The control laws $(U, \varepsilon_1, \Delta_1)$ and $(V, \varepsilon_2, \Delta_2)$ are said to be agreed with respect to the precision parameter if $\varepsilon_1 = \varepsilon_2$. Agreed control laws generate agreed approximate motions, the sequences of which generate agreed limit motions.

A pair of strategies (U, V) generates a nonempty compact (in the metric of the space $C[t_0, \vartheta]$) set $X(t_0, x_0, U, V)$ consisting of limit motions $x[\cdot, t_0, x_0, U, V]$.

Player i chooses his control to maximize the payoff functional

$$I_i = \sigma_i(x(\vartheta)), \quad i = 1, 2 \tag{3.3}$$

where $\sigma_i : R^n \mapsto R^1$ are given continuous functions.

Thus, an non-antagonistic positional (feedback) differential game (NPDG) is defined.

Now we introduce the following definitions [4].

Definition 3.1 A pair of strategies (U^N, V^N) is called a Nash equilibrium solution (NE–solution) of the game, if for any motion $x^*[\cdot] \in X(t_0, x_0, U^N, V^N)$, any moment $\tau \in [t_0, \vartheta]$, and any strategies U and V the following inequalities hold

$$\max_{x[\cdot]} \sigma_1(x[\vartheta, \tau, x^*[\tau], U, V^N]) \le \min_{x[\cdot]} \sigma_1(x[\vartheta, \tau, x^*[\tau], U^N, V^N]), \tag{3.4}$$

$$\max_{x[\cdot]} \sigma_2(x[\vartheta, \tau, x^*[\tau], U^N, V]) \le \min_{x[\cdot]} \sigma_2(x[\vartheta, \tau, x^*[\tau], U^N, V^N]). \tag{3.5}$$

where the operations min are performed over a set of agreed motions, and the operations max by sets of all motions.

Definition 3.2 An NE-solution (U^P, V^P) which is Pareto non-improvable with respect to the values I_1, I_2 (3.3) is called a $P(NE)$-solution.

Now we consider auxiliary zero-sum positional (feedback) differential games Γ_1 and Γ_2. Dynamics of both games is described by the Eq. (3.1). In the game Γ_i Player i maximizes the payoff functional $\sigma_i(x(\vartheta))$ (3.3) and Player $3 - i$ opposes him.

It follows from [10, 11] that both games Γ_1 and Γ_2 have universal saddle points

$$\{u^{(i)}(t, x, \varepsilon), \quad v^{(i)}(t, x, \varepsilon)\}, \quad i = 1, 2 \tag{3.6}$$

and continuous value functions

$$\gamma_1(t, x), \quad \gamma_2(t, x) \tag{3.7}$$

The property of strategies (3.6) to be universal means that they are optimal not only for the fixed initial position $(t_0, x_0) \in G$ but also for any position $(t_*, x_*) \in G$ assumed as initial one.

It is not difficult to see that the value of $\gamma_i(t, x)$ is the guaranteed payoff of the Player i in the position (t, x) of the game.

In [4] the structure of NE- and $P(NE)$-solutions was established. Namely, it was shown that all NE- and $P(NE)$-solutions of the game can be found in the class of pairs of strategies (U, V) each of which generates a unique limit motion (trajectory). The decision strategies that make up such a pair generating the trajectory $x^*(\cdot)$ have the form

$$U^0 = \{u^0(t, x, \varepsilon), \ \beta_1^0(\varepsilon)\}, \quad V^0 = \{v^0(t, x, \varepsilon), \ \beta_2^0(\varepsilon)\}, \tag{3.8}$$

$$u^0(t, x, \varepsilon) = \begin{cases} u^*(t, \varepsilon), & \|x - x^*(t)\| < \varepsilon\varphi(t) \\ u^{(2)}(t, x, \varepsilon), & \|x - x^*(t)\| \ge \varepsilon\varphi(t) \end{cases},$$

$$v^0(t, x, \varepsilon) = \begin{cases} v^*(t, \varepsilon), & \|x - x^*(t)\| < \varepsilon\varphi(t) \\ v^{(1)}(t, x, \varepsilon), & \|x - x^*(t)\| \ge \varepsilon\varphi(t), \end{cases}$$

for all $t \in [t_0, \vartheta]$, $\varepsilon > 0$. In (3.8) we denote by $u^*(t, \varepsilon)$, $v^*(t, \varepsilon)$ families of program controls generating the limit motion $x^*(t)$. The function $\varphi(\cdot)$ and the functions $\beta_1^0(\cdot)$ and $\beta_2^0(\cdot)$ are chosen in such a way that the approximated motions generated by the pair (U^0, V^0) from the initial position (t_0, x_0) do not go beyond the $\varepsilon\varphi(t)$ -neighborhood of the trajectory $x^*(t)$. Functions $u^{(2)}(\cdot, \cdot, \cdot)$ and $v^{(1)}(\cdot, \cdot, \cdot)$ are defined in (3.6). They play the role of punishment strategies for exiting this neighborhood.

Further, for each NE- and $P(NE)$-trajectories $x^*(t)$ the following property holds.

Property 3.1 The point $t = \vartheta$ is the maximum point of the value function $\gamma_i(t, x)$ (3.7) computed along this trajectory, that is,

$$\gamma_i(t, x^*(t)) \leq \gamma_i(\vartheta, x^*(\vartheta)), \ t_0 \leq t \leq \vartheta, \ i = 1, 2 \tag{3.9}$$

3.3 A Non-antagonistic Positional Differential Games with Behavior Types (NPDGwBT): BT-Solution

Now we assume that in addition to the usual normal (nor) type of behavior aimed at maximizing own functionals (3.3), players can use other types of behavior, namely, altruistic, aggressive and paradoxical types [5, 9].

These three types of behavior can be formalized as follows:

Definition 3.3 We say that Player 1 is confined in the current position of the game by *altruistic* (alt) type of behavior if his actions in this position are directed exclusively towards maximizing the functional I_2 (3.3) of Player 2.

Definition 3.4 We say that Player 1 is confined in the current position of the game by *aggressive* (agg) type of behavior if his actions in this position are directed exclusively towards minimizing the functional I_2 (3.3) of Player 2.

Definition 3.5 We will say that Player 1 is confined in the current position of the game by *paradoxical* (par) type of behavior if his actions in this position are directed exclusively towards minimizing own payoff I_1 (3.3).

Similarly, we define the altruistic and aggressive types of Player 2 behavior towards Player 1, as well as the paradoxical type of behavior for Player 2.

Note that the aggressive type of player behavior is actually used in NPDG in the form of punishment strategies contained in the structure of the game's decisions (see, for example, [4]).

The above definitions characterize the extreme types of behavior of players. In reality, however, real individuals behave, as a rule, partly normal, partly altruistic, partly aggressive and partly paradoxical. In other words, mixed types of behavior seem to be more consistent with reality.

If each player is confined to "pure" types of behavior, then in the considered game of two persons with dynamics (3.1) and functionals I_i (3.3) there are 16 possible pairs of types of behavior: (nor, nor), (nor, alt), (nor, agg), (nor, par), (alt, nor), (alt, alt), (alt, agg), (alt, par), (agg, nor), (agg, alt), (agg, agg), (agg, par), (par, nor), (par, alt), (par, agg), (par, par). For the following four pairs (nor, alt), (alt, nor), (agg, par) and (par, agg) the interests of the players coincide and they solve a team problem of control. For the following four pairs (nor, agg), (alt, par), (agg, nor) and (par, alt) players have opposite interests and, therefore, they play a zero-sum differential game. The remaining eight pairs define a non-antagonistic differential games.

The idea of using the players to switch their behavior from one type to another in the course of the game was applied to the game with cooperative dynamics in [9] and for the repeated bimatrix 2×2 game in [6], which allowed to obtain new solutions in these games.

The extension of this approach to non-antagonistic positional differential games leads to new formulation of problems. In particular, it is of interest to see how the player's gains, obtained on Nash solutions, are transformed. The actual task is to minimize the time of "abnormal" behavior, provided that the players' gains are greater than when the players behave normally.

Thus, we assume that players can switch from one type of behavior to another in the course of the game. Such a game will be called a non-antagonistic positional (feedback) differential game with behavior types (NPDGwBT).

In NPDGwBT we assume that simultaneously with the choice of positional strategy, each player also chooses his indicator function defined on the interval $t \in [t_0, \vartheta]$ and taking the value in the set $\{nor, alt, agg, par\}$. We denote the indicator function of Player i by the symbol $\alpha_i : [t_0, \vartheta] \longmapsto \{nor, alt, agg, par\}$, $i = 1, 2$. If the indicator function of some player takes a value, say, alt on some time interval, then this player acts on this interval as an altruist in relation to his partner. Note also that if the indicator functions of both players are identically equal to the value nor on the whole time interval of the game, then we have a classical NPDG.

Thus, in the game NPDGwBT Player 1 controls the choice of a pair of *actions* {position strategy, indicator function} $(U,\ \alpha_1(\cdot))$, and player 2 controls the choice of a pair of actions $(V,\ \alpha_2(\cdot))$.

As mentioned above, for any pair of types of behavior three types of decision making problems can arise: a team problem, a zero-sum game, and a non-antagonistic game. We will assume that the players for each of these three problems are guided by the following Rule 3.1.

Rule 3.1 If on the time interval $(\tau_1, \tau_2) \subset [t_0, \vartheta]$ the player's indicator functions generate a non-antagonistic game, then on this interval players choose one of $P(NE)$-solutions of this game. If a zero-sum game is realized, then as a solution, players choose the saddle point of this game. Finally, if a team problem of control is realized, then players choose one of the pairs of controls such that the value function

$\gamma_i(t, x)$ calculated along the generated trajectory is non-decreasing function, where i is the number of the player whose functional is maximized in team problem.

Generally speaking, the same part of the trajectory can be tracked by several pairs of players' types of behavior, and these pairs may differ from each other by the time of use of abnormal types.

It is natural to introduce the following Rule 3.2.

Rule 3.2 If there are several pairs of types of behavior that track a certain part of the trajectory, then players choose one of them that minimizes the time of using abnormal types of behavior.

We now introduce the definition of the solution of the game NPDGwBT. Note that the set of motions generated by a pair of actions $\{(U, \alpha_1(\cdot)), (V, \alpha_2(\cdot))\}$ coincides with the set of motions generated by the pair (U, V) in the corresponding NPDG.

Definition 3.6 The pair $\{(U^0, \alpha_1^0(\cdot)), (V^0, \alpha_2^0(\cdot))\}$, consistent with Rule 3.1, forms a BT-solution of the game NPDGwBT if there exists a trajectory $x^{BT}(\cdot)$ generated by this pair and there is a $P(NE)$-solution in the corresponding NPDG game generating the trajectory $x^P(\cdot)$ such that the following inequalities are true

$$\sigma_i(x^{BT}(\vartheta)) \geq \sigma_i(x^P(\vartheta)), \quad i = 1, 2, \tag{3.10}$$

where at least one of the inequalities is strict.

Definition 3.7 The BT-solution $\{(U^0, \alpha_1^0(\cdot)), (V^0, \alpha_2^0(\cdot))\}$, which is Pareto non-improvable with respect to the values I_1, I_2 (3.3), is called $P(BT)$-solution of the game NPDGwBT.

Problem 3.1 Find the set of BT-solutions.

Problem 3.2 Find the set of $P(BT)$-solutions.

In the general case, Problems 3.1 and 3.2 have no solutions. However, it is quite expected that the use of abnormal behavior types by players in the game NPDGwBT can in some cases lead to outcomes more preferable for them than in the corresponding game NPDG only with a normal type of behavior. An example of this kind is given in the next section.

3.4 Example

Let equations of dynamics be as follows

$$\dot{x} = u + v, \quad x, u, v \in R^2, \ \|u\| \leq 1, \ \|v\| \leq 1, \ 0 \leq t \leq \vartheta, \ x(0) = x^0, \tag{3.11}$$

where x is the phase vector; u and v are controls of Players 1 and 2, respectively. Let payoff functional of Player i be

$$I_i = \sigma_i(x(\vartheta)) = 18 - \|x(\vartheta) - a^{(i)}\|, \quad i = 1, 2. \tag{3.12}$$

That is, the goal of Player i is to bring vector $x(\vartheta)$ as close as possible to the target point $a^{(i)}$.

Let the following initial conditions and values of parameters be given: $\vartheta = 5.0$, $x_0 = (0, 0)$, $a^{(1)} = (10, 8)$, $a^{(2)} = (-10, 8)$ (Fig. 3.1).

The game has the following phase restrictions. The trajectories of the system (3.11) are forbidden from entering the interior of the set S, which is obtained by removing from the quadrilateral $Oabc$ the line segment Oe. The set S consists of two parts S_1 and S_2, that is, $S = S_1 \cup S_2$.

Coordinates of the points defining the phase constraints:

$$a = (-4.5, 3.6), \; b = (0, 8), \; c = (6.5, 5.2), \; O = (0, 0), \; e = (3.25, 6.6). \tag{3.13}$$

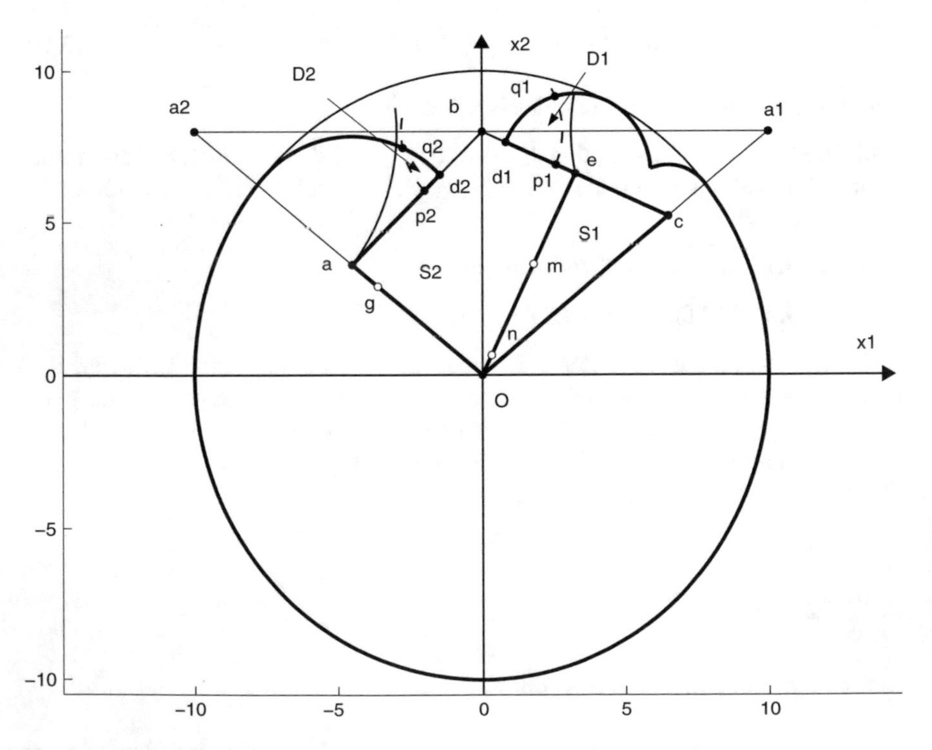

Fig. 3.1 Attainability set

It can be verified that the point a lies on the interval $Oa^{(2)}$, the point c lies on the interval $Oa^{(1)}$, and the point e lies on the interval bc. We also have $|a^{(1)}b| = |a^{(2)}b| = 10$.

Attainability set of the system (3.11) constructed for the moment ϑ consists of points of the circle of radius 10 located not higher than the three-link segment aOc and also bounded by two arcs connecting the large circle with the sides ab and bc of the quadrilateral. The first arc is an arc of the circle with center at the point a and radius $r_1 = 10 - |Oa| = |ad_2|$. The second (composite) arc consists of an arc of the circle with center at the point e and radius $r_2 = 10 - |Oe| = |ed_1|$ and an arc of the circle with center at the point c and radius $r_3 = 10 - |Oc|$ (Fig. 3.1).

Results of approximate calculations: $r_1 = 4.2372$, $r_2 = 2.6432$, $r_3 = 1.6759$, $d_1 = (0.8225, 7.6457)$, $d_2 = (-1.4704, 6.5623)$. In addition, we have: $|Oa^{(1)}| = |Oa^{(2)}| = 12.8062$.

In Fig. 3.1 the dashed lines represent arcs of the circle L with center at the point b and radius $r_4 = |Oa^{(1)}| - |a^{(1)}b| = 12.8062 - 10 = 2.8062$. These arcs intersect the sides ab and bc at the points $p_1 = (2.5773, 6.8898)$ and $p_2 = (-2.0065, 6.0381)$, respectively. By construction, the lengths of the two-links $a^{(1)}bp_2$ and $a^{(2)}bp_1$ are equal to each other and equal to the lengths of the segments $Oa^{(1)}$ and $Oa^{(2)}$.

The value functions $\gamma_1(t,x)$ and $\gamma_2(t,x)$, $0 \le t \le \vartheta$, $x \in R^2 \backslash S$, of the corresponding auxiliary zero-sum games Γ_1 and Γ_2 in this example will be as follows

$$\gamma_i(t,x) = \begin{cases} 18 - \|(x - a^{(i)}\|, & xa^{(i)} \bigcap intS = \emptyset \\ 18 - \rho_S\left(x, a^{(i)}\right), & otherwise \end{cases}, \tag{3.14}$$

where $i = 1, 2$, and $\rho_S\left(x, a^{(i)}\right)$ denotes the smallest of the two distances from the point x to the point $a^{(i)}$, one of which is calculated when the set S is bypassed clockwise and the other counterclockwise.

At first we solve the game NPDG (without abnormal behavior types). One can check that in the game NPDG the trajectory $x(t) \equiv 0$, $t \in [0, 5]$, (stationary point O) is a Nash trajectory. Further, the trajectories constructed along the line Oe, are not Nash ones, since none of them is satisfied condition (3.9). This is also confirmed by the fact that the circle of radius $|a^{(1)}e|$ with the center at the point $a^{(1)}$ has no points in common with the circle L (see Fig. 3.1). Obviously, these are not Nash and all the trajectories that bypass the set S_1 on the right. Are not Nash and all trajectories that bypass the set S_2 on the left, since a circle of radius $|a^{(2)}a|$ with the center at the point $a^{(2)}$ also has no points in common with the circle L. As a result, it turns out that the mentioned trajectory is the only Nash trajectory, and, consequently, the only $P(NE)$-trajectory; the players' gains on it are $I_1 = I_2 = 5.1938$.

Let us now turn to the game NPDGwBT, in which each player during certain periods of time may exhibit altruism and aggression towards other player, and the case of mutual aggression is allowed.

We consider two variants of the game.

Variant I We assume that players 1 and 2 together with a normal type of behavior, can exhibit altruism towards their partner during some time intervals.

Variant II In addition to the assumption of altruism of the players, we assume that each player can act aggressively against another player for some periods of time, and a case of mutual aggression is allowed.

In the attainability set, we find all the points x for which inequalities hold

$$\sigma_i(x) \geq \sigma_i(O),\ i = 1, 2,\ \ \sigma_1(x) + \sigma_2(x) > \sigma_1(O) + \sigma_2(O) \tag{3.15}$$

Such points form two sets D_1 and D_2 (see Fig. 3.1). The set D_1 is bounded by the segment p_1d_1, and also by the arcs p_1q_1 and q_1d_1 of the circles mentioned above. The set D_2 is bounded by the segment p_2d_2, and also by the arcs d_2q_2 and q_2p_2 of the circles mentioned. On the arc p_1q_1, the non-strict inequality (3.15) for i $=$ 2 becomes an equality, and on the arc q_2p_2, the non-strict inequality (3.15) becomes an equality for i $=$ 1. At the remaining points sets D_1 and D_2 , the non-strict inequalities (3.15) for $i \in \{1, 2\}$ are strict.

Now within the framework of Variant I we construct a BT-solution leading to the point $d_2 \in D_2$. Consider the trajectory Oad_2; the players' gains on it are $I_1 = 5.9436$, $I_2 = 9.3501$, that is, each player gains more than on single $P(NE)$-trajectory. How follows from the foregoing, the trajectory Oad_2 is not Nash one. However, if it is possible to construct indicator functions-programs of players that provide motion along this trajectory, then a BT-solution will be constructed.

On the side Oa, we find a point g equidistant from the point $a^{(1)}$ if we go around the set S clockwise, or if we go around S counterclockwise. We obtain $g = (-3.6116, 2.8893)$. Further, if we move along the trajectory Oad_2 with the maximum velocity for $t \in [0, 5]$, then the time to hit the point g will be $t = 2.3125$, and for the point a will be $t = 2.8814$. It can be verified that if we move along this trajectory on the time interval $[0, 2.3125]$ then the function $\gamma_1(t, x)$ (3.14) decreases monotonically and the function $\gamma_2(t, x)$ increases monotonically; for motion on the interval $t \in [2.3125, 2.8814]$, both functions $\gamma_1(t, x)$ and $\gamma_2(t, x)$ increase monotonically; finally, when driving on the remaining interval $t \in [2.8814, 5.0]$, then the function $\gamma_1(t, x)$ increases monotonically, and the function $\gamma_2(t, x)$ decreases.

We check that on the segment Og of the trajectory the pair (alt, nor), which defines a team problem of control, is the only pair of types of behavior that realizes motion on this segment in accordance with Rule 3.1; this is the maximum shift in the direction of the point $a^{(2)}$. In the next segment ga there will already be four pairs of "candidates" (nor, nor), (alt, nor), (nor, alt) and (alt, alt)), but according to Rule 3.2 the last three pairs are discarded; the remaining pair determines a non-antagonistic game, and the motion on this segment will be generated by $P(NE)$-solution of the game. Finally, for the last segment ad_2, the only pair of types of behavior, that generates motion on the segment in accordance with Rule 3.1, there will be a pair (nor, alt) that defines a team problem of control; the motion represents the maximum shift in the direction of the point d_2.

Thus, we have constructed the following indicator function-programs of players

$$\alpha_1^{(1)}(t) = \{alt, t \in [0, 2.3125); \quad nor, t \in [2.3125, 5]\}, \tag{3.16}$$

$$\alpha_2^{(1)}(t) = \{nor, t \in [0, 2.8814); \quad alt, t \in [2.8814, 5]\}. \tag{3.17}$$

We denote by $(U^{(1)}, V^{(1)})$ the pair of players' strategies that generates the limit motion Oad_2 for $t \in [0, 5]$ and is consistent with the constructed indicator functions. Then we obtain the following assertion.

Theorem 3.1 *In Variant I, the pair of actions* $\{(U^{(1)}, \alpha_1^{(1)}(\cdot)), (V^{(1)}, \alpha_2^{(1)}(\cdot))\}$ *(3.16), (3.17) provides the* $P(BT)$*-solution.*

We turn to the Variant II, in which, in addition to assuming the altruism of the players, it assumed that players can use an aggressive type of behavior. We construct a BT-solution, leading to the point $d_1 \in D_1$.

Let us find the point m equidistant from the point $a^{(2)}$ if we go around the set S_2 clockwise, or if we go around S_2 counterclockwise. We also find a point n equidistant from the point $a^{(1)}$ as if we were go around the set S_1 clockwise, or if we go around S_1 counterclockwise. The results of the calculations: $m = (1.7868, 3.6285)$, $n = (0.3190, 0.6478)$.

Consider the trajectory Oed_1; the players' gains on it are $I_1 = 8.8156$, $I_2 = 7.1044$, that is, the gains of both players on this trajectory are greater than the gains on the single $P(NE)$-trajectory. As follows from the above, the trajectory Oed_1 is not Nash one. Therefore, if it is possible to construct indicator functions-programs of players that provide motion along this trajectory, then a BT-solution will be constructed.

First of all find that if we move along the trajectory Oed_1 with the maximum velocity for $t \in [0, 5]$, the time to hit the point n will be $t = 0.3610$, the point m will be $t = 2.0223$, and the point e will be $t = 3.6784$. It is easy to verify that for such a motion along the trajectory Oed_1 on the interval $t \in [0, 0.3610]$, both functions $\gamma_1(t, x)$ and $\gamma_2(t, x)$ (3.14) decrease monotonically; for motion on the interval $t \in [0.3610, 2.0223]$, the function $\gamma_2(t, x)$ continues to decrease, and the function $\gamma_1(t, x)$ increases; for motion on the interval $t \in [1.9135, 3.9620]$, both functions increase; finally, on the remaining interval $t \in [3.9620, 5]$, the function $\gamma_2(t, x)$ continues to increase, and the function $\gamma_1(t, x)$ decreases.

We check that on the segment On of the trajectory, the pair (agg, agg), which determines the non-antagonistic game, is the only pair of types of behaviors that realizes motion on the segment in accordance with Rule 3.1; this is the motion generated by the $P(NE)$-solution, the best for both players. In the next segment nm, two pairs of types of behaviors realize motion on the segment according to Rule 3.1, namely (nor, alt) and (agg, alt); however, according to Rule 3.2, only the pair (nor, alt) remains; it defines a team problem of control in which the motion represents the maximum shift in the direction of point m. There are already four pairs of "candidates" (nor, nor), (alt, nor), (nor, alt) and (alt, alt) on the segment *me*, but according to Rule 3.2 the last three pairs are discarded; the remaining pair defines a non-antagonistic game and the motion on this segment is generated by the $P(NE)$-solution of the game. Finally, for the last segment ed_1, the only pair of types of behaviors is the pair (alt, nor), which defines a team problem of control; the motion represents the maximum shift in the direction of the point d_1.

Thus, we have constructed the following indicator function-programs of players

$$\alpha_1^{(2)}(t) = \{agg, t \in [0, 0.3610); \quad nor, t \in [0.3610, 3.6784); \; alt, t \in [3.6784, 5]\}, \tag{3.18}$$

$$\alpha_2^{(2)}(t) = \{agg, t \in [0, 0.3610); \quad alt, t \in [0.3610, 2.0223); \quad nor, t \in [2.0223, 5]\}. \tag{3.19}$$

We denote by $(U^{(2)}, V^{(2)})$ the pair of players' strategies that generate the limit motion Oed_1 for $t \in [0, 5]$ and is consistent with the constructed indicator functions. Then we obtain the following assertion.

Theorem 3.2 *In Variant II, the pair of actions* $\{(U^{(2)}, \alpha_1^{(2)}(\cdot)), (V^{(2)}, \alpha_2^{(2)}(\cdot))\}$ *(3.18), (3.19) provides the* $P(BT)$*-solution.*

Remark 3.1 It is obvious that Theorem 3.1 is also true for Variant II.

Following the scheme of the proofs of Theorems 3.1 and 3.2 (and also taking into account Remark 3.1), we arrive at the following Theorems.

Theorem 3.3 *In Variant I, the set* D_2 *consists of those and only those points that are endpoints of the trajectories generated by the* BT*-solutions of the game.*

Theorem 3.4 *In Variant II, the sets* D_1 *and* D_2 *consist of those and only those points that are the ends of the trajectories generated by the* BT*-solutions of the game.*

3.5 Conclusion

Realized idea of using the players to switch their behavior from one type to another in the course of the game is somewhat similar to the idea of using trigger strategies [2]. This is indicated by the existence of punishment strategies in the structure (8) of decision strategies. However, there are significant differences. In this paper, we also use more complex switching, namely, from one type of behavior to another, changing the nature of the problem of optimization—from non-antagonistic games to zero-sum games or team problem of control and vice versa. And these switchings are carried out according to pre-selected indicator function-programs.

Each player controls the choice of a pair of actions positional strategy, indicator function. Thus, the possibilities of each player in the general case have expanded (increased) and it is possible to introduce a new concept of a game solution ($P(BT)$-

solution) in which both players increase their payoffs in comparison with the payoffs in Nash equilibrium in the game without switching types of behavior.

For players, it is advantageous to implement $P(BT)$-trajectory; so they will follow the declared indicator function-programs (3.16), (3.17) or (3.18), (3.19).

References

1. Basar, T., Olsder, G.J.: Dynamic Noncooperative Game Theory, 2nd edn. SIAM, New York (1999)
2. Friedman, J.: A noncooperative equilibrium for supergames. Rev. Econ. Stud. **38**, 1–12 (1971)
3. Haurie, A., Krawczyk, J.B., Zaccour, G.: Games and Dynamic Games. World Scientific-Now Publishers Series in Business, vol. 1. World Scientific, Singapore (2012)
4. Kleimenov, A.F.: Nonantagonistic Positional Differential Games. Nauka, Yekaterinburg (1993)
5. Kleimenov, A.F.: On solutions in a nonantagonistic positional differential game. J. Appl. Math. Mech. **61**(5), 717–723 (1997). https://doi.org/10.1016/S0021-8928(97)00094-4
6. Kleimenov, A.F.: An approach to building dynamics for repeated bimatrix 2×2 games involving various behavior types. In: Leitman, G. (ed.) Dynamics and Control. Gordon and Breach, London (1999)
7. Kleimenov, A.F.: Altruistic behavior in a non-antagonistic positional differential game. Math. Theory Games Appl. **7**(4), 40–55 (2015)
8. Kleimenov, A.F.: Application of altruistic and aggressive behavior types in a non-antagonistic positional differential two-person game. Proc. Inst. Math. Mech. Ural Branch RAS **23**(4), 181–191 (2017)
9. Kleimenov, A.F., Kryazhimskii A.V.: Normal behavior, altruism and aggression in cooperative game dynamics. Interim Report IR-98-076. IIASA, Laxenburg (1998)
10. Krasovskii, N.N.: Control of a Dynamic System. Nauka, Moscow (1985)
11. Krasovskii, N.N., Subbotin, A.I.: Game-Theoretical Control Problems. Springer, New York (1988)
12. Petrosyan, L.A., Zenkevich, N.A., Shevkoplyas E.V.: Game Theory. BHV-Petersburg, St. Petersburg (2012)

Chapter 4
Learning in a Game of Strategic Experimentation with Three-Armed Exponential Bandits

Nicolas Klein

Abstract The present article provides some additional results for the two-player game of strategic experimentation with three-armed exponential bandits analyzed in Klein (Games Econ Behav 82:636–657, 2013). Players play replica bandits, with one safe arm and two risky arms, which are known to be of opposite types. It is initially unknown, however, which risky arm is good and which is bad. A good risky arm yields lump sums at exponentially distributed times when pulled. A bad risky arm never yields any payoff. In this article, I give a necessary and sufficient condition for the state of the world eventually to be found out with probability 1 in *any* Markov perfect equilibrium in which at least one player's value function is continuously differentiable. Furthermore, I provide closed-form expressions for the players' value function in a symmetric Markov perfect equilibrium for low and intermediate stakes.

4.1 Introduction

Think of a situation in which agents are initially uncertain about some payoff-relevant aspect of their environment. Yet, they can learn about it over time by exploring different options. Thus, a farmer may not know the yield of a new crop before trying it out. Trying it out implies an opportunity cost, however, as using his field to try the new crop means that he cannot use it to plant a traditional crop, whose yield he already knows. The trade-off he faces is thus between optimally using the information he already has (*exploitation*) and investing resources in order to acquire new information, which will potentially be useful to him in the future (*exploration*).

N. Klein (✉)
Université de Montréal and CIREQ, Département de Sciences Économiques, Montréal, QC, Canada

L. A. Petrosyan et al. (eds.), *Frontiers of Dynamic Games*,
Static & Dynamic Game Theory: Foundations & Applications,
https://doi.org/10.1007/978-3-319-92988-0_4

The so-called *multi-armed bandit model* has become canonical in economics to analyze a decision maker's trade-off between exploration and exploitation.[1]

But now suppose that our farmer has a neighbor and that he can observe the kind of crop planted by his neighbor, as well as its yield. Our farmer would of course prefer that his neighbor experiment with the new crop, as this would allow him to get some information about it without having to bear the (opportunity) cost of producing the information himself. Of course, his neighbor faces precisely the same trade-off, and the informational externality leads to a situation of strategic interaction. Such *strategic* bandit problems have been introduced by Bolton and Harris [2, 3], where players choose between a risky option and a safe one. Here, I use the exponential-bandits variant introduced by Keller et al. [6], and, in particular, adopt the three-armed model of Klein [7].

While in [2, 3] and in [6], the risky option was of the same quality for all players, Klein and Rady [8] introduced negative correlation between players: what was good news to one player was bad news to the other. In [7], I have introduced a setting in which two players have access to *two* risky arms of perfectly negatively correlated types. The comparison of the results in [8] and [7] in particular thus allow for the analysis of the impact of delegating project choice to individual agents.

For the case of perfectly positively correlated two-armed bandits, Keller et al. [6] show that players experiment inefficiently little in equilibrium, as compared to the cooperative benchmark. Indeed, the information players produce is a public good; hence they produce too little of it. Indeed, they both give up on finding out the state of the world too soon (i.e., the *amount* of experimentation is too low) and they learn too slowly (i.e., the *intensity* of experimentation will be inefficiently low). By contrast, Klein and Rady [8] find that, with perfectly negatively correlated two-armed bandits, the *amount* of experimentation is always at the efficient level. Furthermore, there exists an efficient equilibrium if and only if the stakes at play are *below* a certain threshold. By contrast, in [7], I show that, when both agents have access to two perfectly negatively correlated risky arms, there exists an efficient equilibrium if and only if the stakes at play *exceed* a certain threshold. In the present article, I provide closed-form expressions for the players' value function in a symmetric Markov perfect equilibrium for the cases in which there does not exist an efficient equilibrium. Furthermore, I give a necessary and sufficient condition for learning to be complete, i.e. for the state of the world to be found out with probability 1, in *any* Markov perfect equilibrium in which at least one player's value function is continuously differentiable.

The rest of this article is organised as follows. Section 4.2 explains the model setup; Sect. 4.3 analyzes conditions under which complete learning will prevail; Sect. 4.4 analyzes equilibrium for low and intermediate stakes, while Sect. 4.5 concludes. Formal proofs are collected in Sect. 4.6.

[1]The multi-armed bandit model was first introduced by Thompson [10] and Robbins [9], and subsequently analyzed, amongst others, by Bradt et al. [4] and Bellman [1]. Gittins and Jones [5] provided the famous Gittins-index characterization of an optimal policy.

4.2 Model Setup

The setup is as in [7]: There are two agents playing a three-armed bandit in continuous time each. One arm is safe in that it yields a known flow payoff of $s > 0$ when pulled; the other two arms, A and B, are risky in that they can be either good or bad. It is known that exactly one of the two risky arms is good and that the same risky arm is good for both players. Which between arms A and B is good and which is bad is initially unknown. The good risky arm yields lump sums $h > 0$ at exponentially distributed times with parameter $\lambda > 0$, when it is pulled. The bad risky arm always yields 0. The parameters λ, s and h, are common knowledge among the players. I assume that $g := \lambda h > s > 0$.

More specifically, either player $i \in \{1, 2\}$ can decide in continuous time how to distribute a unit endowment flow over the three arms of his bandit; i.e., at each instant $t \in \mathscr{R}_+$, he chooses $(k_{i,A}, k_{i,B}) \in \{(a, b) \in [0, 1]^2 : a + b \leq 1\}$, where $k_{i,A}(t)$ $(k_{i,B}(t))$ denotes the fraction of the unit endowment flow player i devotes to arm A (B) at instant t.

Players start out from a common prior $p_0 \in (0, 1)$ that it is their risky arms A that are good. As everyone's action choices, as well as the outcomes of these action choices, are perfectly publicly observable, there is no private information at any time. Thus, players will share a common posterior belief that it is their risky arms A that are good at all times $t \geq 0$. We shall denoted by p_t this belief at instant t. As only a good risky arm can ever yield a lump-sum payoff, $p_\tau = 1$ $(p_\tau = 0)$ at all times $\tau > t$ if either player has received a lump sum from arm A (B) at time t. If no such breakthrough has occurred yet by time t, the belief satisfies

$$p_t = \frac{p_0 e^{-\lambda \int_0^t (k_{1,A}(\tau)+k_{2,A}(\tau))\,d\tau}}{p_0 e^{-\lambda \int_0^t (k_{1,A}(\tau)+k_{2,A}(\tau))\,d\tau} + (1-p_0) e^{-\lambda \int_0^t (k_{1,B}(\tau)+k_{2,B}(\tau))\,d\tau}}. \tag{4.1}$$

Following much of the literature, I focus on Markov perfect equilibria with the common posterior belief p_t as the state variable (which I shall sometimes simply refer to as *equilibrium*). A Markov strategy for player i is a time-invariant, piecewise continuous, function $(k_{i,A}, k_{i,B}) : [0, 1] \rightarrow \{(a, b) \in [0, 1]^2 : a + b \leq 1\}$, $p_t \mapsto (k_{i,A}, k_{i,B})(p_t)$. As in [8], a *pair* of Markov strategies is said to be *admissible* if there exists a solution to the corresponding law of motion of beliefs (derived from Bayes' rule) that coincides with the limit of the unique discrete-time solution. An inadmissible strategy pair is assumed to give both players a payoff of $-\infty$.

Players discount payoffs at the common discount rate $r > 0$. An admissible strategy pair $((k_{1,A}, k_{1,B}), (k_{2,A}, k_{2,B}))$ induces a payoff function u_i for players $i \in \{1, 2\}$, which is given by

$$u_i(p) = \mathscr{E}\left[\int_0^\infty re^{-rt}\left\{(k_{i,A}(p_t)p_t + k_{i,B}(p_t)(1-p_t))g \right.\right.$$
$$\left.\left. +[1 - k_{i,A}(p_t) - k_{i,B}(p_t)]s\right\} dt \,\middle|\, p_0 = p\right], \tag{4.2}$$

where the expectation is taken with respect to the process of beliefs $\{p_t\}_{t\in\mathscr{R}_+}$. Player i's objective is to maximize u_i. As one can see immediately from player i's objective, the other player's actions impact u_i only via the players' common belief process $\{p_t\}_{t\in\mathscr{R}_+}$; i.e., ours is a game of purely informational externalities.

I say that the stakes are *high* if $\frac{g}{s} \geq \frac{4(r+\lambda)}{2r+3\lambda}$; they are *intermediate* if $\frac{2r+\lambda}{r+\lambda} < \frac{g}{s} < \frac{4(r+\lambda)}{2r+3\lambda}$; they are *low* if $\frac{g}{s} \leq \frac{2r+\lambda}{r+\lambda}$, and *very low* if $\frac{g}{s} < \frac{2(r+\lambda)}{r+2\lambda}$. It is immediate to verify that the stakes are low if and only if $p_1^* := \frac{rs}{(r+\lambda)(g-s)+rs} \geq \frac{1}{2}$; they are very low if and only if $p_2^* := \frac{rs}{(r+2\lambda)(g-s)+rs} \geq \frac{1}{2}$.

Klein [7, Section 4] shows that the utilitarian planner's solution has a bang-bang structure.[2] If the stakes at play are not very low, the planner would always use the risky arm that looks momentarily the most promising; he would never use the safe arm. This means that learning will be complete, i.e. the true state of the world will be found out with probability 1. If the stakes are very low, by contrast, the planner would use the safe arm for all beliefs in $[1 - p_2^*, p_2^*]$ and the risky arm that looks momentarily the most promising for all other beliefs. Thus, learning will be incomplete in this case. A single player acting in isolation would optimally pursue the same policy, with p_1^* replacing p_2^*, and "low stakes" replacing "very low stakes," in the previous statements.

4.3 Complete Learning

As already mentioned in the introduction, Keller et al. [6] identified two dimensions of inefficiency in their model: On the one hand, players give up on finding out about the true state of the world too soon, i.e. the experimentation *amount* is inefficiently small. On the other hand, players also learn too slowly, i.e. the experimentation *intensity* is inefficiently low. If one were merely to focus on the long-run properties of learning, only the former effect would be of interest. Keller et al. [6] show that, because of the informational externalities, all experimentation stops at the single-agent cutoff belief in any equilibrium; the efficient cutoff belief would be more pessimistic, though, as it takes into account that the information a player generates benefits the other players also.[3] Furthermore, learning is always incomplete, i.e.

[2]The utilitarian planner maximizes the sum of the players' utilities. The solution to this problem is the policy the players would want to commit to at the outset of the game if they had commitment power. It thus constitutes a natural efficient benchmark against which to compare our equilibria.

[3]By contrast, Bolton and Harris [2] identified an *encouragement effect* in their model. It makes players experiment at beliefs that are more pessimistic than their single-agent cutoffs. This is because they will receive good news with some probability, which will make the other players more optimistic also. This then induces them to provide more experimentation, from which the first player then benefits in turn. With fully revealing breakthroughs as in [6, 8], or this model, however, a player could not care less what others might do after a breakthrough, as there will not be anything left to learn. Therefore, there is no encouragement effect in these models.

there is a positive probability that the truth will never be found out.[4] In [8], however, the *amount* of experimentation is always at the efficient level.[5] This is because both players cannot be exceedingly pessimistic at the same time. Indeed, as soon as players' single-agent cutoffs overlap, at any possible belief at least one of them is loath to give up completely, although players may not be experimenting with the enthusiasm required by efficiency. In particular, learning will be complete in any equilibrium if and only if efficiency so requires.

This section will show that which of these effects prevails here depends on the stakes at play: If stakes are so high that the single-agent cutoffs overlap, players would not be willing ever completely to give up on finding out the true state of the world even if they were by themselves. Yet, since all a player's partner is doing is to provide him some additional information for free, a player should be expected to do at least as well as if he were by himself. Hence, the Klein and Rady [8] effect obtains if players' single-agent cutoffs overlap, and, in any equilibrium (in which at least one player's value function is smooth),[6] the true state of the world will eventually be found out with probability 1 (i.e. learning will be *complete*), as efficiency requires. In the opposite case, however, the informational externality identified by Keller et al. [6] carries the day, and, as we will see in the next section, there exists an equilibrium entailing an inefficiently low amount of experimentation. For some parameters, this implies incomplete equilibrium learning while efficiency calls for complete learning.

To state the next lemma, I write u_1^* for the value function of a single agent operating a bandit with only a safe arm and a risky arm A, while I denote by u_2^* the value function of a single agent operating a bandit with only a safe arm and a risky arm B. It is straightforward to verify that $u_2^*(p) = u_1^*(1-p)$ for all p and that[7]

$$u_1^*(p) = \begin{cases} s & \text{if } p \leq p_1^*, \\ g\left[p + \frac{\lambda p_1^*}{\lambda p_1^* + r}(1-p)\left(\frac{\Omega(p)}{\Omega(p_1^*)}\right)^{\frac{r}{\lambda}}\right] & \text{if } p > p_1^* \end{cases}, \tag{4.3}$$

[4]The efficient solution in [6] also implies incomplete learning.

[5]For perfect negative correlation, this is true in any equilibrium; for general negative correlation, there always exists an equilibrium with this property.

[6]The technical requirement that at least one player's value function be continuously differentiable is needed on account of complications pertaining to the admissibility of strategies. I use it in the proof of Lemma 4.1 to establish that the safe payoff s constitutes a lower bound on the player's equilibrium value. However, by e.g. insisting on playing $(1, 0)$ at a single belief $\hat{p}$ while playing $(0, 0)$ everywhere else in a neighborhood of $\hat{p}$, a player could e.g. force the other player to play $(0, 1)$ at $\hat{p}$ for mere admissibility reasons. Thus, both players' *equilibrium* value functions might be pushed below s at certain beliefs $\hat{p}$. For the purposes of this section, I rule out such implausible behavior by restricting attention to equilibria in which at least one player's value function is smooth.

[7]See Prop.3.1 in [6].

where $\Omega(p) := \frac{1-p}{p}$ denotes the odds ratio. The following lemma tells us that u_1^* and u_2^* are both lower bounds on a player's value in *any* equilibrium, provided his value is smooth.

Lemma 4.1 (Lower Bound on Equilibrium Payoffs) *Let $u \in C^1$ be a player's equilibrium value function. Then, $u(p) \geq \max\{u_1^*(p), u_2^*(p)\}$ for all $p \in [0, 1]$.*

The intuition for this result is very straightforward. Indeed, there are only informational externalities, no payoff externalities, in our model. Thus, intuitively, a player can only benefit from any information his opponent provides him for free; therefore, he should be expected to do at least as well as if he were by himself, forgoing the use of one of his risky arms to boot.

Now, if $\frac{g}{s} > \frac{2r+\lambda}{r+\lambda}$, then $p_1^* < \frac{1}{2} < 1 - p_1^*$, so at any belief p, we have that $u_1^*(p) > s$ or $u_2^*(p) > s$ or both. Thus, there cannot exist a p such that $(k_{1,A}, k_{1,B})(p) = (k_{2,A}, k_{2,B})(p) = (0, 0)$ be mutually best responses as this would mean $u_1(p) = u_2(p) = s$. This proves the following proposition:

Proposition 4.1 (Complete Learning) *If $\frac{g}{s} > \frac{2r+\lambda}{r+\lambda}$, learning will be complete in any Markov perfect equilibrium in which at least one player's value function is continuously differentiable.*

It is the same threshold $\frac{2r+\lambda}{r+\lambda}$ above which complete learning is efficient, and prevails in any equilibrium, in the perfectly negatively correlated two-armed bandit case.[8] In our setting, however, complete learning is efficient for a larger set of parameters, as we saw in Sect. 4.2. In the following section, I shall proceed to a more thorough analysis of the strategic problem.

4.4 Equilibrium Payoff Functions

In [7], I have shown that there exists an efficient equilibrium in this model if and only if the stakes are high. The purpose of this section is to construct a symmetric equilibrium for those parameter values for which there does not exist an efficient equilibrium. I define symmetry in keeping with [2] as well as [6]:

Definition 4.1 An equilibrium is said to be *symmetric* if equilibrium strategies $((k_{1,A}, k_{1,B}), (k_{2,A}, k_{2,B}))$ satisfy $(k_{1,A}, k_{1,B})(p) = (k_{2,A}, k_{2,B})(p)$ for all $p \in [0, 1]$.

As a matter of course, in any symmetric equilibrium, $u_1(p) = u_2(p)$ for all $p \in [0, 1]$. I shall denote the players' common value function by u. By the same token, I shall write $k_{1,A} = k_{2,A} = k_A$ and $k_{1,B} = k_{2,B} = k_B$.

[8] See Proposition 8 in [8].

4.4.1 Low Stakes

Recall that the stakes are low if, and only if, the single-agent cutoffs for the two risky arms do not overlap. It can be shown that in this case there exists an equilibrium that is essentially two copies of the Keller et al. [6] symmetric equilibrium (see their Proposition 5.1), mirrored at the $p = \frac{1}{2}$ axis.

Proposition 4.2 (Symmetric MPE for Low Stakes) *If $\frac{g}{s} \leq \frac{2r+\lambda}{r+\lambda}$, there exists a symmetric equilibrium where both players exclusively use the safe arm on $[1 - p_1^*, p_1^*]$, the risky arm A above the belief $\hat{p} > p_1^*$, and the risky arm B at beliefs below $1 - \hat{p}$, where $\hat{p}$ is defined implicitly by*

$$\Omega(p^m)^{-1} - \Omega(\hat{p})^{-1} = \frac{r+\lambda}{\lambda}\left[\frac{1}{1-\hat{p}} - \frac{1}{1-p_1^*} - \Omega(p_1^*)^{-1}\ln\left(\frac{\Omega(p_1^*)}{\Omega(\hat{p})}\right)\right]. \tag{4.4}$$

In $[p_1^*, \hat{p}]$, the fraction $k_A(p) = \frac{u(p)-s}{c_A(p)}$ is allocated to risky arm A, while $1 - k_A(p)$ is allocated to the safe arm; in $[1 - \hat{p}, 1 - p_1^*]$, the fraction $k_B(p) = \frac{u(p)-s}{c_B(p)}$ is allocated to risky arm B, while $1 - k_B(p)$ is allocated to the safe arm.

Let $V_h(p) := pg + C_h(1-p)\Omega(p)^{\frac{r}{2\lambda}}$, and $V_l(p) := (1-p)g + C_l p\Omega(p)^{-\frac{r}{2\lambda}}$. Then, the players' value function is continuously differentiable, and given by $u(p) = W(p)$ if $1 - \hat{p} \leq p \leq \hat{p}$, where $W(p)$ is defined by

$$W(p) := \begin{cases} s + \frac{r}{\lambda}s\left[\Omega(p_1^*)^{-1}\left(1 - \frac{p}{p_1^*}\right) - p\ln\left(\frac{\Omega(p)}{\Omega(p_1^*)}\right)\right] & \text{if } 1-\hat{p} < p < 1-p_1^* \\ s \text{ if } 1 - p_1^* \leq p \leq p_1^* & \\ s + \frac{r}{\lambda}s\left[\Omega(p_1^*)\left(1 - \frac{1-p}{1-p_1^*}\right) - (1-p)\ln\left(\frac{\Omega(p_1^*)}{\Omega(p)}\right)\right] & \text{if } p_1^* < p < \hat{p} \end{cases}; \tag{4.5}$$

$u(p) = V_h(p)$ if $\hat{p} \leq p$, while $u(p) = V_l(p)$ if $p \leq 1 - \hat{p}$, where the constants of integration C_h and C_l are determined by $V_h(\hat{p}) = W(\hat{p})$ and $V_l(1-\hat{p}) = W(1-\hat{p})$, respectively.

Thus, in this equilibrium, even though either player knows that one of his risky arms is good, whenever the uncertainty is greatest, the safe option is attractive to the point that he cannot be bothered to find out which one it is. When players are relatively certain which risky arm is good, they invest all their resources in that arm. When the uncertainty is of medium intensity, the equilibrium has the flavor of a mixed-strategy equilibrium, with players devoting a uniquely determined fraction of their resources to the risky arm they deem more likely to be good, with the rest being invested in the safe option. As a matter of fact, the experimentation intensity decreases continuously from $k_A(\hat{p}) = 1$ to $k_A(p_1^*) = 0$ (from $k_B(1 - \hat{p}) = 1$ to $k_B(1 - p_1^*) = 0$). Intuitively, the situation is very much reminiscent of the classical Battle of the Sexes game: If one's partner experiments, one would like to free-ride on his efforts; if one's partner plays safe, though, one would rather do the experimentation oneself than give up on finding out the truth. On the relevant range of beliefs it is the case that as players become more optimistic, they have to raise their experimentation intensities in order to increase free-riding incentives for their partner. This is necessary to keep their partner indifferent, and hence willing to mix, over both options.

Having seen that for $\frac{g}{s} \leq \frac{2r+\lambda}{r+\lambda}$, there exists an equilibrium with smooth value functions that implies incomplete learning, we are now in a position to strengthen our result on the long-run properties of equilibrium learning:

Corollary 4.1 (Complete Learning) *Learning will be complete in any Markov perfect equilibrium in which at least one player's value function is smooth, if and only if* $\frac{g}{s} > \frac{2r+\lambda}{r+\lambda}$.

For perfect negative correlation, Klein and Rady [8] found that with the possible exception of the knife-edge case $\frac{g}{s} = \frac{2r+\lambda}{r+\lambda}$, learning was going to be complete in any equilibrium if and only if complete learning was efficient. While the proposition pertains to the exact same parameter set on which complete learning prevails in [8], we here find by contrast that if $\frac{2(r+\lambda)}{r+2\lambda} < \frac{g}{s} \leq \frac{2r+\lambda}{r+\lambda}$, efficiency uniquely calls for complete learning, yet there exists an equilibrium entailing incomplete learning. This is because with three-armed bandits information is more valuable to the utilitarian planner, as in case of a success he gets the full payoff of a good risky arm. With negatively correlated two-armed bandits, however, the planner cannot shift resources between the two types of risky arm; thus, his payoff in case of a success is just $\frac{g+s}{2}$.

4.4.2 Intermediate Stakes

For intermediate stakes, the equilibrium I construct is essentially of the same structure as the previous one: It is symmetric and it requires players to mix on some interval of beliefs. However, there does not exist an interval where both players play safe, so that players will always eventually find out the true state of the world, even though they do so inefficiently slowly.

Proposition 4.3 (Symmetric MPE for Intermediate Stakes) *If* $\frac{2r+\lambda}{r+\lambda} < \frac{g}{s} < \frac{4(r+\lambda)}{2r+3\lambda}$, *there exists a symmetric equilibrium. Let* $\check{p} := \frac{\lambda+r}{\lambda}(2p^m - 1)$, *and* $\mathscr{W}(p)$ *be defined by*

$$\mathscr{W}(p) := \begin{cases} s + \frac{r+\lambda}{\lambda}(g-s) - \frac{r}{\lambda}ps\,(2 + \ln(\Omega(p))) & \text{if } p \leq \frac{1}{2} \\ s + \frac{r+\lambda}{\lambda}(g-s) - \frac{r}{\lambda}(1-p)s\,(2 - \ln(\Omega(p))) & \text{if } p \geq \frac{1}{2} \end{cases} \tag{4.6}$$

Now, let $p_1^\dagger > \frac{1}{2}$ *and* $p_2^\dagger > \frac{1}{2}$ *be defined by* $\mathscr{W}(p_1^\dagger) = \frac{\lambda+r(1-p_1^\dagger)}{\lambda+r}g$ *and* $\mathscr{W}(p_2^\dagger) = 2s - p_2^\dagger g$, *respectively. Then, let* $p^\dagger := p_1^\dagger$ *if* $p_1^\dagger \geq \check{p}$; *otherwise, let* $p^\dagger := p_2^\dagger$.

In equilibrium, both players will exclusively use their risky arm A in $[p^\dagger, 1]$, *and their risky arm B in* $[0, 1-p^\dagger]$. *In* $]\frac{1}{2}, p^\dagger]$, *the fraction* $k_A(p) = \frac{\mathscr{W}(p)-s}{c_A(p)}$ *is allocated to risky arm A, while* $1 - k_A(p)$ *is allocated to the safe arm; in* $[p^\dagger, \frac{1}{2}[$*, the fraction* $k_B(p) = \frac{\mathscr{W}(p)-s}{c_B(p)}$ *is allocated to risky arm B, while* $1 - k_B(p)$ *is allocated to the safe arm. At* $p = \frac{1}{2}$*, a fraction of* $k_A(\frac{1}{2}) = k_B(\frac{1}{2}) = \frac{(\lambda+r)g-(2r+\lambda)s}{\lambda(2s-g)}$ *is allocated to either risky arm, with the rest being allocated to the safe arm.*

Let $V_h(p) := pg + C_h(1-p)\Omega(p)^{\frac{r}{2\lambda}}$*, and* $V_l(p) := (1-p)g + C_l p\Omega(p)^{-\frac{r}{2\lambda}}$*. Then, the players' value function is continuously differentiable, and given by* $u(p) = \mathscr{W}(p)$ *in* $[1-p^\dagger, p^\dagger]$*, by* $u(p) = V_h(p)$ *in* $[p^\dagger, 1]$*, and* $u(p) = V_l(p)$ *in* $[0, 1-p^\dagger]$*, with the constants of integration* C_h *and* C_l *being determined by* $V_h(p^\dagger) = \mathscr{W}(p^\dagger)$ *and* $V_l(1-p^\dagger) = \mathscr{W}(1-p^\dagger)$*.*

Thus, no matter what initial prior belief players start out from, there is a positive probability that beliefs will end up at $p = \frac{1}{2}$, and hence they will try the risky project that looked initially less auspicious. Therefore, in contrast to the equilibrium for low stakes, there is a positive value attached to the option of having access to the second risky project.

4.5 Conclusion

I have analyzed a game of strategic experimentation with three-armed bandits, where the two risky arms are perfectly negatively correlated. In [7], I have shown that there exists an efficient equilibrium if and only if the stakes are high. Here, we have seen that any equilibrium in which at least one player's value is smooth involves complete learning if stakes are not low. If stakes are intermediate in size, all equilibria are inefficient, though they involve complete learning (provided both players' value functions are not kinked), as required by efficiency. If the stakes are low, all equilibria are inefficient, and there exists an equilibrium implying an inefficiently low amount of experimentation. In particular, if the stakes are low but not very low, there exists an equilibrium that involves incomplete learning while efficiency requires complete learning; if the stakes are very low, the efficient solution also implies incomplete learning.

From an economic point of view, the reason for the prevalence of free-riding in Markov perfect equilibrium when the types of the risky arms are perfectly positively correlated is as follows. If a player deviates by providing less effort than he is supposed to, the other players will be more optimistic than they should be as a result, and hence more willing to pick up the deviating player's slack. This makes players more inclined to free-ride. However, if players' risky arms are negatively correlated as in [8], it is impossible for both of them to be very pessimistic about their respective projects at the same time, and free-riding only appears if the players' respective single-agent cut-offs overlap. Otherwise, i.e., if the stakes are low, there exists an efficient equilibrium. By contrast, in our setting, there exists an efficient equilibrium if and only if the stakes are high [7], i.e. if and only if both players are always sufficiently optimistic about one of their projects. Otherwise, the positive correlation between players makes incentives for free-riding reappear.

4.6 Proofs

This section collects the proofs of our results. We note that player i's Bellman equation is given by (see [7])

$$u_i(p) = s + k_{j,A} B_A(p, u_i) + k_{j,B} B_B(p, u_i) + \max_{\{(k_{i,A},k_{i,B})\in[0,1]^2: k_{i,A}+k_{i,B}\leq 1\}} \left\{k_{i,A}\left[B_A(p,u_i) - c_A(p)\right] + k_{i,B}\left[B_B(p,u_i) - c_B(p)\right]\right\}, \tag{4.7}$$

where $\{j\} = \{1,2\}\backslash\{i\}$, $B_A(p,u) := \frac{\lambda}{r} p[g - u(p) - (1-p)u'(p)]$ and $B_B(p,u) := \frac{\lambda}{r}(1-p)[g - u(p) - pu'(p)]$ measure the learning benefit from playing arm A and arm B, respectively, while $c_A(p) := s - pg$ and $c_B(p) := s - (1-p)g$ measure the appertaining myopic opportunity cost of doing so. A myopic player (i.e. a player whose discount rate $r \to \infty$) would use risky arm A (B) if and only if $c_A(p) > 0$ ($c_B(p) > 0$), i.e., if and only if $p > p^m := \frac{s}{g}$ ($p < 1 - p^m$).

Furthermore, we note for future reference (see Appendix A in [7]) that, on any open interval of beliefs on which $((1,0),(1,0))$ is played, both players' value functions satisfy the ODE

$$2\lambda p(1-p)u'(p) + (2\lambda p + r)u(p) = (2\lambda + r)pg. \tag{4.8}$$

On any open interval of beliefs at which a player is indifferent between his safe arm and his risky arm A, his value function satisfies the ODE

$$\lambda p(1-p)u'(p) + \lambda p u(p) = (\lambda + r)pg - rs. \tag{4.9}$$

4.6.1 Proof of Lemma 4.1

In a first step, I show that s is a lower bound on u. Assume to the contrary that there exists a belief $p^\dagger \in]0,1[$ such that $u(p^\dagger) < s$. Then, since u is continuously differentiable and $u(0) = u(1) = g > s$, there exists a belief $\tilde{p} \in]0,1[$ such that $u(\tilde{p}) < s$ and $u'(\tilde{p}) = 0$. I write B_A and B_B for $B_A(p,u)$ and $B_B(p,u)$, respectively, suppressing arguments whenever this is convenient. Moreover, I define $\hat{B}_A(p) := \frac{\lambda}{r} p(g - s) > 0$ and $\hat{B}_B(p) := \frac{\lambda}{r}(1-p)(g-s) > 0$, while denoting by $(k_{j,A}, k_{j,B})$ the other player's action at $\tilde{p}$ in the equilibrium underlying the value function u. Now, at $\tilde{p}$, $u < s$ immediately implies $B_A = \frac{\lambda}{r}\tilde{p}(g - u) > \hat{B}_A$ and $B_B = \frac{\lambda}{r}(1-\tilde{p})(g-u) > \hat{B}_B$, and we have that

$$u - s \geq k_{j,A}(B_A - \hat{B}_A) + k_{j,B}(B_B - \hat{B}_B) = (k_{j,A}\tilde{p} + k_{j,B}(1-\tilde{p}))(s - u) \geq 0, \tag{4.10}$$

a contradiction to $u < s$.[9] Thus, we have already shown that u_1^* bounds u from below at all beliefs $p \leq p_1^*$.

Now, suppose there exists a belief $p > p_1^*$ at which $u < u_1^*$. I now write $B_A^* := \frac{\lambda}{r}p[g - u_1^* - (1-p)(u_1^*)'(p)] = u_1^* - pg$ and $B_B^* := \frac{\lambda}{r}(1-p)[g - u_1^* + p(u_1^*)'(p)]$. Since $B_A^* + B_B^* = \frac{\lambda}{r}(g - u_1^*)$, and hence $B_B^* = \frac{\lambda}{r}(g - u_1^*) - (u_1^* - pg)$, we have that $B_B^* \geq 0$ if and only if $u_1^* \leq \frac{\lambda + rp}{\lambda + r}g =: w_1(p)$. Let $\tilde{p}$ be defined by $w_1(\tilde{p}) = s$; it is straightforward to show that $\tilde{p} < p_1^*$. Noting furthermore that $u_1^*(p_1^*) = s$, $w_1(1) = u_1^*(1) = g$, and that w_1 is linear whereas u_1^* is strictly convex in p, we conclude that $u_1^* < w_1$ and hence $B_B^* > 0$ on $[p_1^*, 1[$. Moreover, since $B_A^* \geq 0$ (see [6]), we have $u_1^* = pg + B_A^* \leq pg + k_{j,B}B_B^* + (1 + k_{j,A})B_A^*$ on $[p^*, 1]$, for any $(k_{j,A}, k_{j,B})$.

Since s is a lower bound on u, by continuity, $u(p) < u_1^*(p)$ implies the existence of a belief strictly greater than p_1^* where $u < u_1^*$ and $u_1' \leq (u_1^*)'$. This immediately yields $B_A > B_A^* > c_A$, as well as

$$u - u_1^* \geq pg + k_{j,B}B_B + (1 + k_{j,A})B_A - [pg + (1 + k_{j,A})B_A^* + k_{j,B}B_B^*] \tag{4.11}$$

$$= k_{j,B}(B_A + B_B - B_A^* - B_B^*) + (1 + k_{j,A} - k_{j,B})(B_A - B_A^*) \tag{4.12}$$

$$= k_{j,B}\frac{\lambda}{r}(u_1^* - u_1) + (1 + k_{j,A} - k_{j,B})(B_A - B_A^*) > 0, \tag{4.13}$$

a contradiction.[10]

An analogous argument applies for u_2^*. □

4.6.2 *Proof of Proposition 4.2*

First, I show that $\hat{p}$ as defined in the proposition indeed exists and is unique in $]p_1^*, 1[$. It is immediate to verify that the left-hand side of the defining equation is decreasing, while the right-hand side is increasing in $\hat{p}$. Moreover, for $\hat{p} = p_1^*$, the left-hand side is strictly positive, while the right-hand side is zero. Now, for $\hat{p} \uparrow 1$, the left-hand side tends to $-\infty$, while the right-hand side is positive. The claim thus follows by continuity.

[9]Strictly speaking, the first inequality relies on the admissibility of the action (0, 0) at $\tilde{p}$. However, even if (0, 0) should not be admissible at $\tilde{p}$, my definition of strategies still guarantees the existence of a neighborhood of $\tilde{p}$ in which (0, 0) is admissible everywhere except at $\tilde{p}$. Hence, by continuous differentiability of u, there exists a belief $\tilde{\tilde{p}}$ in this neighborhood at which the same contradiction can be derived.

[10]Again, strictly speaking, the first inequality relies on the admissibility of the action (1, 0) at the belief in question, and my previous remark applies.

The proposed policies imply a well-defined law of motion for the posterior belief. It is immediate to verify that the function u satisfies value matching and smooth pasting at p_1^* and $1 - p_1^*$. To show that it is continuously differentiable, it remains to be shown that smooth pasting is satisfied at $\hat{p}$ and $1 - \hat{p}$. From the appertaining ODEs, we have that

$$\lambda \hat{p}(1 - \hat{p})u'(\hat{p}-) + \lambda \hat{p}u(\hat{p}) = (\lambda + r)\hat{p}g - rs \tag{4.14}$$

and

$$2\lambda \hat{p}(1 - \hat{p})u'(\hat{p}+) + (2\lambda \hat{p} + r)u(\hat{p}) = (2\lambda + r)\hat{p}g, \tag{4.15}$$

where I write $u'(\hat{p}-) := \lim_{p\uparrow\hat{p}} u'(p)$ and $u'(\hat{p}+) := \lim_{p\downarrow\hat{p}} u'(p)$. Now, $u'(\hat{p}-) = u'(\hat{p}+)$ if and only if $u(\hat{p}) = 2s - \hat{p}g$. Now, algebra shows that indeed $W(\hat{p}) = 2s - \hat{p}g$. By symmetry, we can thus conclude that $W(1-\hat{p}) = 2s-(1-\hat{p})g$ and that u is continuously differentiable. Furthermore, it is strictly decreasing on $]0, 1 - p_1^*[$ and strictly increasing on $]p_1^*, 1[$. Moreover, $u = s + 2B_B - c_B$ on $[0, 1 - \hat{p}]$, $u = s + k_B B_B$ on $[1 - \hat{p}, 1 - p_1^*]$, $u = s$ on $[1 - p_1^*, p_1^*]$, $u = s + k_A B_A$ on $[p_1^*, \hat{p}]$ and $u = s + 2B_A - c_A$ on $[\hat{p}, 1]$, which shows that u is indeed the players' payoff function from $((k_A, k_B), (k_A, k_B))$.

Consider first the interval $]1 - p_1^*, p_1^*[$. It has to be shown that $B_A - c_A < 0$ and $B_B - c_B < 0$. On $]1 - p_1^*, p_1^*[$, we have that $u = s$ and $u' = 0$, and therefore $B_A - c_A = \frac{\lambda+r}{r} pg - \frac{\lambda p + r}{r} s$. This is strictly negative if and only if $p < p_1^*$. By the same token, $B_B - c_B = \frac{\lambda+r}{r}(1 - p)g - \frac{\lambda(1-p)+r}{r} s$. This is strictly negative if and only if $p > 1 - p_1^*$.

Now, consider the interval $]p_1^*, \hat{p}[$. Here, $B_A = c_A$ by construction, as k_A is determined by the indifference condition and symmetry. It remains to be shown that $B_B \leq c_B$ here. Using the relevant differential equation, I find that $B_B = \frac{\lambda}{r}(g - u) + pg - s$. This is less than $c_B = s - (1 - p)g$ if and only if $u \geq \frac{\lambda+r}{\lambda} g - \frac{2r}{\lambda} s$. Yet, $\frac{\lambda+r}{\lambda} g - \frac{2r}{\lambda} s \leq s$ if and only if $\frac{g}{s} \leq \frac{2r+\lambda}{r+\lambda}$, so that the relevant inequality is satisfied. The interval $]1 - \hat{p}, 1 - p_1^*[$ is treated in an analogous way.

Finally, consider the interval $]\hat{p}, 1[$. Plugging in the relevant differential equation yields $B_A - B_B = u - pg - \frac{\lambda}{r}(g - u)$. This exceeds $c_A - c_B = (1 - 2p)g$ if and only if $u \geq \frac{\lambda+r(1-p)}{\lambda+r} g$. At $\hat{p}$, the indifference condition gives us $k_A(\hat{p}) = 1$, which implies $u(\hat{p}) = 2s - \hat{p}g$. Since $p \mapsto \frac{\lambda+r(1-p)}{\lambda+r} g$ is decreasing and u is increasing, it is sufficient for us to show that $u(\hat{p}) \geq \frac{\lambda+r(1-\hat{p})}{\lambda+r} g$, which is equivalent to $\hat{p} \leq \frac{\lambda+r}{\lambda}(2p^m - 1)$. From the indifference condition for the experimentation intensity $\tilde{k}_A(p) := \frac{u(p)-s}{c_A(p)}$, we see that $\tilde{k}_A$ is strictly increasing on $]p_1^*, p^m[$, and that $\lim_{p\uparrow p^m} \tilde{k}_A(p) = +\infty$; hence $\hat{p} < p^m$. Therefore, it is sufficient to show that $p^m \leq \frac{\lambda+r}{\lambda}(2p^m - 1)$, which is equivalent to $\frac{g}{s} \leq \frac{2r+\lambda}{r+\lambda}$. □

4.6.3 Proof of Proposition 4.3

The proposed policies imply a well-defined law of motion for the posterior belief. The function u is strictly decreasing on $]0, \frac{1}{2}[$ and strictly increasing on $]\frac{1}{2}, 1[$. Furthermore, as $\lim_{p\uparrow\frac{1}{2}} u'(p) = \lim_{p\downarrow\frac{1}{2}} u'(p) = 0$, the function u is continuously differentiable. Moreover, $u = s + 2B_B - c_B$ on $[0, 1 - p^\dagger]$, $u = s + k_B B_B$ on $[1 - p^\dagger, \frac{1}{2}]$, $u = s + k_A B_A$ on $[\frac{1}{2}, p^\dagger]$ and $u = s + 2B_A - c_A$ on $[p^\dagger, 1]$, which shows that u is indeed the players' payoff function from $((k_A, k_B), (k_A, k_B))$.

To establish existence and uniqueness of $p^\dagger$, note that $p \mapsto \frac{\lambda+r(1-p)}{\lambda+r} g$ and $p \mapsto 2s - pg$ are strictly decreasing in p, whereas $\mathscr{W}$ is strictly increasing in p on $]\frac{1}{2}, 1[$. Now, $\mathscr{W}(\frac{1}{2}) = \frac{r+\lambda}{\lambda} g - \frac{2r}{\lambda} s$. This is strictly less than $\frac{\lambda+\frac{r}{2}}{\lambda+r} g$ and $2s - \frac{g}{2}$ whenever $\frac{g}{s} < \frac{4(r+\lambda)}{2r+3\lambda}$. Moreover, $\mathscr{W}(\frac{1}{2})$ strictly exceeds $\frac{\lambda+r(1-p^m)}{\lambda+r} g = g - \frac{r}{r+\lambda} s$ and $2s - p^m g = s$ whenever $\frac{g}{s} > \frac{2r+\lambda}{r+\lambda}$. Thus, I have established uniqueness and existence of $p^\dagger$ and that $p^\dagger \in]\frac{1}{2}, p^m[$.

By construction, $u > \max\{\frac{\lambda+r(1-p)}{\lambda+r} g, 2s - pg\}$ in $]p^\dagger, 1]$, which, by Lemma A.1 in [7], implies that $((1, 0), (1, 0))$ are mutually best responses in this region; by the same token, $u > \max\{\frac{\lambda+rp}{\lambda+r} g, 2s - (1 - p)g\}$ in $[0, 1 - p^\dagger[$, which, by Lemma A.1 in [7], implies that $((0, 1), (0, 1))$ are mutually best responses in that region.

Now, consider the interval $]\frac{1}{2}, p^\dagger]$. Here, $B_A = c_A$ by construction, so all that remains to be shown is $B_B \leq c_B$. By plugging in the indifference condition for u', I get $B_B = \frac{\lambda}{r}(g - u) + pg - s$. This is less than $c_B = s - (1 - p)g$ if and only if $u \geq \frac{\lambda+r}{\lambda} g - \frac{2r}{\lambda} s = \mathscr{W}(\frac{1}{2}) = u(\frac{1}{2})$, which is satisfied by the monotonicity properties of u. An analogous argument establishes $B_A \leq c_A$ on $[1 - p^\dagger, \frac{1}{2}[$. □

References

1. Bellman, R.: A problem in the sequential design of experiments. Sankhya Indian J. Stat. (1933–1960) **16**(3/4), 221–229 (1956)
2. Bolton, P., Harris, C.: Strategic experimentation. Econometrica **67**, 349–374 (1999)
3. Bolton, P., Harris, C.: Strategic experimentation: the Undiscounted case. In: Hammond, P.J., Myles, G.D. (eds.) Incentives, Organizations and Public Economics – Papers in Honour of Sir James Mirrlees, pp. 53–68. Oxford University Press, Oxford (2000)
4. Bradt, R., Johnson, S., Karlin, S.: On sequential designs for maximizing the sum of n observations. Ann. Math. Stat. **27**, 1060–1074 (1956)
5. Gittins, J., Jones, D.: A dynamic allocation index for the sequential design of experiments. In: Progress in Statistics, European Meeting of Statisticians, 1972, vol. 1, pp. 241–266. North-Holland, Amsterdam (1974)
6. Keller G., Rady, S., Cripps, M.: Strategic experimentation with exponential bandits. Econometrica **73**, 39–68 (2005)
7. Klein, N.: Strategic learning in teams. Games Econ. Behav. **82**, 636–657 (2013)
8. Klein, N., Rady, S.: Negatively correlated bandits. Rev. Econ. Stud. **78**, 693–732 (2011)

9. Robbins, H.: Some aspects of the sequential design of experiments. Bull. Am. Math. Soc. **58**, 527–535 (1952)
10. Thompson, W.: On the likelihood that one unknown probability exceeds another in view of the evidence of two samples. Biometrika **25**, 285–294 (1933)

Chapter 5
Solution for a System of Hamilton–Jacobi Equations of Special Type and a Link with Nash Equilibrium

Ekaterina A. Kolpakova

Abstract The paper is concerned with systems of Hamilton–Jacobi PDEs of the special type. This type of systems of Hamilton–Jacobi PDEs is closely related with a bilevel optimal control problem. The paper aims to construct equilibria in this bilevel optimal control problem using the generalized solution for the system of the Hamilton–Jacobi PDEs. We introduce the definition of the solution for the system of the Hamilton–Jacobi PDEs in a class of multivalued functions. The notion of the generalized solution is based on the notions of minimax solution and M-solution to Hamilton–Jacobi equations proposed by Subbotin. We prove the existence theorem for the solution of the system of the Hamilton–Jacobi PDEs.

5.1 Introduction

The paper deals with a differential game, the dynamics of the game is entirely defined by the policy of the first player. The payoff functional of the first player is also determined by the control of the first player and the payoff functional of the second player depends on control of both players. Actually we investigate a bilevel optimal control problem. In considerable problem Nash equilibrium coincides with Stackelberg equilibrium [1, 5]. We restrict our attention to the case when the players use open-loop strategies and examine this problem applying the solution of the system of Hamilton–Jacobi equations.

The solution for a strongly coupled system of the Hamilton–Jacobi equations is open mathematical problem. For the general case there is no existence theorems. Furthermore the system of Hamilton–Jacobi equations is connected with the system of the quasilinear first order PDEs. The systems of quasilinear PDEs (the system

E. A. Kolpakova (✉)
Krasovskii Institute of Mathematics and Mechanics UrB RAS, Yekaterinburg, Russia

L. A. Petrosyan et al. (eds.), *Frontiers of Dynamic Games*,
Static & Dynamic Game Theory: Foundations & Applications,
https://doi.org/10.1007/978-3-319-92988-0_5

of conservation laws) describe many physical processes. If we differentiate the system of Hamilton–Jacobi equations w.r.t. phase variable x, then we obtain a system of quasilinear equations. The existence theorems for a generalized solution are obtained only for initial values with a small total variation [3, 8]. Using this link Bressan and Shen [3] constructed Nash strategies in the feedback for some non-zero sum two players differential game on the line. The authors do not solve the system of Hamilton–Jacobi equations, but they solve the corresponding strictly hyperbolic system of quasilinear PDEs. This way can be applied only in the case of the scalar phase variable and the hyperbolic system of quasilinear equations. Analogous constructions for a differential game with simple motions were described in [4].

As we mentioned above the theory of the system of Hamilton–Jacobi equations is open mathematical problem, at the same time the theory of generalized solution for the single Hamilton–Jacobi equation is well-developed. Subbotin proposed the notion of minimax solution, he proved the existence and uniqueness theorems [13]. Crandall et al. introduced the viscosity approach [6]. Moreover Subbotin proved the equivalence of these approaches.

In the paper we consider the systems of Hamilton–Jacobi equations where the first equation of the system does not depend on the solution of the second equation, and the second equation depends on partial derivatives of the solution for the first equation. This implies that we can solve the equations of the system sequentially. This system is connected with a bilevel optimal control problem [16]. Using the minimax/viscosity approach we obtain the solution of the first equation of the system. Further we substitute the derivative of the minimax/viscosity solution in the second equation. The second equation is solved in the framework of M-solutions [9].

Our main result is the following. We show that the solution for the system of Hamilton–Jacobi equations of special type belongs to a class of multivalued map. We construct this multivalued solution and connects with a Nash equilibrium in a bilevel optimal control problem.

5.2 Bilevel Optimal Control Problem

A bilevel optimal control problem is a particle case of two-level differential games. Let us consider the bilevel optimal control problem with dynamics

$$\dot{x} = f(t, x, u), \ x(t_0) = x_0, \ u \in U \subset \mathbb{R}^n. \tag{5.1}$$

Here $t \in [0, T]$, $x \in \mathbb{R}^n$. The players maximize payoff functionals I_1, I_2:

$$I_1(u(\cdot)) = \sigma_1(x(T)) + \int_{t_0}^{T} g_1(t, x(t), u(t))dt,$$

$$I_2(u(\cdot), v(\cdot)) = \sigma_2(x(T)) + \int_{t_0}^{T} g_2(t, x(t), u(t), v(t))dt.$$

Here u and v are controls of the players. Assume that $U, V \subset R^n$ are compact sets. Denote the set of all measurable controls of the first player by $\tilde{U}$:

$$\tilde{U} = \{u : [t_0, T] \to U, u \text{ are measurable functions}\},$$

and the set of all measurable controls of the second player by $\tilde{V}$:

$$\tilde{V} = \{v : [t_0, T] \to V, v \text{ are measurable functions}\}.$$

From [7, 12] it follows that the payoffs of the players satisfy the system of the Hamilton–Jacobi equations:

$$\frac{\partial c}{\partial t} + H_1(t, x, p) = 0, \;\; c(T, x) = \sigma_1(x); \tag{5.2}$$

$$\frac{\partial w}{\partial t} + H_2(t, x, p, q) = 0, \;\; w(T, x) = \sigma_2(x), \tag{5.3}$$

under condition

$$H_1(t, x, p) = \max_{u \in U} \langle f(t, x, u), p\rangle + g_1(t, x, u)$$

$$= \langle f(t, x, u^*(t, x, p)), p\rangle + g_1(t, x, u^*(t, x, p)),$$

$$H_2(t, x, q) = \langle f\Big(t, x, u^*\Big(t, x, \frac{\partial c(t, x)}{\partial x}\Big)\Big), q\rangle + \max_{v \in V} g_2\Big(t, x, u^*\Big(t, x, \frac{\partial c(t, x)}{\partial x}\Big), v\Big).$$

Here

$$u^*(t, x, p) \in \arg\max_{u \in U} \langle f(t, x, u), p\rangle + g_1(t, x, u), \tag{5.4}$$

$p = \frac{\partial c}{\partial x}, q = \frac{\partial w}{\partial x}$.

Further we shall assume that

A1. the function $H_1 : [0, T] \times \mathbb{R}^n \times \mathbb{R}^n \to \mathbb{R}$ is continuously differentiable, H_1 satisfies sublinear condition w.r.t. x, p, the function H_1 is strongly convex w.r.t. p for any $(t, x) \in [0, T] \times \mathbb{R}^n$.
A2. the function σ_1 is Lipschitz continuous.
A3. the function $H_2 : [0, T] \times \mathbb{R}^n \times \mathbb{R}^n \times \mathbb{R}^n \to \mathbb{R}$ is continuously differentiable, H_2 satisfies sublinear condition w.r.t. x, p, q, the function H_2 is strongly convex w.r.t. q for any $(t, x) \in [0, T] \times \mathbb{R}^n$.
A4. the function σ_2 is continuously differentiable.

From assumptions $A1$, $A3$ we get

$$g_1(t, x, p) = H_1^*\Big(t, x, \frac{\partial H_1(t, x, p)}{\partial p}\Big),$$

$$g_2(t, x, p, q) = H_2^*\Big(t, x, \frac{\partial H_1}{\partial p}, \frac{\partial H_2(t, x, p, q)}{\partial q}\Big).$$

Here H_1^*, H_2^* are conjugate functions to H_1, H_2, $\frac{\partial H_1}{\partial p} = \Big(\frac{\partial H_1}{\partial p_1}, \dots, \frac{\partial H_1}{\partial p_n}\Big)$. Hence g_1, g_2 are continuous functions w.r.t all variables. Since condition $A1$ holds a measurable function (5.4) $u^* : (t, x, p) \to U$ is well-defined.

Let us introduce the mapping

$$\begin{aligned}(t_0, x_0) \to \xi(t_0, x_0) = \{\xi \in \mathbb{R}^n : \tilde{x}(t_0, \xi) = x_0, \tilde{x}(T, \xi) = \xi, \\ \tilde{s}(T, \xi) = D_x\sigma_1(\xi),\ \tilde{z}(T, \xi) = \sigma_1(\xi),\ \tilde{z}(t_0, \xi) = c(t_0, x_0)\}\end{aligned} \tag{5.5}$$

Here $(\tilde{x}(\cdot), \tilde{s}(\cdot), \tilde{z}(\cdot))$ is the unique and extendable solution of the characteristic system for Bellman equation (5.2):

$$\dot{\tilde{x}} = \frac{\partial H_1(t, \tilde{x}, \tilde{s})}{\partial \tilde{s}},\ \dot{\tilde{s}} = -\frac{\partial H_1(t, \tilde{x}, \tilde{s})}{\partial \tilde{x}},\ \dot{\tilde{z}} = \langle \frac{\partial H_1(t, \tilde{x}, \tilde{s})}{\partial \tilde{s}}, \tilde{s}\rangle - H_1(t, \tilde{x}, \tilde{s})$$

with a boundary condition

$$\tilde{x}(T, \xi) = \xi,\ \tilde{s}(T, \xi) = D_x\sigma_1(\xi),\ \tilde{z}(T, \xi) = \sigma_1(\xi),\ \xi \in \mathbb{R}^n.$$

It follows from [11, 15] that for any point $(t_0, x_0) \in [0, T] \times \mathbb{R}^n$ assumption $A1$ guarantees the existence of optimal open-loop control $u^0(\cdot; t_0, x_0)$ satisfying the relation

$$\max_{u(\cdot) \in \tilde{U}} I_1(u(\cdot)) = I_1(u^0(\cdot; t_0, x_0)) = c(t_0, x_0).$$

Pontryagin's Maximum principle implies that the optimal open-loop control $u^0(\cdot; t_0, x_0)$ of the first player for the initial point $(t_0, x_0) \in [0, T] \times \mathbb{R}^n$ is defined by the rule

$$u^0(t; t_0, x_0) \in \arg\max_{u \in U}[\langle \tilde{s}(t, \xi_0), f(t, \tilde{x}(t, \xi_0), u)\rangle + g_1(t, \tilde{x}(t, \xi_0), u)], \ \forall t \in [t_0, T] \tag{5.6}$$

Here $(\tilde{x}(\cdot), \tilde{s}(\cdot), \tilde{z}(\cdot))$ is the solution of the characteristic system for problem (5.2) for any $t \in [t_0, T]$, for any $\xi_0 \in \xi(t_0, x_0)$ defining by (5.5).

We determine the set of optimal open-loop controls of the first player

$$U^0(t_0, x_0) = \Big\{u(\cdot) : [t_0, T] \to U \text{are measurable functions, satisfying (5.6)}\Big\}.$$

Remark 5.1 Equivalently the first player's control can be considered in feedback strategies [14]. In this case the optimal feedback is given by

$$u(t, x) \in \arg\max_{u \in U}\left[\frac{dc(t, x)}{d(1, f(t, x, u))} + g_1(t, x, u)\right],$$

where c is the solution of Cauchy problem (5.2), $\frac{dc(t,x)}{d(1,f(t,x,u))}$ is the derivative of c at the point (t, x) in the direction $(1, f(t, x, u))$.

5.3 The Solution of the System of the Hamilton–Jacobi Equations

In this section we will focus on solution of system of Hamilton–Jacobi equations (5.2), (5.3). We begin with definition of a minimax/viscosity solution of Cauchy problem (5.2).

Definition 5.1 The continuous function $c : [0, T] \times \mathbb{R}^n \to \mathbb{R}$ is said to be the minimax/viscosity solution if $c(T, x) = \sigma_2(x)$, $x \in \mathbb{R}^n$ and the following inequalities hold for any $(t, x) \in (0, T) \times \mathbb{R}^n$

$$\alpha + H_1(t, x, \beta) \le 0, \quad (\alpha, \beta) \in D^-c(t, x),$$

$$\alpha + H_1(t, x, \beta) \ge 0, \quad (\alpha, \beta) \in D^+c(t, x).$$

Here $D^-c(t, x)$ and $D^+c(t, x)$ are sub- and superdifferentials of function c at a point (t, x).

It is known from [13] that under conditions $A1$–$A3$ there exists the unique minimax solution $c(\cdot, \cdot)$ in problem (5.2).

We recall the properties of the minimax solution of problem (5.2) under conditions $A1$,A_2 from [13, 14]:

1. the minimax solution $c(\cdot,\cdot)$ is a locally Lipschitz function;
2. the superdifferential of the minimax solution $D^+c(t,x) \neq \emptyset$ for any point $(t,x) \in [0,T] \times \mathbb{R}^n$.

We solve the system of Hamilton–Jacobi equations sequentially. The minimax solution of the first equation (5.2) is a Lipschitz continuous function. Thus the partial derivative of the minimax solution can be discontinuous w.r.t. x. We substitute the superdifferential $D_x^+c(\cdot,\cdot)$ of function c for p in the second equation (5.3), therefore we obtain the multivalued Hamiltonian

$$\tilde{H}(t,x,q) = H_2(t,x,D_x c(t,x),q). \tag{5.7}$$

Hence, we have the Hamilton–Jacobi equation with the multivalued Hamiltonian:

$$\frac{\partial w}{\partial t} + \tilde{H}(t,x,q) = 0,\ w(T,x) = \sigma_2(x). \tag{5.8}$$

A.I. Subbotin proposed the notion of M-solution for Cauchy problem (5.8) with the multivalued Hamiltonian relative to x.

Consider the differential inclusion

$$(\dot{x},\dot{z}) \in E(t,x,q),\ E(t,x,q) = \{(f,g) : f \in \frac{\partial H_2(t,x,p,q)}{\partial q},\ p \in D^+c(t,x),$$
$$\langle f,q\rangle - g \in [H_{2*}(t,x,q), H_2^*(t,x,q)], q \in \mathbb{R}^n\}. \tag{5.9}$$

Here $\frac{\partial H_2(t,x,p,q)}{\partial q} = \left(\frac{\partial H_2(t,x,p,q)}{\partial q_1},\ldots,\frac{\partial H_2(t,x,p,q)}{\partial q_n}\right)$,

$$H_{2*}(t,x,q) = \lim \inf_{(\tau,\xi)\to(t,x)} \tilde{H}(\tau,\xi,q),\ H_2^*(t,x,q) = \lim \sup_{(\tau,\xi)\to(t,x)} \tilde{H}(\tau,\xi,q). \tag{5.10}$$

It follows from [10] that differential inclusion (5.9) is an admissible characteristical inclusion. Recall some definitions and theorem from the work [9].

Definition 5.2 The closed set $W \subset [0,T] \times \mathbb{R}^n \rightrightarrows \mathbb{R}$ is viable w.r.t. differential inclusion (5.9), if for any point $(t_0,x_0,z_0) \in W$ there exist $\tau > 0$ and a trajectory $(x(\cdot),z(\cdot))$ of admissible differential inclusion (5.9) such that $(x(0),z(0)) = (x_0,z_0)$, $(t,x(t),z(t)) \in W$ for any $t \in [0,\tau]$.

Definition 5.3 The closed maximal set $W \subset [0,T] \times \mathbb{R}^n \rightrightarrows \mathbb{R}$ is called the M-solution of Cauchy problem for Hamilton–Jacobi equation (5.8), if W is viable w.r.t. admissible differential inclusion (5.9) and satisfies the condition

$$(T,x,z) \in W \Rightarrow z = \sigma_2(x)\ \forall\, x \in \mathbb{R}^n.$$

Definition 5.4 The closed set $W \subset [0, T] \times \mathbb{R}^n \times \mathbb{R}$ is said to be the epi-solution (hypo-solution) of problem (5.8) if W is viable w.r.t. admissible differential inclusion (5.9) and satisfies the condition

$$(T, x, z) \in W \Rightarrow z \geq \sigma_2(x)((T, x, z) \in W \Rightarrow z \leq \sigma_2(x)) \; \forall \, x \in \mathbb{R}^n.$$

We introduce the definition for a generalized solution of the system of the Hamilton–Jacobi equations.

Definition 5.5 The multivalued map (c, w), where $c(\cdot, \cdot) : [0, T] \times \mathbb{R}^n \to \mathbb{R}$, $w : [0, T] \times \mathbb{R}^n \rightrightarrows \mathbb{R}$ is called a generalized solution of Cauchy problem for the system of Hamilton–Jacobi equations (5.2), (5.3), if the function $c(\cdot, \cdot)$ is the minimax solution of problem (5.2), the map $w(\cdot, \cdot)$ is the M-solution of problem (5.8).

Theorem 5.1 ([10]) *Let $w : [0, T] \times \mathbb{R}^n \to \mathbb{R}$ be a multivalued map and gr w is closed set. Suppose that $w(t, x)$ is not empty for $t \in [0, T], x \in \mathbb{R}^n$ and put*

$$w_*(t, x) = \min_{z \in w(t,x)} z > -\infty, \; w^*(t, x) = \max_{z \in w(t,x)} z < \infty.$$

The map w is the M-solution of problem (5.8) iff epi w_* and hypo w^* are the M-solutions of problem (5.8).

Given $t \in [t_0, T], x \in \mathbb{R}^n, u \in U$ let

$$(t, x, u) \to Q(t, x, u) = \arg\max_{v \in V} g_2(t, x, u, v). \tag{5.11}$$

be the set of optimal controls of the second player. Consider the map $\Gamma(u(\cdot)) : \tilde{U} \to \mathbb{R}$ given by the following rule

$$u(\cdot) \to \sigma_2(x[T; t_0, x_0]) + \int_{t_0}^{T} g_2(t, x[t; t_0, x_0], u(t), Q(t, x[t; t_0, x_0], u(t)))dt, \tag{5.12}$$

$u(\cdot) \in U^0(t_0, x_0)$, the function $x[\cdot; t_0, x_0]$ is a solution of the problem

$$\dot{x} = f(t, x, u(t)), \; u(\cdot) \in U^0(t_0, x_0), x(t_0) = x_0. \tag{5.13}$$

Put

$$w(t_0, x_0) = \bigcup_{u(\cdot) \in U^0(t_0, x_0)} \Gamma(u) \tag{5.14}$$

Lemma 5.1 *Map (5.14) is compact-valued.*

Proof Let us choose $w_i = \Gamma(u_i(\cdot)) \in w(t_0, x_0)$. We show that if $w_i \to w_0$, $i \to \infty$, then $w_0 \in w(t_0, x_0)$.

Let us define the set generalized controls

$$\Lambda = \{\mu : [t_0, T] \times U \to [0, +\infty) \text{ is measurable },$$

$\forall\, [\tau_1, \tau_2] \subset [0, T] \quad \mu([\tau_1, \tau_2] \times U) = \tau_2 - \tau_1, \}$. Here λ is Lebesgue measure on $[0, T]$. Hence the trajectory $x(\cdot)$ under control μ has the form

$$x(t) = x_0 + \int\limits_{[t_0,t]\times U} f(\tau, x(\tau), u)\mu(d(\tau, u)).$$

In this case the first player's outcome is

$$I_1(\mu) = \sigma_1(x(T)) + \int\limits_{[t_0,t]\times U} g_1(\tau, x(\tau), u)\mu(d(\tau, u)).$$

We consider the set of generalized optimal controls

$$M_{t_0} = \{\mu \in \Lambda : \mu \text{ maximizes } I_1(\mu)\}.$$

It is known from [15] that the set M_{t_0} is a compact metric set. Now we show the link between $U^0(t_0, x_0)$ and M_{t_0}. If $u(\cdot) \in U^0(t_0, x_0)$ then there exists $\mu_{u(\cdot)} \in M_{t_0}$ such that

$$\forall\, \varphi \in C([0, T] \times U) \qquad \int\limits_{[0,T]\times U} \varphi(t, u)\mu_{u(\cdot)}(d(t, u)) = \int\limits_0^T \varphi(t, u(t))dt.$$

Hence from $u_i \in U^0(t_0, x_0)$ we obtain $\mu_i = \mu_{u_i(\cdot)} \in M_{t_0}$. Consider $\mu_i \to \mu^*$ as $i \to \infty$. Since M_{t_0} is a closed set we get $\mu^* \in M_{t_0}$. Let us construct $u^* \in U^0(t_0, x_0)$ such that $\mu^* = \mu_{u^*(\cdot)}$.

We have

$$\lim_{i\to\infty} w_i = \lim_{i\to\infty} \Gamma(u_i(\cdot)) = \Gamma(u^*(\cdot)) = w_0.$$

Hence $w_0 = \Gamma(u^*(\cdot)) \in w(t_0, x_0)$. Since $\Gamma(u)$ is bounded on the set $U^0(t_0, x_0)$ it follows that $w(t, x_0)$ is bounded.

We prove the following theorem.

Theorem 5.2 *If conditions A1–A4 hold, then the multivalued map w, defining (5.14) is the M-solution of problem (5.8).*

Proof Put

$$w^*(t_0, x_0) = \max_{y \in w(t_0, x_0)} y,$$

where w is defined by (5.14). Let us show that hypograph w^* is viable w.r.t. differential inclusion (5.9).

We fix the position $(t_0, x_0) \in [0, T] \times \mathbb{R}^n$. Choose $(t_0, x_0, z_0) \in$ hypo w^*, $z_0 \leq w^*(t_0, x_0)$. If assumptions $A1$–$A3$ are true, then in the optimal control problem with payoff functional I_1 there exists an optimal open-loop control u^* in the class of measurable functions. And control u^* generates the trajectory ξ:

$$\dot{\xi} = f(t, \xi, u^*(t)), \quad \xi(t_0) = x_0.$$

The choice of point z_0 and Bellman's optimality principle yield the equality

$$z_0 \leq w^*(t_0, x_0) = w^*(t, \xi(t))$$

$$+\int_{t_0}^{t} g_2(\tau, \xi[\tau; t_0, x_0], u^*(\tau), Q(\tau, \xi[\tau; t_0, x_0], u^*(\tau))d\tau.$$

Further we have

$$z_0 - \int_{t_0}^{t} g_2(\tau, \xi[\tau; t_0, x_0], u^*(\tau), Q(\tau, \xi[\tau; t_0, x_0], u^*(\tau)))d\tau \leq w^*(t, \xi(t))$$

for any $t \in [t_0, T]$. Note that

$$z(t) = z_0 - \int_{t_0}^{t} g_2(\tau, \xi[\tau; t_0, x_0], u^*(\tau), Q(\tau, \xi[\tau; t_0, x_0], u^*(\tau)))d\tau,$$

hence the trajectory $(\xi(\cdot), z(\cdot))$ satisfies to differential inclusion (5.9). From definition of the Hamiltonian H_2 it follows that

$$g = \dot{z} = -g_2(t, \xi(t), u^*(t), Q(t, \xi(t), u^*(t))), \ \langle f(t, \xi(t), u^*(t), p\rangle - g$$

$$= \langle f(t, \xi(t), u^*(t), p\rangle + g_2(t, \xi(t), u^*(t), Q(t, \xi(t), u^*(t))) \in$$

$$[H_{2*}(t, \xi(t), p), H_2^*(t, \xi(t), p)].$$

Hence $(t, \xi(t), z(t)) \in$ hypo $w^*(t, \xi(t)), t \in [t_0, T]$. Therefore hypo w^* is a closed set, satisfying the definition of the hypo-solution.

Put

$$w_*(t_0, x_0) = \min_{y \in w(t_0, x_0)} y,$$

where w is defined by (5.14). We choose a point $(t_0, x_0, z_0) \in \text{epi}\, w_*$, $z_0 \geq w_*(t_0, x_0)$. Let us consider the optimal trajectory $\xi(\cdot)$ of dynamical system (5.1), generated by control u_* and satisfying to initial condition $\xi(t_0) = x_0$. Since $\xi(\cdot)$ is the optimal trajectory we have

$$w_*(t, \xi(t)) + \int_{t_0}^{t} g_2(\tau, \xi[\tau; t_0, x_0], u_*(\tau), Q(\tau, \xi[\tau; t_0, x_0], u_*(\tau)))d\tau$$

$= w_*(t_0, x_0) \leq z_0$. Therefore

$$w_*(t, \xi(t)) \leq z_0$$

$$- \int_{t_0}^{t} g_2(\tau, \xi[\tau; t_0, x_0], u_*(\tau), Q(\tau, \xi[\tau; t_0, x_0], u_*(\tau)))d\tau = z(t),$$

that is the trajectory $(\xi(\cdot), z(\cdot))$ lies in the epigraph w_*. We show that $z(\cdot)$ is a solution of differential inclusion (5.9). Really

$$g = \dot{z} = -g_2(t, \xi(t), u_*(t), Q(t, \xi(t), u_*(t))), \ \langle f(t, \xi(t), u_*(t), p\rangle - g$$

$$= \langle f(t, \xi(t), u_*(t), p\rangle + g_2(t, \xi(t), u_*(t), Q(t, \xi(t), u_*(t))) \in$$

$$[H_{2*}(t, \xi(t), p), H_2^*(t, \xi(t), p)].$$

Consequently epi w_* is a closed set, satisfying the definition of the epi-solution.

Using Theorem 5.1 we obtain epi $w_* \bigcap$ hypo w^* is the M-solution of problem (5.8). We note that epi $w_*(T, x) \bigcap$ hypo $w^*(T, x) = \sigma_2(x)$, $x \in \mathbb{R}^n$.

Remark 5.2 We have proved that multivalued map (5.14) is the M-solution of problem (5.8). From definition 5.3 the M-solution is maximal-valued. Let us assume that there exist two M-solutions W and W' of problem (5.8). Then we have inclusions $W \subseteq W'$ and $W' \subseteq W$. Hence $W = W'$ and the M-solution is unique.

5.4 Design of Nash Equilibrium

Let us recall the definition of a Nash equilibrium in program strategies.

Definition 5.6 ([2]) A couple of strategies $(\bar{u}(\cdot), \bar{v}(\cdot))$ is a Nash equilibrium in two persons differential game if following inequalities hold for any $u(\cdot) \in \tilde{U}$, $v(\cdot) \in \tilde{V}$

$$\sigma_1(\bar{x}(T)) + \int_{t_0}^{T} g_1(t, \bar{x}(t), \bar{u}(t))dt \geq \sigma_1(x^{[1]}(T)) + \int_{t_0}^{T} g_1(t, x^{[1]}(t), u(t))dt,$$

$$\sigma_2(\bar{x}(T)) + \int_{t_0}^{T} g_2(t, \bar{x}(t), \bar{u}(t), \bar{v}(t))dt \geq \sigma_2(\bar{x}(T)) + \int_{t_0}^{T} g_2(t, \bar{x}(t), \bar{u}(t), v(t))dt,$$

$t \in [t_0, T]$, where

$$\dot{\bar{x}}(t) = f(t, \bar{x}(t), \bar{u}(t)),\ \dot{x}^{[1]}(t) = f(t, x^{[1]}(t), u(t)),\ \bar{x}(t_0) = x^{[1]}(t_0) = x_0.$$

Let us define the control $\bar{u}(\cdot)$ by formula (5.6). The control $\bar{u}(\cdot)$ maximizes the functional I_1 for optimal control problem (5.1), and therefore the first inequality holds in Definition 5.6.

Let $\bar{v}(\cdot)$ be given by

$$\bar{v}(t) \in \arg\max_{v \in V}\{g_2(t, \bar{x}(t), \bar{u}(t), v)\}, t \in [t_0, T], \tag{5.15}$$

where $\bar{x}(\cdot)$ is a solution of problem $\dot{\bar{x}}(t) = f(t, \bar{x}(t), \bar{u}(t))$, $\bar{x}(t_0) = x_0$. Since g_2 is a continuous function w.r.t. all variables, $\bar{x}(\cdot)$ is a differentiable function and $\bar{u}(\cdot)$ is measurable function we see that $g_2(\cdot, \bar{x}(\cdot), \bar{u}(\cdot), v)$ is a measurable function w.r.t. t and multivalued map $G(t) = \{g_2(t, \bar{x}(t), \bar{u}(t), v) : v \in V\}, t \in [t_0, T]$ is measurable w.r.t. t. The multivalued map $Gm(t) = \max_{v \in V} g_2(t, \bar{x}(t), \bar{u}(t), v)$ is upper semicontinuous therefore this map is measurable w.r.t t. Using this fact and Casteing's theorem [15], we get the map

$$\arg\max_{v \in V} g_2(\cdot, x(\cdot), \bar{u}(\cdot), v) : [t_0, T] \rightrightarrows V$$

is measurable. Hence from Neiman–Aumann–Casteing's theorem [15] the measurable multivalued map has a measurable selector $\bar{v}(\cdot) : [t_0, T] \to \mathbb{R}^n$.

By the definition $\bar{v}$ (5.15) the second inequality for integral parts holds in Definition 5.6.

Hence the couple of strategies $(\bar{u}, \bar{v})$ provides a Nash equilibrium. The first player solves the optimal control problem and the payoff does not depend on behavior of the second player. Choosing the control $\bar{u}(\cdot; t_0, x_0)$, the first player will obtain a payoff $c(t_0, x_0)$. We shall show how the choice of the control of the first player influences on the payoff of the second player.

Remark 5.3 Let us fix the point $(t_0, x_0) \in [0, T] \times \mathbb{R}^n$. Let (c, w) be the generalized solution of problem (5.2), (5.3), $\alpha \in w(t_0, x_0)$, then there exists a couple of Nash equilibrium strategies (u^*, v^*):

$$u^*(t) \in \arg \max_{u(\cdot)\in U^0(t_0,x_0)} \Gamma(u), \;\; v^*(t) = Q(t, x[t; t_0, x_0]), u^*(t)),$$

Γ is defined by (5.12), $x^*[\cdot; t_0, x_0]$ satisfies (5.1). From $A3$ we can use arg max instead of arg sup. The corresponding payoffs of players at the point $(t_0, x_0) \in [0, T] \times \mathbb{R}^n$ equal to $(c(t_0, x_0), \alpha)$.

5.5 Example

Let us consider the optimal control problem

$$\dot{x} = u, \;\; x(t_0) = x_0,$$

$x \in \mathbb{R}$, $t \in [0, T]$, $|u| \leq 1$, $|v| \leq 1$. Leader maximizes the payoff functional

$$I_1(u(\cdot)) = |x(T)| - \int_{t_0}^{T} \frac{u^2}{2} dt \to \max,$$

and the follower maximizes payoff functional

$$I_2(u(\cdot), v(\cdot)) = x(T) - \int_{t_0}^{T} v^2 + uvdt \to \max.$$

The system of Hamilton–Jacobi equations has the form

$$\frac{\partial c}{\partial t} + \max_{u\in U}[pu - \frac{u^2}{2}] = 0, \;\; c(T, x) = |x|,$$

$$\frac{\partial w}{\partial t} + qu^0(t, x, p) + \max_{v\in V}[-v^2 - u^0(t, x, p)v] = 0, w(T, x) = x,$$

$x \in \mathbb{R}$, $t \in [0, T]$, $p = \frac{\partial c}{\partial x}$, $q = \frac{\partial w}{\partial x}$. Using formula (5.4) we obtain

$$u^0(t, x, p) = \begin{cases} p, \text{ if } |p| \leq 1, \\ 1, \text{ if } p > 1, \\ -1, \text{ if } p < -1. \end{cases}$$

Lax–Hopf formula yields the solution of the first Hamilton–Jacobi equation

$$c(t, x) = |x| - 1/2(t - T).$$

Now by formula (5.6) the open-loop control of the leader

$$u^0(t; t_0, x_0) = \begin{cases} 1, & \text{if } x_0 > 0, \\ -1, & \text{if } x_0 < 0, \\ \{-1, 1\}, & \text{if } x_0 = 0. \end{cases}$$

Applying (5.11) we construct the map Q

$$Q(u) = -\frac{u}{2}.$$

Hence the open-loop control of the follower

$$v^0(t; t_0, x_0) = \begin{cases} -\frac{1}{2}, & \text{if } x_0 > 0, \\ \frac{1}{2}, & \text{if } x_0 < 0, \\ \left[-\frac{1}{2}, \frac{1}{2}\right], & \text{if } x_0 = 0. \end{cases}$$

Further we construct M-solution of the second equation

$$w(t, x) = \begin{cases} x + \frac{3}{4}(t - T), & \text{if } x < 0, \\ x - \frac{5}{4}(t - T), & \text{if } x > 0, \\ \left\{x + \frac{3}{4}(t - T), x - \frac{5}{4}(t - T)\right\}, & \text{if } x = 0. \end{cases}$$

We see that the solution of the second Hamilton–Jacobi equation is multivalued under $x = 0$.

The payoffs of the players at the point $(t_0, x_0) \in [0, T] \times \mathbb{R}^n$ equal to $(|x_0| - 1/2(t_0 - T), \alpha)$, where $\alpha \in w(t_0, x_0)$.

Acknowledgements This work was supported by the Russian Fond of Fundamental Researches under grant No. 17-01-00074.

References

1. Averboukh, Y., Baklanov A.: Stackelberg solutions of differential games in the class of nonanticipative strategies. Dyn. Games Appl. **4**(1), 1–9 (2014)
2. Basar, T., Olsder, G.J.: Dynamic Noncooperative Game Theory. SIAM, Philadelphia (1999)
3. Bressan, A., Shen, W.: Semi-cooperative strategies for differential games. Int. J. Game Theory **32**, 1–33 (2004)
4. Cardaliaguet, P., Plaskacz, S.: Existence and uniqueness of a Nash equilibrium feedback for a simple nonzero-sum differential game. Int. J. Game Theory **32**, 33–71 (2003)
5. Chen, C.I., Cruz, J.B., Jr.: Stackelberg solution for two-person games with biased information patterns. IEEE Trans. Autom. Control **6**, 791–798 (1972)
6. Crandall, M.G., Ishii, H., Lions, P.-L.: User's guide to viscosity solutions of second order partial differential equations. Bull. Am. Math. Soc. (N.S.) **27**, 1–67 (1992)
7. Friedman, A.: Differential Games. Wiley, Hoboken (1971)
8. Glimm, J.: Solutions in the large for nonlinear systems of equations. Commun. Pure Appl. Math. **18**, 697–715 (1965)
9. Lakhtin, A.S., Subbotin, A.I.: Multivalued solutions of first-order partial differential equations. Sb. Math. **189**(6), 849–873 (1998)
10. Lakhtin, A.S., Subbotin, A.I.: The minimax and viscosity solutions in discontinuous partial differential equations of the first order. Dokl. Akad. Nauk **359**, 452–455 (1998)
11. Pontryagin, L.S., Boltyanskii, V.G., Gamkrelidze, R.V., Mishchenko, E.F.: The Mathematical Theory of Optimal Processes. Pergamon Press, New York (1964)
12. Starr, A.W., Ho, Y.C.: Non-zero sum differential games. J. Optim. Theory Appl. **3**(3), 184–206 (1969)
13. Subbotin, A.I.: Generalized Solutions of First-Order PDEs: The Dynamical Optimization Perspectives. Birkhauser, Boston (1995)
14. Subbotina, N.N.: The method of characteristics for Hamilton–Jacobi equations and applications to dynamical optimization. J. Math. Sci. **135**(3), 2955–3091 (2006)
15. Warga, J.: Optimal Control of Differential and Functional Equations. Academic Press, New York (1972)
16. Ye, J.J.: Optimal strategies for bilevel dynamic problems. SIAM J. Control Optim. **35**(2), 512–531 (1997)

Chapter 6
The Impact of Discounted Indices on Equilibrium Strategies of Players in Dynamical Bimatrix Games

Nikolay Krasovskii and Alexander Tarasyev

Abstract The paper deals with construction of solutions in dynamical bimatrix games. It is assumed that integral payoffs are discounted on the infinite time interval. The dynamics of the game is subject to the system of differential equations describing the behavior of players. The problem of construction of equilibrium trajectories is analyzed in the framework of the minimax approach proposed by N. N. Krasovskii and A. I. Subbotin in the differential games theory. The concept of dynamical Nash equilibrium developed by A. F. Kleimenov is applied to design the structure of the game solution. For obtaining constructive control strategies of players, the maximum principle of L. S. Pontryagin is used in conjunction with the generalized method of characteristics for Hamilton–Jacobi equations. The impact of the discount index is indicated for equilibrium strategies of the game.

6.1 Introduction

The dynamical bimatrix game with discounted integral payoff functionals is considered on the infinite horizon. Usually the discount parameter appears to be very uncertain value which reflects subjective components in economic and financial models. In this case models with discounted indices require an implementation of sensitivity analysis for solutions with respect to changing of the discount parameter. In the paper we build optimal control strategies based on Krasovskii minimax approach [10, 11], using constructions of Pontryagin maximum principle [21] and Subbotin technique of method of characteristics for generalized (minimax)

N. Krasovskii (✉)
Krasovskii Institute of Mathematics and Mechanics UrB RAS, Yekaterinburg, Russia
e-mail: n.a.krasovskii@imm.uran.ru

A. Tarasyev
Krasovskii Institute of Mathematics and Mechanics UrB RAS, Yekaterinburg, Russia

Ural Federal University, Yekaterinburg, Russia
e-mail: tam@imm.uran.ru

L. A. Petrosyan et al. (eds.), *Frontiers of Dynamic Games*,
Static & Dynamic Game Theory: Foundations & Applications,
https://doi.org/10.1007/978-3-319-92988-0_6

solutions of Hamilton-Jacobi equations [22, 23]. Basing on constructed optimal control strategies we simulate equilibrium trajectories for dynamical bimatrix game in the framework of Kleimenov approach [8]. It is important to note that in considered statement we can obtain analytical solutions for control strategies depending explicitly on uncertain discount parameter. That allows to implement the sensitivity analysis of equilibrium trajectories with respect to changing of discount parameter and determine the asymptotical behavior of solutions when the discount parameter converges to zero. It is shown that control strategies and equilibrium solutions asymptotically converge to the solution of dynamical bimatrix game with average integral payoff functional considered in papers by Arnold [1].

It is worth to note that we use dynamical constructions and methods of evolutionary games analysis proposed in the paper [18]. To explain the dynamics of players' interaction we use elements of evolutionary games models [2, 5, 6, 25, 27]. For the analysis of shifting equilibrium trajectories from competitive static Nash equilibrium to the points of cooperative Pareto maximum we consider ideas and constructions of cooperative dynamical games [20]. The dynamics of bimatrix game can be interpreted as a generalization of Kolmogorov's equations for probabilities of states [9], which are widely used in Markov processes, stochastic models of mathematical economics and queuing theory. The generalization is understood in the sense that parameters of the dynamics are not fixed a priori and appear to be control parameters and are constructed by the feedback principle in the framework of control theory and differential games theory.

The solution of dynamical bimatrix games is based on construction of positional strategies that maximize own payoffs at any behavior of competing players, which means "guaranteeing" strategies [10, 11, 19]. The construction of solutions on the infinite horizon is divided into fragments with a finite horizon for which Pontryagin maximum principle is used [21] in accordance with constructions of positional differential games theory [11]. More precisely, elements of maximum principle are considered in the aggregate with the method of characteristics for Hamilton-Jacobi equations [12, 22, 24, 26]. The optimal trajectory in each time interval is constructed from pieces of characteristics while switching moments from one characteristic to another are determined by maximum principle. In this method switching moments and points generate switching lines in the phase space which determine the synthesis of optimal positional strategies. Let us note that analogous methods for construction of positional strategies are used in papers [7, 13–17].

In the framework of proposed approach we consider the model of competition on financial markets which is described by dynamical bimatrix game. For this game we construct switching curves for optimal control strategies and synthesize equilibrium trajectories for various values of the discount parameter. We analyze the qualitative behavior of equilibrium trajectories and demonstrate that equilibrium trajectories of dynamical bimatrix game provide better results than static Nash equilibrium. Results of the sensitivity analysis for obtained solutions are demonstrated. This analysis shows that switching curves of optimal control strategies for the series of the discount parameter values have the convergence property by the parameter. We provide calculations confirming the fact that equilibrium trajectories in the problem with discounting converge to the equilibrium trajectory in the problem with average integral payoff functional.

6.2 Model Dynamics

The system of differential equations which defines the dynamics of behavior for two players is investigated

$$\begin{aligned} \dot{x}(t) &= -x(t) + u(t), \quad x(t_0) = x_0, \\ \dot{y}(t) &= -y(t) + v(t), \quad y(t_0) = y_0. \end{aligned} \tag{6.1}$$

The parameter $x = x(t)$, $0 \leq x \leq 1$, means the probability that first player holds to the first strategy (respectively, $(1-x)$ is the probability that he holds to the second strategy). The parameter $y = y(t)$, $0 \leq y \leq 1$, is the probability of choosing the first strategy by the second player (respectively, $(1 - y)$ is the probability that he holds to the second strategy). Control parameters $u = u(t)$ and $v = v(t)$ satisfy conditions $0 \leq u \leq 1$, $0 \leq v \leq 1$, and can be interpreted as signals, that recommend change of strategies by players. For example, value $u = 0$ $(v = 0)$ corresponds to the signal: "change the first strategy to the second one". The value $u = 1$ $(v = 1)$ corresponds to the signal: "change the second strategy to the first one". The value $u = x$ $(v = y)$ corresponds to the signal: "keep the previous strategy".

It is worth to note, that the basis for the dynamics (6.1) and its properties were examined in papers [18, 25]. This dynamics generalizes Kolmogorov's differential equations for probabilities of states [9]. Such generalization assumes that coefficients of incoming and outgoing streams inside coalitions of players are not fixed a priori and can be constructed as positional strategies in the controlled process.

6.3 Local Payoff Functions

Let us assume that the payoff of the first player is described my the matrix $A = a_{ij}$, and the payoff of the second player is described by the matrix $B = b_{ij}$

$$A = \begin{pmatrix} a_{11} & a_{12} \\ a_{21} & a_{22} \end{pmatrix}, \quad B = \begin{pmatrix} b_{11} & b_{12} \\ b_{21} & b_{22} \end{pmatrix}.$$

Local payoff functions of the players in the time period t, $t \in [t_0, +\infty)$ are determined by the mathematical expectation of payoffs, given by corresponding matrices A and B in the bimatrix game, and can be interpreted as "local" interests of the players

$$g_A(x(t), y(t)) = C_A x(t)y(t) - \alpha_1 x(t) - \alpha_2 y(t) + a_{22},$$

$$g_B(x(t), y(t)) = C_B x(t)y(t) - \beta_1 x(t) - \beta_2 y(t) + b_{22}.$$

Here parameters C_A, α_1, α_2 and C_B, β_1, β_2 are determined according to the classical theory of bimatrix games (see [27])

$$C_A = a_{11} - a_{12} - a_{21} + a_{22}, \quad D_A = a_{11}a_{22} - a_{12}a_{21},$$
$$\alpha_1 = a_{22} - a_{12}, \quad \alpha_2 = a_{22} - a_{21},$$

$$C_B = b_{11} - b_{12} - b_{21} + b_{22}, \quad D_B = b_{11}b_{22} - b_{12}b_{21},$$
$$\beta_1 = b_{22} - b_{12}, \quad \beta_2 = b_{22} - b_{21}.$$

6.4 Nash Equilibrium in the Differential Game with Discounted Functionals

In this section we consider the non-zero sum differential game for two players with discounted payoff functionals on the infinite horizon

$$JD_A^{\infty} = [JD_A^-, JD_A^+], \tag{6.2}$$
$$JD_A^- = JD_A^-(x(\cdot), y(\cdot)) = \liminf_{T\to\infty} \int_{t_0}^{T} e^{-\lambda(t-t_0)} g_A(x(t), y(t))\, dt,$$
$$JD_A^+ = JD_A^+(x(\cdot), y(\cdot)) = \limsup_{T\to\infty} \int_{t_0}^{T} e^{-\lambda(t-t_0)} g_A(x(t), y(t))\, dt,$$

defined on the trajectories $(x(\cdot), y(\cdot))$ of the system (6.1).

Payoff functionals of the second player JD_B^{∞}, JD_B^-, JD_B^+ are determined analogously by replacement of the function $g_A(x, y)$ by the function $g_B(x, y)$.

Discounted functionals (6.2) are traditional for the problems of evolutionary economics and economic growth [6, 12], and are related to the idea of depreciation of financial funds in time. In the problems of optimal guaranteed control such functionals were considered in the paper [25]. Unlike payoff functionals optimized in each period, discounted functionals admit the possibility of loss in some periods in order to win in other periods and obtain better integral result in all periods. This fact allows the system to stay longer in favorable domains where values of local payoffs for the players are strictly better than values of static Nash equilibrium.

Let us introduce the notion of dynamical Nash equilibrium for the evolutionary game with the dynamics (6.1) and discounted payoff functionals (6.2) in the context of constructions of non-antagonistic positional differential games [8, 11, 18]. Let us define the dynamical Nash equilibrium in the class of positional strategies (feedbacks) $U = u(t, x, y, \varepsilon)$, $V = v(t, x, y, \varepsilon)$.

Definition 6.1 The dynamical Nash equilibria (U^0, V^0), $U^0 = u^0(t, x, y, \varepsilon)$, $V^0 = v^0(t, x, y, \varepsilon)$ from the class of controls by the feedback principle $U =$

$u(t, x, y, \varepsilon)$, $V = v(t, x, y, \varepsilon)$ for the given problem is determined by inequalities

$$JD_A^-(x^0(\cdot), y^0(\cdot)) \geq JD_A^+(x_1(\cdot), y_1(\cdot)) - \varepsilon,$$
$$JD_B^-(x^0(\cdot), y^0(\cdot)) \geq JD_B^+(x_2(\cdot), y_2(\cdot)) - \varepsilon,$$

$$(x^0(\cdot), y^0(\cdot)) \in X(x_0, y_0, U^0, V^0), \quad (x_1(\cdot), y_1(\cdot)) \in X(x_0, y_0, U, V^0),$$
$$(x_2(\cdot), y_2(\cdot)) \in X(x_0, y_0, U^0, V).$$

Here symbol X stands for the set of trajectories, starting from initial point and generated by corresponding postional strategies is the sense of the paper [11].

6.5 Auxiliary Zero-Sum Games

For the construction of desired equilibrium feedbacks U^0, V^0 we use the approach [8]. In accordance with this approach we construct the equilibrium using optimal feedbacks for differential games $\Gamma_A = \Gamma_A^- \cup \Gamma_A^+$ and $\Gamma_B = \Gamma_B^- \cup \Gamma_B^+$ with payoffs JD_A^∞ and JD_B^∞ (6.2). In the gamed Γ_A the first player maximizes the functional $JD_A^-(x(\cdot), y(\cdot))$ with the guarantee using the feedback $U = u(t, x, y, \varepsilon)$, and the second player oppositely provides the minimization of the functional $JD_A^+(x(\cdot), y(\cdot))$ using the feedback $V = v(t, x, y, \varepsilon)$. Vice versa, in the game Γ_B the second player maximizes the functional $JD_B^-(x(\cdot), y(\cdot))$ with the guarantee, and the first player maximizes the functional $JD_B^+(x(\cdot), y(\cdot))$.

Let us introduce following notations. By $u_A^0 = u_A^0(t, x, y, \varepsilon)$ and $v_B^0 = v_B^0(t, x, y, \varepsilon)$ we denote feedbacks that solve the problem of guaranteed maximization for payoff functionals JD_A^- and JD_B^- respectively. Let us note, that these feedbacks represent the guaranteed maximization of players' payoffs in the long run and can be named "positive". By $u_B^0 = u_B^0(t, x, y, \varepsilon)$ and $v_A^0 = v_A^0(t, x, y, \varepsilon)$ we denote feedbacks mostly favorable for opposite players, namely, those, that minimize payoff functionals JD_B^+, JD_A^+ of the opposite players. Let us call them "punishing".

Let us note, that inflexible solutions of selected problems can be obtained in the framework of the classical bimatrix games theory. Let us propose for definiteness, (this proposition is given for illustration without loss of generality for the solution), that the following relations corresponding to the almost antagonistic structure of bimatrix game hold for the parameters of matrices A and B,

$$\begin{aligned} &C_A > 0, && C_B < 0, \\ &0 < x_A = \frac{\alpha_2}{C_A} < 1, && 0 < x_B = \frac{\beta_2}{C_B} < 1, \\ &0 < y_A = \frac{\alpha_1}{C_A} < 1, && 0 < y_B = \frac{\beta_1}{C_B} < 1. \end{aligned} \tag{6.3}$$

The following proposition is fair.

Lemma 6.1 *Differential games Γ_A^-, Γ_A^+ have equal values*

$$w_A^- = w_A^+ = w_A = \frac{D_A}{C_A},$$

and differential games Γ_B^-, Γ_B^+ have equal values

$$w_B^- = w_B^+ = w_B = \frac{D_B}{C_B}$$

for any initial position $(x_0, y_0) \in [0, 1] \times [1, 0]$. These values, for example, can be guaranteed by "positive" feedbacks u_A^{cl}, v_B^{cl} corresponding to classical solutions x_A, y_B

$$u_A^0 = u_A^{cl} = u_A^{cl}(x, y) = \begin{cases} 0, & x_A < x \leq 1, \\ 1, & 0 \leq x < x_A, \\ [0, 1], & x = x_A. \end{cases}$$

$$v_B^0 = v_B^{cl} = v_B^{cl}(x, y) = \begin{cases} 0, & y_B < y \leq 1, \\ 1, & 0 \leq y < y_B, \\ [0, 1], & y = y_B. \end{cases}$$

"Punishing" feedbacks are determined by formulas

$$u_B^0 = u_B^{cl} = u_B^{cl}(x, y) = \begin{cases} 0, & x_B < x \leq 1, \\ 1, & 0 \leq x < x_B, \\ [0, 1], & x = x_B, \end{cases}$$

$$v_A^0 = v_A^{cl} = v_A^{cl}(x, y) = \begin{cases} 0, & y_A < y \leq 1, \\ 1, & 0 \leq y < y_A, \\ [0, 1], & y = y_A, \end{cases}$$

and correspond to classical solutions x_B, y_A (6.3), which generate the static Nash equilibrium $NE = (x_B, y_A)$.

The proof of this proposition can me obtained by the direct substitution of shown strategies to corresponding payoff functionals (6.2).

Remark 6.1 Values of payoff functions $g_A(x, y)$, $g_B(x, y)$ coincide at points (x_A, y_B), (x_B, y_A)

$$g_A(x_A, y_B) = g_A(x_B, y_A) = w_A, \qquad g_B(x_A, y_B) = g_B(x_B, y_A) = w_B.$$

The point $NE = (x_B, y_A)$ is the "mutually punishing" Nash equilibrium, and the point (x_A, y_B) does not possess equilibrium properties in the corresponding static game.

6.6 Construction of the Dynamical Nash Equilibrium

Let us construct the pair of feedbacks, which consist the Nash equilibrium. For this let us combine "positive" feedbacks u_A^0, v_B^0 and "punishing" feedbacks u_B^0, v_A^0.

Let us choose the initial position $(x_0, y_0) \in [0, 1]\times[0, 1]$ and accuracy parameter $\varepsilon > 0$. Let us choose the trajectory $(x^0(\cdot), y^0(\cdot)) \in X(x_0, y_0, U_A^0(\cdot), v_B^0(\cdot))$, generated by "positive" $u_A^0 = U_A^0(t, x, y, \varepsilon)$ and $v_B^0 = v_B^0(t, x, y, \varepsilon)$. Let us choose $T_\varepsilon > 0$ such that

$$\begin{array}{rcl} g_A(x^0(t), y^0(t)) & > & JD_A^-(x^0(\cdot), y^0(\cdot)) - \varepsilon, \\ g_B(x^0(t), y^0(t)) & > & JD_B^-(x^0(\cdot), y^0(\cdot)) - \varepsilon, \\ & t \in [T_\varepsilon, +\infty]. & \end{array}$$

Let us denote by $u_A^\varepsilon(t)$: $[0, T_\varepsilon) \to [0, 1]$, $v_B^\varepsilon(t)$: $[0, T_\varepsilon) \to [0, 1]$ step-by-step implementation of strategies u_A^0, v_B^0 such that the corresponding step-by-step trajectory $(x_\varepsilon(\cdot), y_\varepsilon(\cdot))$ satisfies the condition

$$\max_{t\in[0,T_\varepsilon]} \|(x^0(t), y^0(t)) - (x_\varepsilon(t), y_\varepsilon(t))\| < \varepsilon.$$

From the results of the paper [8] the next proposition follows.

Lemma 6.2 *The pair of feedbacks $U^0 = u^0(t, x, y, \varepsilon)$, $V^0 = v^0(t, x, y, \varepsilon)$, combines together "positive" feedbacks u_A^0, v_B^0 and "punishing" feedbacks u_B^0, v_A^0 according to relations*

$$U^0 = u^0(t, x, y, \varepsilon) = \begin{cases} u_A^\varepsilon(t), & \|(x, y) - (x_\varepsilon(t), y_\varepsilon(t))\| < \varepsilon, \\ u_B^0(x, y), & otherwise, \end{cases}$$

$$V^0 = v^0(t, x, y, \varepsilon) = \begin{cases} v_B^\varepsilon(t), & \|(x, y) - (x_\varepsilon(t), y_\varepsilon(t))\| < \varepsilon, \\ v_A^0(x, y), & otherwise \end{cases}$$

is the dynamical ε-Nash equilibrium.

Below we construct flexible "positive" feedbacks that generate trajectories $(x^{fl}(\cdot), y^{fl}(\cdot))$, which reduce to "better" positions than the inflexible dynamical equilibrium (x_B, y_A), (x_A, y_B) by both criteria $JD_A^\infty(x^{fl}(\cdot), y^{fl}(\cdot)) \geq v_A$, $JD_B^\infty(x^{fl}(\cdot), y^{fl}(\cdot)) \geq v_B$.

6.7 Two-Step Optimal Control Problems

For the construction of "positive" feedbacks $u_A^0 = u_A^{fl}(x, y)$, $v_B^0 = v_B^{fl}(x, y)$ we consider in this section the auxiliary two-step optimal control problem with discounted payoff functional for the first player in the situation, when actions of the second player are most unfavorable. Namely, let us analyze the optimal control problem for the dynamical system (6.1)

$$\begin{aligned} \dot{x}(t) &= -x(t) + u(t), \quad x(0) = x_0, \\ \dot{y}(t) &= -y(t) + v(t), \quad y(0) = y_0. \end{aligned} \tag{6.4}$$

with the payoff functional

$$JD_A^f = \int_0^{T_f} e^{-\lambda t} g_A(x(t), y(t))\, dt. \tag{6.5}$$

Here without loss of generality let us consider that $t_0 = 0$, $T = T_f$, and terminal moment of time $T_f = T_f(x_0, y_0)$ we determine later.

Without loss of generality, we assume that the value of the static game equals to zero

$$w_A = \frac{D_A}{C_A} = 0, \tag{6.6}$$

and next conditions hold

$$C_A > 0, \quad 0 < x_A = \frac{\alpha_2}{C_A} < 1, \quad 0 < y_A = \frac{\alpha_1}{C_A} < 1. \tag{6.7}$$

Let us consider the case when initial conditions (x_0, y_0) of the system (6.4) satisfy relations

$$x_0 = x_A, \quad y_0 > y_A. \tag{6.8}$$

Let us assume that actions of the second player are mostly unfavorable for the first player. For trajectories of the system (6.4), which start from initial positions (x_0, y_0) (6.8), these actions $v_A^0 = v_A^{cl}(x, y)$ are determined by the relation

$$v_A^{cl}(x, y) \equiv 0.$$

Optimal actions $u_A^0 = u_A^{fl}(x, y)$ of the first player according to the payoff functional JD_A^f (6.5) in this situation can be presented as the two-step impulse control: it equals one from the initial time moment $t_0 = 0$ till the moment of optimal switch s

and then equals to zero till the final time moment T_f

$$u_A^0(t) = u_A^{fl}(x(t), y(t)) = \begin{cases} 1, & \text{if} \quad t_0 \leq t < s, \\ 0, & \text{if} \quad s \leq t < T_f. \end{cases}$$

Here the parameter s is the optimization parameter. The final time moment T_f is determined by the following condition. The trajectory $(x(\cdot), y(\cdot))$ of the system (6.4), which starts from the line where $x(t_0) = x_A$, returns to this line when $x(T_f) = x_A$.

Let us consider two aggregates of characteristics. The first onc is described by the system of differential equations with the value on the control parameter $u = 1$

$$\begin{aligned} \dot{x}(t) &= -x(t) + 1, \\ \dot{y}(t) &= -y(t), \end{aligned} \tag{6.9}$$

solutions of which are determined by the Cauchy formula

$$x(t) = (x_0 - 1)e^{-t} + 1, \quad y(t) = y_0 e^{-t}. \tag{6.10}$$

Here initial positions (x_0, y_0) satisfy conditions (6.8) and time parameter t satisfies the inequality $0 \leq t < s$.

The second aggregate of characteristics is given by the system of differential equations with the value of the control parameter $u = 0$

$$\begin{aligned} \dot{x}(t) &= -x(t), \\ \dot{y}(t) &= -y(t), \end{aligned} \tag{6.11}$$

solutions of which are determined by the Cauchy formula

$$x(t) = x_1 e^{-t}, \quad y(t) = y_1 e^{-t}. \tag{6.12}$$

Here initial positions $(x_1, y_1) = (x_1(s), y_1(s))$ are determined by relations

$$x_1 = x_1(s) = (x_0 - 1)e^{-s} + 1, \quad y_1 = y_1(s) = y_0 e^{-s}, \tag{6.13}$$

and the time parameter t satisfies the inequality $0 \leq t < p$. Here the final time moment $p = p(s)$ and the final position $(x_2, y_2) = (x_2(s), y_2(s))$ of the whole trajectory $(x(\cdot), y(\cdot))$ is given by formulas

$$x_1 e^{-p} = x_A, \quad p = p(s) = \ln \frac{x_1(s)}{x_A}, \quad x_2 = x_A, \quad y_2 = y_1 e^{-p}. \tag{6.14}$$

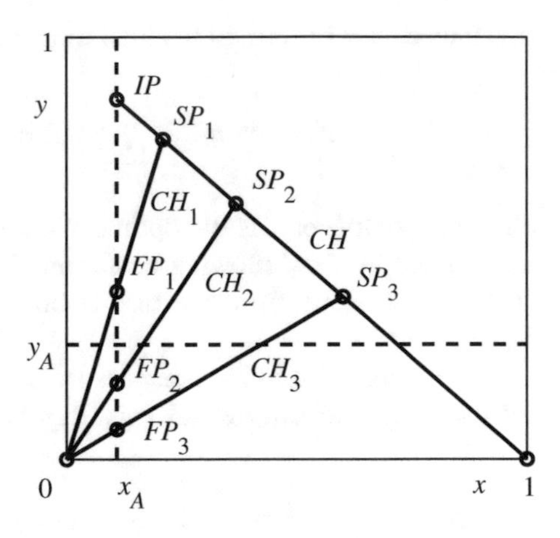

Fig. 6.1 Families of characteristics and switching points in the two-step optimal control problem

The optimal control problem is to find such moment of time s and the corresponding switching point $(x_1, y_1) = (x_1(s), y_1(s))$ on the trajectory $(x(\cdot), y(\cdot))$, where the integral $I = I(s)$ reaches the maximum value

$$I(s) = I_1(s) + I_2(s), \tag{6.15}$$

$$I_1(s) = \int_0^s e^{-\lambda t}(C_A((x_0 - 1)e^{-t} + 1)y_0 e^{-t} - \alpha_1((x_0 - 1)e^{-t} + 1)$$

$$-\alpha_2 y_0 e^{-t} + a_{22})\, dt,$$

$$I_2(s) = e^{-\lambda s} \int_0^{p(s)} e^{-\lambda t}(C_A x_1(s) y_1(s) e^{-2t} - \alpha_1 x_1(s) e^{-t} - \alpha_2 y_1(s) e^{-t} + a_{22})\, dt.$$

On the Fig. 6.1 we depict the initial position IP, chosen on the line $x = x_A$ when $y > y_A$, the characteristic CH, oriented on the vertex $(1, 0)$ of the unit square, characteristics CH_1, CH_2, CH_3, oriented on the vertex $(0, 0)$ of the unit square, switching points SP_1, SP_2, SP_3 of the motion along characteristics and final points of the motion FP_1, FP_2, FP_3, located of the line $x = x_A$.

6.8 The Solution of the Two-Step Optimal Control Problem

We obtain the solution of the two-step optimal control problem (6.9)–(6.15), by calculating the derivative dI/ds, presenting it as the function of optimal switching points $(x, y) = (x_1, y_1)$, equating this derivative to zero $dI/ds = 0$ and finding the equation $F(x, y) = 0$ for the curve, that consist from optimal switching points (x, y).

Sufficient maximum conditions in such construction are obtained from the fact that the integral $I(s)$ has the property of monotonic increase by the variable s in the initial period, because the integrand $g_A(x, y)$ is positive, $g_A(x, y) > w_A = 0$, in the domain $x > x_A$, $y > y_A$. In the finite period the integral $I(s)$ strictly monotonically decreases by the variable s, because the integrand $g_A(x, y)$ is negative, $g_A(x, y) < w_A = 0$, in the domain $x > x_A$, $y < y_A$.

Firstly let us calculate integrals I_1, I_2

$$
\begin{aligned}
I_1 = I_1(s) = C_A(x_0 - 1)y_0 \frac{(1 - e^{-(\lambda+2)s})}{(\lambda + 2)} + C_A y_0 \frac{(1 - e^{-(\lambda+1)s})}{(\lambda + 1)} \\
-\alpha_1(x_0 - 1)\frac{(1 - e^{-(\lambda+1)s})}{(\lambda + 1)} - \alpha_1 \frac{(1 - e^{-\lambda s})}{\lambda} \\
-\alpha_2 y_0 \frac{(1 - e^{-(\lambda+1)s})}{(\lambda + 1)} + a_{22} \frac{(1 - e^{-\lambda s})}{\lambda}.
\end{aligned}
$$

$$
\begin{aligned}
I_2 = I_2(s) = e^{-\lambda s} C_A x_1(s) y_1(s) \frac{(1 - e^{-(\lambda+2)p(s)})}{(\lambda + 2)} \\
-e^{-\lambda s}\alpha_1 x_1(s) \frac{(1 - e^{-(\lambda+1)p(s)})}{(\lambda + 1)} \\
-e^{-\lambda s}\alpha_2 y_1(s) \frac{(1 - e^{-(\lambda+1)p(s)})}{(\lambda + 1)} \\
+e^{-\lambda s} a_{22} \frac{(1 - e^{-\lambda p(s)})}{\lambda}.
\end{aligned}
$$

Let us calculate derivatives dI_1/ds, dI_2/ds and present them as functions of optimal switching points $(x, y) = (x_1, y_1)$

$$
\begin{aligned}
\frac{dI_1}{ds} &= C_A(x_0 - 1)y_0 e^{-2s} e^{-\lambda s} + C_A y_0 e^{-s} e^{-\lambda s} \\
&\quad -\alpha_1(x_0 - 1)e^{-s} e^{-\lambda s} - \alpha_1 e^{-\lambda s} - \alpha_2 y_0 e^{-s} e^{-\lambda s} + a_{22} e^{-\lambda s} \\
&= e^{-\lambda s}(C_A x y - \alpha_1 x - \alpha_2 y + a_{22}).
\end{aligned}
$$

While calculating the derivative dI_2/ds let us take into the account next expressions for derivatives dx/ds, dy/ds, dp/ds and the exponent e^{-p} as functions of variables (x, y):

$$
\frac{dx}{ds} = 1 - x, \quad \frac{dy}{ds} = -y, \quad \frac{dp}{ds} = \frac{1 - x}{x}, \quad e^{-p} = \frac{\alpha_2}{C_A x}.
$$

Let us introduce the new derivative $q = e^{-p}$ and obtain the expression for dI_2/ds

$$\frac{dI_2}{ds} = e^{-\lambda s}\Big(- \lambda C_A xy\frac{(1-q^{(\lambda+2)})}{(\lambda+2)} + C_A(1-x)y\frac{(1-q^{(\lambda+2)})}{(\lambda+2)}$$
$$-C_A xy\frac{(1-q^{(\lambda+2)})}{(\lambda+2)} + C_A(1-x)yq^{(\lambda+2)}$$
$$+\lambda\alpha_1 x\frac{(1-q^{(\lambda+1)})}{(\lambda+1)} - \alpha_1(1-x)\frac{(1-q^{(\lambda+1)})}{(\lambda+1)} - \alpha_1(1-x)q^{(\lambda+1)}$$
$$+\lambda\alpha_2 y\frac{(1-q^{(\lambda+1)})}{(\lambda+1)} + \alpha_2 y\frac{(1-q^{(\lambda+1)})}{(\lambda+1)} - \alpha_2 y\frac{(1-x)}{x}q^{(\lambda+1)}$$
$$+a_{22}\frac{(1-x)}{x}q^{\lambda} - a_{22}(1-q^{\lambda})\Big).$$

Let us summarize derivatives dI_1/ds and dI_2/ds, equalize the expression to zero and express y by x in the following form:

$$y = \Big(\alpha_1 x - \lambda\alpha_1 x\frac{(1-q^{(\lambda+1)})}{(\lambda+1)} + \alpha_1(1-x)\frac{(1-q^{(\lambda+1)})}{(\lambda+1)} + \alpha_1(1-x)q^{(\lambda+1)}$$
$$-a_{22}\frac{(1-x)}{x}q^{\lambda} + a_{22}(1-q^{\lambda}) - a_{22}\Big)\Big/$$
$$\Big(C_A x - \lambda C_A x\frac{(1-q^{(\lambda+2)})}{(\lambda+2)} + C_A(1-2x)\frac{(1-q^{(\lambda+2)})}{(\lambda+2)} + C_A(1-x)q^{(\lambda+2)}$$
$$+\lambda\alpha_2\frac{(1-q^{(\lambda+1)})}{(\lambda+1)} + \alpha_2\frac{(1-q^{(\lambda+1)})}{(\lambda+1)} - \alpha_2\frac{(1-x)}{x}q^{(\lambda+1)} - \alpha_2\Big).$$

Simplifying the expression we obtain the formula:

$$y = \Big(\alpha_1\frac{(1-q^{(\lambda+1)})}{(\lambda+1)} + \alpha_1 q^{(\lambda+1)} - a_{22}\frac{1}{x}q^{\lambda}\Big)\Big/$$
$$\Big(C_A\frac{(1-q^{(\lambda+2)})}{(\lambda+2)} + C_A q^{(\lambda+2)} - \alpha_2\frac{1}{x}q^{(\lambda+1)}\Big).$$

Taking into the account the fact that $w_A = 0$ (6.6), we obtain $a_{22} = (\alpha_1\alpha_2)/C_A$. By substitution of this relation and the expression $q = \alpha_2/(C_A x)$ to previous formula we obtain:

$$y = \Big(\alpha_1\Big(1 - \Big(\frac{\alpha_2}{C_A x}\Big)^{(\lambda+1)}\Big)(\lambda+2)\Big)\Big/\Big(C_A\Big(1 - \Big(\frac{\alpha_2}{C_A x}\Big)^{(\lambda+2)}\Big)(\lambda+1)\Big).$$

Multiplying both parts on the expression by $x^{(\lambda+2)}$, we obtain:

$$y = \left(\alpha_1\left(x^{(\lambda+1)} - \left(\frac{\alpha_2}{C_A}\right)^{(\lambda+1)}\right)(\lambda+2)x\right) \Big/ \left(C_A\left(x^{(\lambda+2)} - \left(\frac{\alpha_2}{C_A}\right)^{(\lambda+2)}\right)(\lambda+1)\right).$$

Taking into the account relations $x_A = \alpha_2/C_A$ and $y_A = \alpha_1/C_A$ (6.7), we obtain the final expression for the switching curve $M_A^1(\lambda)$:

$$y = \frac{(\lambda+2)\left(x^{(\lambda+1)} - x_A^{(\lambda+1)}\right)y_A x}{(\lambda+1)\left(x^{(\lambda+2)} - x_A^{(\lambda+2)}\right)}.$$

To construct the final switching curve $M_A(\lambda)$ for the optimal strategy of the first player in the game with the discounted functional in the case $C_A > 0$, we add to the curve $M_A^1(\lambda)$ the similar curve $M_A^2(\lambda)$ in the domain, where $x \le y_A$ and $y \le y_A$

$$M_A(\lambda) = M_A^1(\lambda) \cup M_A^2(\lambda), \tag{6.16}$$

$$M_A^1(\lambda) = \Bigg\{(x, y) \in [0, 1] \times [0, 1]:$$

$$y = \frac{(\lambda+2)\left(x^{(\lambda+1)} - x_A^{(\lambda+1)}\right)y_A x}{(\lambda+1)\left(x^{(\lambda+2)} - x_A^{(\lambda+2)}\right)}, \ x \ge x_A, \ y \ge y_A\Bigg\},$$

$$M_A^2(\lambda) = \Bigg\{(x, y) \in [0, 1] \times [0, 1]:$$

$$y = -\frac{(\lambda+2)\left((1-x)^{(\lambda+1)} - (1-x_A)^{(\lambda+1)}\right)(1-y_A)(1-x)}{(\lambda+1)\left((1-x)^{(\lambda+2)} - (1-x_A)^{(\lambda+2)}\right)} + 1,$$

$$x \le x_A, \ y \le y_A\Bigg\}.$$

In the case when $C_A < 0$, curves $M_A(\lambda)$, $M_A^1(\lambda)$ and$M_A^2(\lambda)$ are described by formulas

$$M_A(\lambda) = M_A^1(\lambda) \cup M_A^2(\lambda), \tag{6.17}$$

$$M_A^1(\lambda) = \Bigg\{(x, y) \in [0, 1] \times [0, 1]:$$

$$y = \frac{(\lambda+2)\left((1-x)^{(\lambda+1)} - (1-x_A)^{(\lambda+1)}\right)y_A(1-x)}{(\lambda+1)\left((1-x)^{(\lambda+2)} - (1-x_A)^{(\lambda+2)}\right)},$$

$$x \le x_A, \ y \ge y_A\Bigg\},$$

$$M_A^2(\lambda) = \Big\{(x, y) \in [0, 1] \times [0, 1]:$$

$$y = -\frac{(\lambda + 2)\left(x^{(\lambda+1)} - x_A^{(\lambda+1)}\right)(1 - y_A)x}{(\lambda + 1)\left(x^{(\lambda+2)} - x_A^{(\lambda+2)}\right)} + 1, \ x \geq x_A, \ y \leq y_A\Big\}.$$

The curve $M_A(\lambda)$ divides the unit square $[0, 1] \times [0, 1]$ into two parts: the upper part

$$D_A^u \supset \{(x, y) : \quad x = x_A, \quad y > y_A\}$$

and the lower part

$$D_A^l \supset \{(x, y) : \quad x = x_A, \quad y < y_A\}.$$

The "positive" feedback u_A^{fl} has the following structure

$$u_A^{fl} = u_A^{fl}(x, y) = \begin{cases} \max\{0, -sgn(C_A)\}, & if \ (x, y) \in D_A^u, \\ \max\{0, sgn(C_A)\}, & if \ (x, y) \in D_A^l, \\ [0, 1], & if \ (x, y) \in M_A(\lambda). \end{cases} \tag{6.18}$$

On the Fig. 6.2 we show switching curves $M_A^1(\lambda)$, $M_A^2(\lambda)$ for the first player. Directions of velocities $\dot{x}$ are depicted by horizontal (left and right) arrows.

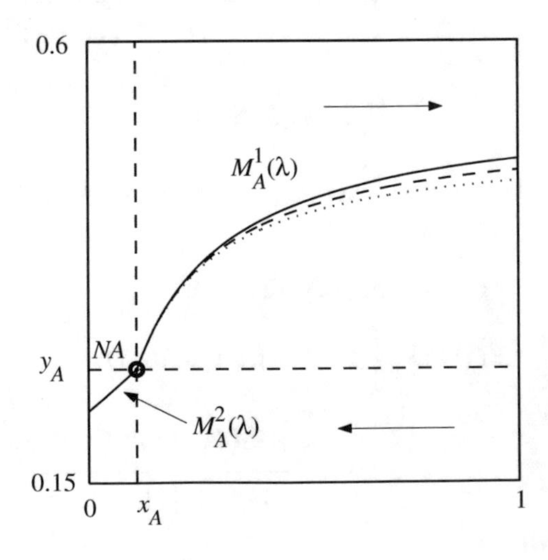

Fig. 6.2 Switching curves $M_A^1(\lambda)$, $M_A^2(\lambda)$ for the first player in the problem with discounted payoff functionals

For the second player one can get similar switching curves $M_B(\lambda)$ for the optimal control problem with the discounted functional, corresponding to the matrix B. More precisely, in the case when $C_B > 0$, the switching curve $M_B(\lambda)$ is given by relations

$$M_B(\lambda) = M_B^1(\lambda) \cup M_B^2(\lambda), \tag{6.19}$$

$$M_B^1(\lambda) = \Big\{(x, y) \in [0, 1] \times [0, 1]:$$

$$x = \frac{(\lambda + 2)\Big(y^{(\lambda+1)} - y_B^{(\lambda+1)}\Big)x_B y}{(\lambda + 1)\Big(y^{(\lambda+2)} - y_B^{(\lambda+2)}\Big)}, \ x \geq x_B, \ y \geq y_B\Big\},$$

$$M_B^2(\lambda) = \Big\{(x, y) \in [0, 1] \times [0, 1]:$$

$$x = -\frac{(\lambda + 2)\Big((1 - y)^{(\lambda+1)} - (1 - y_B)^{(\lambda+1)}\Big)(1 - x_B)(1 - y)}{(\lambda + 1)\Big((1 - y)^{(\lambda+2)} - (1 - y_B)^{(\lambda+2)}\Big)} + 1,$$

$$x \leq x_B, \ y \leq y_B\Big\}.$$

In the case when the parameter C_B is negative $C_B < 0$, curves $M_B(\lambda)$, $M_B^1(\lambda)$ and $M_B^2(\lambda)$ are determined by formulas

$$M_B(\lambda) = M_B^1(\lambda) \cup M_B^2(\lambda), \tag{6.20}$$

$$M_B^1(\lambda) = \Big\{(x, y) \in [0, 1] \times [0, 1]:$$

$$x = \frac{(\lambda + 2)\Big((1 - y)^{(\lambda+1)} - (1 - y_B)^{(\lambda+1)}\Big)x_B(1 - y)}{(\lambda + 1)\Big((1 - y)^{(\lambda+2)} - (1 - y_B)^{(\lambda+2)}\Big)},$$

$$x \geq x_B, \ y \leq y_B\Big\},$$

$$M_B^2(\lambda) = \Big\{(x, y) \in [0, 1] \times [0, 1]:$$

$$x = -\frac{(\lambda + 2)\Big(y^{(\lambda+1)} - y_B^{(\lambda+1)}\Big)(1 - x_B)y}{(\lambda + 1)\Big(y^{(\lambda+2)} - y_B^{(\lambda+2)}\Big)} + 1, \ x \leq x_B, \ y \geq y_B\Big\}.$$

The curve $M_B(\lambda)$ divide the unit square $[0, 1] \times [0, 1]$ into two parts: the left part

$$D_B^l \supset \{(x, y) : \quad x < x_B, \quad y = y_B\}$$

and the right part

$$D_B^r \supset \{(x, y) : \quad x > x_B, \quad y = y_B\}.$$

The "positive" feedback v_B^{fl} has the following structure

$$v_B^{fl} = v_B^{fl}(x, y) = \begin{cases} \max\{0, -sgn(C_B)\}, & if\ (x, y) \in D_B^l, \\ \max\{0, sgn(C_B)\}, & if\ (x, y) \in D_B^r, \\ [0, 1], & if\ (x, y) \in M_B(\lambda). \end{cases} \tag{6.21}$$

Remark 6.2 Let us note that in papers by Arnold [1] average integral payoff functionals were considered

$$\frac{1}{(T - t_0)} \int_{t_0}^{T} g_A(x(t), y(t))\, dt. \tag{6.22}$$

In the paper [16] switching curves for optimal control strategies of players in the game with average integral functionals were obtained. For example, for the first player in the case when $C_A > 0$ the switching curve in the domain $x \geq x_A$, $y \geq y_A$ is described by relations

$$y = \frac{2\alpha_1 x}{C_A x + \alpha_2}. \tag{6.23}$$

The asymptotical analysis of solutions (6.16) for the game with discounted payoff functionals shows, that according to L'Hospital's rule, when the discount parameter λ tends to zero, the relation for switching curves (6.16) of the control strategy for the first player converges to switching curves (6.23) in the game with average integral payoff functionals (6.22).

On the Fig. 6.2. the solid line shows the switching curve of control strategies for the first player in the game with average integral payoff functionals, which is asymptotically approximated by solutions of the game with discounted functionals when $\lambda \downarrow 0$. The dashed line and the dotted line show switching curves of control strategies for the first player in the game with discounted payoff functionals with values of the discount parameter $\lambda = 0.1$ and $\lambda = 0.2$, respectively.

On the Fig. 6.3 we show switching curves $M_B^1(\lambda)$, $M_B^2(\lambda)$ for the second player. Directions of velocities $\dot{y}$ are depicted by vertical (up and down) arrows.

It is worth to clarify the asymptotical behavior of switching curves for optimal control when discount parameters can be infinitely large. In this case, one can check that switching curve $M_A(\lambda)$ for optimal control in the problem with discounted

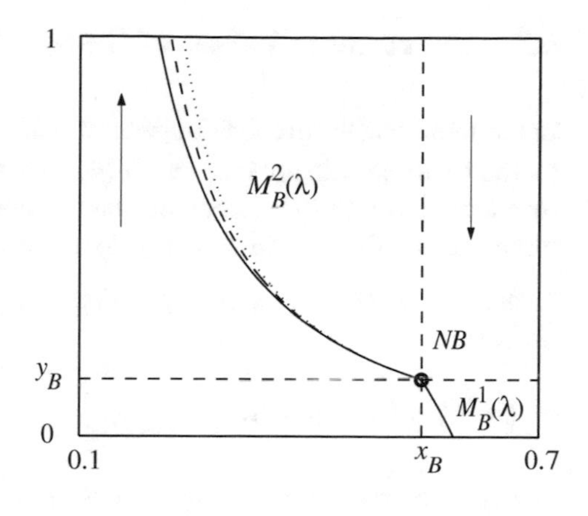

Fig. 6.3 Switching curves $M_B^1(\lambda)$, $M_B^2(\lambda)$ for the second player in the problem with discounted payoff functionals

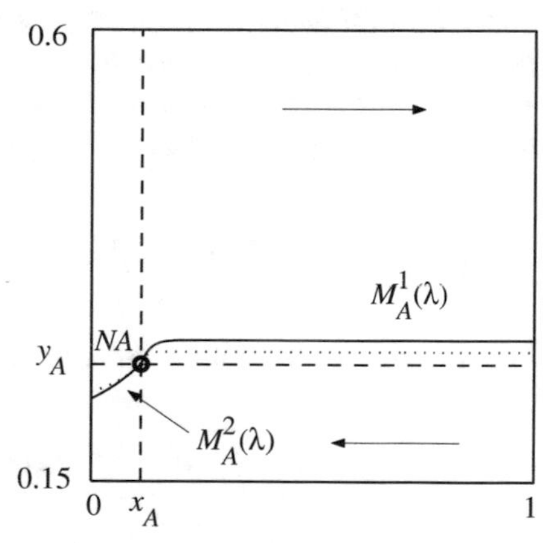

Fig. 6.4 Asymptotical behavior of switching curve $M_A(\lambda)$ for the first player in the problem with discounted payoff functionals

integral payoffs describing long-term interests of players converge to the switching line $y = y_A$ generated by the short-run payoff function $g_A(x, y)$ when the discount parameter λ tends to infinity. Such behavior of the switching curve $M_A(\lambda)$ is shown on the Fig. 6.4.

$$y = \frac{(\lambda+2)\left(x^{(\lambda+1)} - x_A^{(\lambda+1)}\right) y_A x}{(\lambda+1)\left(x^{(\lambda+2)} - x_A^{(\lambda+2)}\right)}$$

$$= \left(1 + \frac{1}{(\lambda+1)}\right) \frac{\left(1 - \left(\frac{x_A}{x}\right)^{(\lambda+1)}\right)}{\left(1 - \left(\frac{x_A}{x}\right)^{(\lambda+2)}\right)} y_A \to y_A, \quad \text{when } \lambda \to +\infty.$$

6.9 Guaranteed Values of Discounted Payoffs

Let us formulate the proposition, which confirms, that the "positive" optimal control by the feedback principle $u_A^{fl}(x, y)$ (6.18) with the switching cure M_A, defined by formulas (6.16), (6.17), guarantee that the value of discounted payoff functionals is more or equal than the value w_A (6.6) of the static matrix game.

Theorem 6.1 *For any initial position* $(x_0, y_0) \in [0, 1] \times [0, 1]$ *and for any trajectory*

$$(x^{fl}(\cdot), y^{fl}(\cdot)) \in X(x_0, y_0, u_A^{fl}), \quad x^{fl}(t_0) = x_0, \quad y^{fl}(t_0) = y_0, \quad t_0 = 0,$$

generated by the optimal control by the feedback principle $u_A^{fl} = u_A^{fl}(x, y)$ *there exists the final moment of time* $t_* \in [0, T_A]$ *such that in this moment of time the trajectory* $(x^{fl}(\cdot), y^{fl}(\cdot))$ *reaches the line where* $x = x_A$, *namely* $x^{fl}(t_*) = x_A$. *Then, according to the construction of the optimal control, that maximizes the integral (6.15) by the feedback principle* u_A^{fl}, *the following estimate holds*

$$\int_{t_*}^{T} e^{-\lambda(t-t_*)} g_A(x(t), y(t))\, dt \geq \frac{w_A}{\lambda}\big(1 - e^{-\lambda(T-t_*)}\big), \quad \forall T \geq t_*. \tag{6.24}$$

In particular, this inequality remains valid when time T *tends to infinity*

$$\liminf_{T \to +\infty} \lambda \int_{t_*}^{T} e^{-\lambda(t-t_*)} g_A(x^{fl}(t), y^{fl}(t))\, dt \geq w_A. \tag{6.25}$$

Inequalities (6.24), (6.25) mean, that the value of the discounted functional is not worse, than the value w_A *(6.6) of the static matrix game.*

The analogous result is fair for trajectories, which generated by the optimal control v_B^{fl} *(6.21), that corresponds to the switching curve* M_B *(6.19), (6.20).*

Proof The result of the theorem follows from the fact that the value of the payoff functional (6.5) is maximum on the constructed broken line. In particular, it is more or equal, than the value of this functional on the trajectory which stays on the segment $x = x_A$ (see Fig. 6.1) with the control $u(t) = x_A$. The value of the payoff functional on such trajectory is following

$$\int_{t_*}^{T} e^{-\lambda(t-t_*)} w_A\, dt = \frac{w_A}{\lambda}\big(1 - e^{-\lambda(T-t_*)}\big).$$

These arguments imply the required relation (6.24), which in the limit transition provides the relation (6.25). □

Remark 6.3 Let us consider the acceptable trajectory $(x^{fl}_{AB}(\cdot), y^{fl}_{AB}(\cdot))$, generated by "positive" feedbacks u^{fl}_A (6.18), v^{fl}_B (6.21). Then in accordance with the Theorem 6.1, next inequalities take place

$$\liminf_{T\to+\infty} \lambda \int_{t_*}^{T} e^{-\lambda(t-t_*)} g_A(x^{fl}_{AB}(t), y^{fl}_{AB}(t))\, dt \geq w_A$$

$$\liminf_{T\to+\infty} \lambda \int_{t_*}^{T} e^{-\lambda(t-t_*)} g_B(x^{fl}_{AB}(t), y^{fl}_{AB}(t))\, dt \geq w_B$$

and, hence, the acceptable trajectory $(x^{fl}_{AB}(\cdot), y^{fl}_{AB}(\cdot))$ provides the better result for both players, than trajectories, convergent to points of the static Nash equilibrium, in which corresponding payoffs are equal to values w_A and w_B.

6.10 Equilibrium Trajectories in the Game with Discounted Payoffs

Let us consider payoff matrices of players on the financial market, which reflect the data of investigated markets of stocks [3] and bonds [4] in USA. The matrix A corresponds to the behavior of traders, which play on increase of the course and are called "bulls". The matrix B corresponds to the behavior of traders, which play on the depreciation of the course and are called "bears". Parameters of matrices represent rate of return for stocks and bonds, expressed in the form of interest rates,

$$A = \begin{pmatrix} 10 & 0 \\ 1.75 & 3 \end{pmatrix}, \quad B = \begin{pmatrix} -5 & 3 \\ 10 & 0.5 \end{pmatrix}. \tag{6.26}$$

Characteristic parameters of static games are given at the following levels [27]

$$C_A = a_{11} - a_{12} - a_{21} + a_{22} = 11.25,$$

$$\alpha_1 = a_{22} - a_{12} = 3, \quad \alpha_2 = a_{22} - a_{21} = 1.25,$$

$$x_A = \frac{\alpha_2}{C_A} = 0.11, \quad y_A = \frac{\alpha_1}{C_A} = 0.27;$$

$$C_B = b_{11} - b_{12} - b_{21} + b_{22} = -17.5,$$

$$\beta_1 = b_{22} - b_{12} = -2.5, \quad \beta_2 = b_{22} - b_{21} = -9.5,$$

$$x_B = \frac{\beta_2}{C_B} = 0.54, \quad y_B = \frac{\beta_1}{C_B} = 0.14.$$

On the Fig. 6.5 we present broken lines of players' best replies, saddle points NA, NB in static antagonistic games, the point of the Nash equilibrium NE in the static bimatrix game.

Let us note, that players of the coalition of "bulls" gain in the case of upward trend of markets, when players of both coalitions invest in the same market. And players of the coalition of "bears" make profit from investments in the case of downward trend of markets when players of the coalition of "bulls" move their investments from one market to another.

For the game of coalitions of "bulls" and "bears" we construct switching curves $M_A(\lambda)$, $M_B(\lambda)$ and provide calculations of equilibrium trajectories of the market dynamics with the value of the discount parameter $\lambda = 0.1$.

This calculations are presented on the Fig. 6.6. Here we show saddle points NA, NB in static antagonistic games, the point of the Nash equilibrium NE in the static bimatrix game, switching lines for players' controls $M_A(\lambda) = M_A^1(\lambda) \bigcup M_A^2(\lambda)$ and $M_B(\lambda) = M_B^1(\lambda) \bigcup M_B^2(\lambda)$ in the dynamical bimatrix game with discounted payoff functionals for matrices A, B (6.26). The field of velocities of players is depicted by arrows.

The field of directions generates equilibrium trajectories, one of which is presented on the Fig. 6.6. This trajectory $TR(\lambda) = (x_{AB}^{fl}(\cdot), y_{AB}^{fl}(\cdot))$ starts from the initial position $IP = (0.1, 0.9)$ and moves along the characteristic in the direction of the vertex $(1, 1)$ of the unit square $[0, 1] \times [0, 1]$ with control signals $u = 1$, $v = 1$. Then it crosses the switching line $M_B(\lambda)$, and the second coalition switches the control v from 1 to 0. Then, the trajectory $TR(\lambda)$ moves in the direction of the vertex $(1, 0)$ until it reaches the switching line $M_A(\lambda)$. Here players of the first coalition change the control signal u from 1 to 0. After that the movement of the trajectory is directed along the characteristic to the vertex $(0, 0)$. Then the trajectory crosses the line $M_B(\lambda)$, on which the sliding mode arises, during which the switch of controls of the second coalition occurs, and the trajectory $TR(\lambda)$ converge to

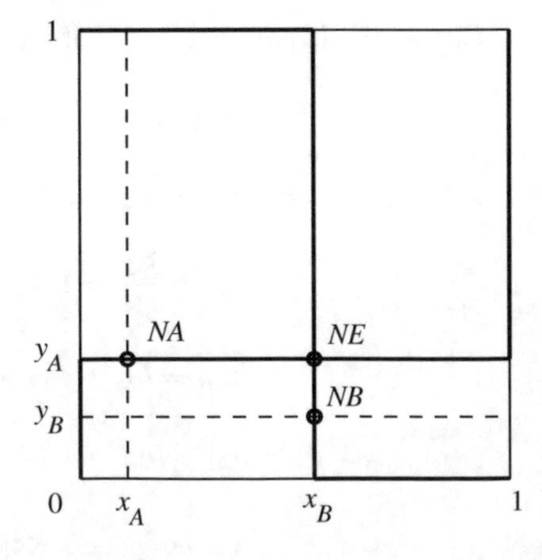

Fig. 6.5 Saddle points NA, NB and the point of the Nash equilibrium NE

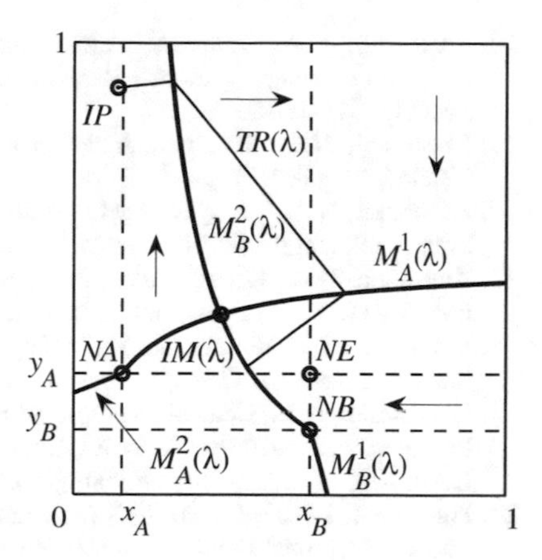

Fig. 6.6 The equilibrium trajectory in the game with discounted payoffs

the point $IM(\lambda) = M_A(\lambda) \bigcap M_B(\lambda)$ of the intersection of switching lines $M_A(\lambda)$, $M_B(\lambda)$.

Acknowledgements The paper is supported by the Project 18-1-1-10 "Development of the concept of feedback control, minimax approach and singular perturbations in the theory of differential equations" of the Integrated Program of UrB RAS

References

1. Arnold, V.I.: Optimization in mean and phase transitions in controlled dynamical systems. Funct. Anal. Appl. **36**, 83–92 (2002). https://doi.org/10.1023/A:1015655005114
2. Basar, T., Olsder, G.J.: Dynamic Noncooperative Game Theory. Academic Press, London (1982)
3. CNN Money [Electronic resource]. http://money.cnn.com/
4. Forex Market [Electronic resource]. http://www.fxstreet.ru.com/
5. Friedman, D.: Evolutionary games in economics. Econometrica **59**(3), 637–666 (1991)
6. Intriligator, M.: Mathematical Optimization and Economic Theory. Prentice-Hall, New York (1971)
7. Klaassen, G., Kryazhimskii, A.V., Tarasyev, A.M.: Multiequilibrium game of timing and competition of gas pipeline projects. J. Optim. Theory Appl. **120**(1), 147–179 (2004)
8. Kleimenov, A.F.: Nonantagonistic Positional Differential Games. Nauka, Yekaterinburg (1993)
9. Kolmogorov, A.N.: On analytical methods in probability theory. Uspekhi Mat. Nauk **5**, 5–41 (1938)
10. Krasovskii, A.N., Krasovskii, N.N.: Control Under Lack of Information. Birkhauser, Boston (1995)
11. Krasovskii, N.N., Subbotin, A.I.: Game-Theoretical Control Problems. Springer, New York (1988)
12. Krasovskii, A.A., Taras'ev, A.M.: Dynamic optimization of investments in the economic growth models. Autom. Remote Control **68**(10), 1765–1777 (2007)
13. Krasovskii, N.A., Tarasyev, A.M.: Search for maximum points of a vector criterion based on decomposition properties. Proc. Steklov Inst. Math. **269**, 174 (2010). https://doi.org/10.1134/S0081543810060155

14. Krasovskii, N.A., Tarasyev, A.M.: Decomposition algorithm of searching equilibria in a dynamic game. Autom. Remote Control **76**, 185 (2015). https://doi.org/10.1134/S0005117915100136
15. Krasovskii, N.A., Tarasyev, A.M.: Equilibrium Solutions in Dynamical Games. UrGAU, Yekaterinburg (2015)
16. Krasovskii, N.A., Tarasyev, A.M.: Equilibrium trajectories in dynamical bimatrix games with average integral payoff functionals. Math. Game Theory Appl. **8**(2), 58–90 (2016)
17. Krasovskii, N.A., Kryazhimskiy, A.V., Tarasyev, A.M.: Hamilton–Jacobi equations in evolutionary games. Proc. Inst. Math. Mech. UrB RAS **20**(3), 114–131 (2014)
18. Kryazhimskii, A.V., Osipov, Yu.S.: On differential-evolutionary games. Proc. Steklov Inst. Math. **211**, 234–261 (1995)
19. Kurzhanskii, A.B.: Control and Observation Under Uncertainty. Nauka, Moscow (1977)
20. Petrosjan, L.A., Zenkevich, N.A.: Conditions for sustainable cooperation. Autom. Remote Control **76**, 84 (2015). https://doi.org/10.1134/S0005117915100148
21. Pontryagin, L.S., Boltyanskii, V.G., Gamkrelidze, R.V., Mischenko, E.F.: The Mathematical Theory of Optimal Processes. Interscience Publishers, New York (1962)
22. Subbotin, A.I.: Minimax Inequalities and Hamilton–Jacobi Equations. Nauka, Moscow (1991)
23. Subbotin, A.I., Tarasyev, A.M.: Conjugate derivatives of the value function of a differential game. Dokl. AN SSSR **283**(3), 559–564 (1985)
24. Subbotina, N.N.: The Cauchy method of characteristics and generalized solutions of Hamilton–Jacobi–Bellman equations. Dokl. Acad. Nauk SSSR **320**(3), 556–561 (1991)
25. Tarasyev, A.M.: A Differential model for a 2×2-evolutionary game dynamics. IIASA Working Paper, Laxenburg, Austria, WP-94-063 (1994). http://pure.iiasa.ac.at/4148/1/WP-94-063.pdf
26. Ushakov, V.N., Uspenskii, A.A., Lebedev, P.D.: Geometry of singular curves of a class of time-optimal problems. Vestn. Sankt-Peterburgsk. Univ. **10**(3), 157–167 (2013)
27. Vorobyev, N.N.: Game Theory for Economists and System Scientists. Nauka, Moscow (1985)

Chapter 7
On Control Reconstruction Problems for Dynamic Systems Linear in Controls

Evgeniy Krupennikov

Abstract In differential games the a posteriori analysis of motions, namely, trajectories of the dynamics and the analysis of the players' controls generating these trajectories are very important. This paper is devoted to solving problems of reconstruction of trajectories and controls in differential games using known history of inaccurate measurements of a realized trajectory. A new method for solving reconstruction problems is suggested and justified for a class of differential games with dynamics, linear in controls and non-linear in state coordinates. This method relies on necessary optimality conditions in auxiliary variational problems. An illustrating example is exposed.

7.1 Introduction

This paper is devoted to solving inverse problems of reconstruction of players' trajectories and controls in differential games, using known inaccurate measurements of the realized trajectories. The a posteriori analysis is an important part of the decision making in the future. Inverse problems may occur in many areas such as economics, engineering, medicine and many others that involve the task of reconstruction of the players' controls by known inaccurate trajectory measurements.

The inverse problems have been studied by many authors. The approach suggested by Osipov and Kryazhimskii [6, 7] is one of the closest to the material of this paper. The method suggested by them reconstructs the controls by using a regularized (a variation of Tikhonov regularization [12]) procedure of control with

E. Krupennikov (✉)
Krasovskii Institute of Mathematics and Mechanics UrB RAS, Yekaterinburg, Russia

Ural Federal University, Yekaterinburg, Russia
e-mail: krupennikov@imm.uran.ru

L. A. Petrosyan et al. (eds.), *Frontiers of Dynamic Games*,
Static & Dynamic Game Theory: Foundations & Applications,
https://doi.org/10.1007/978-3-319-92988-0_7

a guide. This procedure allows to reconstruct the controls on-line. It is originated from the works of Krasovskii's school on the theory of optimal feedback [3, 4].

Another method for solving dynamic reconstruction problems by known history of inaccurate measurements has been suggested by Subbotina et al. [10]. It is based on a method, which use necessary optimality conditions for auxiliary optimal control problems [9]. This method has been also developed in [5, 8, 10, 11]. A modification of this approach is presented in this paper. It relies on necessary optimality conditions in an auxiliary variational problem on extremum for an integral functional. The functional is a variation of a Tikhonov regularizator.

In this paper the suggested method is justified for a special class of differential games with dynamics linear in controls and non-linear in state coordinates. Results of simulation are exposed.

7.2 Dynamics

We consider a differential game with dynamics of the form

$$\dot{x}(t) = G(x(t), t)u(t), \quad x(\cdot) : [0, T] \to R^n, \quad u(\cdot) : [0, T] \to R^n, \quad t \in [0, T]. \tag{7.1}$$

Here $G(x, t)$ is an $n \times n$ matrix with elements $g_{ij}(x, t) : R^n \times [0, T] \to R$, $i = 1, ..., n$, $j = 1, ..., n$ that have continuous derivatives

$$\frac{\partial g_{ij}(x, t)}{\partial t}, \quad \frac{\partial g_{ij}(x, t)}{\partial x_k}, \quad i = 1, ..., n, \ j = 1, ..., n, \ k = 1, ..., n,$$
$$x \in R^n, \ t \in [0, T].$$

In (7.1) $x_i(t)$ is the state of the ith player, while $u_i(t)$ is the control of the ith player, restricted by constraints

$$|u_i(t)| \leq \overline{U} < \infty, \quad i = 1, \ldots, n, \quad t \in [0, T]. \tag{7.2}$$

We consider piecewise continuous controls with finite number of points of discontinuity.

7.3 Input Data

It is supposed that some base trajectory $x^*(\cdot) : [0, T] \to R^n$ of system (7.1) has been realized on the interval $t \in [0, T]$. Let $u^*(\cdot) : [0, T] \to R^n$ be the piecewise continuous control satisfying constrains (7.2) that generated this trajectory.

We assume that measurements $y^\delta(\cdot, \delta) = y^\delta(\cdot) : [0, T] \to R^n$ of the base trajectory $x^*(t)$ are known and they are twice continuously differentiable functions that determine $x^*(t)$ with the known accuracy $\delta > 0$, i.e.

$$|y_i^\delta(t) - x_i^*(t)| \le \delta, \quad i = 1, ..., n, \quad t \in [0, T]. \tag{7.3}$$

7.4 Hypotheses

We introduce two hypotheses on the input data.

Hypothesis 7.1 *There exist such compact set $\Psi \subset R^n$, such constant $r > 0$ and such constants $\underline{\omega} > 0$, $\overline{\omega} > 0$, $\omega' > 0$ that*

$$\begin{gathered}
\Psi \supset \{x \in R^n : |x_i - x_i^*(t)| \le r \quad \forall t \in [0, T]\}, \\
0 < \underline{\omega}^2 \le |det G(x, t)| \le \overline{\omega}^2, \quad \left|\frac{\partial g_{ij}(x, t)}{\partial t}\right| \le \omega', \quad \left|\frac{\partial g_{ij}(x, t)}{\partial x_k}\right| \le \omega', \\
i = 1, \dots, n, \quad j = 1, \dots, n, \quad k = 1, \dots, n, \quad x \in \Psi, \quad t \in [0, T].
\end{gathered} \tag{7.4}$$

Let's introduce the following constants

$$\begin{gathered}
R_1 = \frac{\pi \overline{\omega}}{\underline{\omega}^3}, \quad R_2 = \frac{\pi \overline{\omega}^2}{\underline{\omega}^4}(\overline{\omega}^2 + \omega'), \quad R_3 = \frac{\pi n T \overline{\omega}^4 \omega'}{\underline{\omega}^2}, \\
R_4 = \frac{T \pi \overline{\omega}^2}{2\underline{\omega}^4} + 4\frac{\pi \overline{\omega}^2}{\underline{\omega}} + 2\frac{R_1 \pi \overline{\omega}^2}{\underline{\omega}}\left(\ln \frac{0.5 T \overline{\omega}}{\underline{\omega}^2} + 1\right) + R_2 + 1, \\
R_w = \max\{\frac{1 + \underline{\omega}}{\underline{\omega}}(R_4 + 2R_3), (1 + \underline{\omega})(R_4 + 2R_3)\},
\end{gathered} \tag{7.5}$$

which will be used in Hypothesis 7.2 and Theorem 7.1.

Hypothesis 7.2 *There exist such constants $\delta_0 \in (0, \min\{0.5r, \frac{1}{R_w}\}]$ and $\overline{Y} > 0$ that for any $\delta \in (0, \delta_0]$*

$$|y_i^\delta(t)| \le \overline{Y}, \quad |\dot{y}_i^\delta(t)| \le \overline{Y}, \quad \left|\dot{x}_i^*(t)\right| \le \overline{Y}, \quad t \in [0, T], \quad i = 1, \dots, n \tag{7.6}$$

and for any $\delta \in (0, \delta_0]$ exists such compact $\Omega^\delta \subset [0, T]$ with measure $\mu\Omega^\delta = \beta^\delta \overset{\delta \to 0}{\longrightarrow}$ that

$$|\ddot{y}_i^\delta(t)| \le \overline{Y}, \; T \in [0, T] \setminus \Omega^\delta, \quad \max_{t \in \Omega^\delta} |\ddot{y}_i^\delta(t)| \beta^\delta \le \overline{Y}, \quad i = 1, \dots, n. \tag{7.7}$$

Remark 7.1 Conditions (7.6) reflect the fact that the right hand sides of Eq. (7.1) are restricted.

Remark 7.2 In Hypothesis 7.2 the constant $\overline{Y}$ is unified for all inequalities to simplify the further calculations and explanations.

Remark 7.3 Hypothesis 7.2 allows the functions $\dot{y}^{\delta}(\cdot)$ to be able to approximate piecewise continuous functions $\dot{x}^{*}(\cdot) = g(x^{*}(\cdot), \cdot)u^{*}(\cdot)$.

7.5 Problem Statement

Let's consider the following reconstruction problem: for a given $\delta \in (0, \delta_0]$ and a given measurement function $y^{\delta}(\cdot)$ fulfilling estimates (7.3) and Hypothesis 7.2 to find a function $u(\cdot, \delta) = u^{\delta}(\cdot) : [0, T] \to R^n$ that satisfies the following conditions:

1. The function $u^{\delta}(\cdot)$ belongs to the set of admissible controls, i.e. the set of piecewise continuous functions with finite number of points of discontinuity satisfying constraints (7.2);
2. The control $u^{\delta}(\cdot)$ generates trajectory $x(\cdot, \delta) = x^{\delta}(\cdot) : [0, T] \to R^n$ of system (7.1) with boundary condition $x^{\delta}(T) = y^{\delta}(T)$. In other words, there exists a unique solution $x^{\delta}(\cdot) : [0, T] \to R^n$ of the system

$$\dot{x}^{\delta}(t) = G(x^{\delta}(t), t)u^{\delta}(t), \quad t \in [0, T]$$

   that satisfy the boundary condition $x^{\delta}(T) = y^{\delta}(T)$.
3. Functions $x^{\delta}(\cdot)$ and $u^{\delta}(\cdot)$ satisfy conditions

$$\lim_{\delta \to 0} \|x_i^{\delta}(\cdot) - x_i^{*}(\cdot)\|_{C_{[0,T]}} = 0, \quad \lim_{\delta \to 0} \|u_i^{\delta}(\cdot) - u_i^{*}(\cdot)\|_{L_{2,[0,T]}} = 0, \quad i = 1, ..., n. \tag{7.8}$$

Hereinafter

$$\|f(\cdot)\|_{C_{[0,T]}} = \max_{t \in [0,T]} |f(t)|, \quad f(\cdot) : [0, T] \to R$$

is the norm in the space of continuous functions C and

$$\|f(\cdot)\|_{L_{2,[0,T]}} = \sqrt{\int_0^T \sum_{i=1}^{n} f_i^2(\tau) d\tau}, \quad f(\cdot) : [0, T] \to R^n$$

is the norm in space L_2.

7.6 A Solution of the Inverse Problem

7.6.1 Auxiliary Problem

To solve the inverse problem in Sect. 7.5, we introduce an auxiliary variational problem (AVP) for fixed parameters $\delta \in (0, \delta_0]$, $\alpha > 0$ and a given measurement function $y^\delta(\cdot)$ satisfying estimates (7.3) and Hypothesis 7.2.

We consider the set of pairs of continuously differentiable functions $F_{xu} = \{\{x(\cdot), u(\cdot)\} : x(\cdot) : [0, T] \to R^n,\ u(\cdot) : [0, T] \to R^n\}$ that satisfy differential equations (7.1) and the following boundary conditions

$$x(T) = y^\delta(T),\ u(T) = G^{-1}(y^\delta(T), T)\dot{y}^\delta(T). \tag{7.9}$$

Hereinafter G^{-1} is the inverse matrix for non degenerate matrix G. Let us remark that due to Hypothesis 7.1, the inverse matrix $G^{-1}(y^\delta(T), T)$ exists.

AVP is to find a pair of functions $x(\cdot, \delta, \alpha) = x^{\delta,\alpha}(\cdot) : [0, T] \to R^n$ and $u(\cdot, \delta, \alpha) = u^{\delta,\alpha}(\cdot) : [0, T] \to R^n$ such that $\{x^{\delta,\alpha}(\cdot), u^{\delta,\alpha}(\cdot)\} \in F_{xu}$ and such that they provide an extremum for the integral functional

$$I(x(\cdot), u(\cdot)) = \int_0^T \left[-\frac{\|x(t) - y^\delta(t)\|^2}{2} + \frac{\alpha^2\|u(t)\|^2}{2} \right] dt. \tag{7.10}$$

Here α is a small regularising parameter [12] and $\|f\| = \sqrt{\sum_{i=1}^{n} f_i^2}$, $f \in R^n$ is Euclidean norm in R^n.

7.6.2 Necessary Optimality Conditions in the AVP

We can write the necessary optimality conditions for the AVP (7.1), (7.10), (7.9) in Lagrange form [14]. Lagrangian for the AVP has the form

$$L(x, u, \dot{x}, \lambda(t), t) = -\frac{\|x - y^\delta(t)\|^2}{2} + \frac{\alpha^2\|u\|^2}{2} + \sum_{i=1}^{n} \left[\lambda_i(t) \sum_{j=1}^{n} \left[\dot{x}_i - g_{ij}(x, t)u_j\right] \right],$$

where $\lambda(t) : [0, T] \to R^n$ is the Lagrange multipliers vector.

The $2n$ corresponding Euler equations are

$$\begin{gathered}\dot{\lambda}_i(t) + (x_i(t) - y_i^\delta(t)) + \sum_{j=1}^{n}\left[\lambda_j(t)\sum_{k=1}^{n} u_k(t)\frac{\partial g_{jk}}{\partial x_i}(x(t), t)\right] = 0, \\ -\alpha^2 u_i(t) + \sum_{j=1}^{n}\left[\lambda_j(t) g_{ji}(x(t), t)\right] = 0, \quad i = 1, \ldots, n.\end{gathered} \tag{7.11}$$

The first n equations in (7.11) can be rewritten in vector form:

$$\dot{\lambda}_i(t) + (x_i(t) - y_i^\delta(t)) + \langle \lambda_j(t), \frac{\partial G}{\partial x_i}(x(t), t)u(t)\rangle = \overrightarrow{0}, \quad i = 1, \ldots, n. \tag{7.12}$$

Hereinafter $\langle a, b\rangle$ means the scalar product of vectors $a \in R^n$, $b \in R^n$ and $\frac{\partial G}{\partial x_i}(x(t), t)$ is a matrix with elements $\frac{\partial g_{jk}}{\partial x_i}(x(t), t)$, $j = 1, ..., n$, $k = 1, ..., n$.

The last n equations in (7.11) define the relations between the controls $u_i(t)$ and the Lagrange multipliers $\lambda_i(t)$, $i = 1, \ldots, n$:

$$u(t) = \frac{1}{\alpha^2} G^T(x(t), t)\lambda(t). \tag{7.13}$$

Hereinafter G^T means transpose of a matrix G.

We can substitute equations (7.13) into (7.12) and (7.1) to rewrite them in the form of Hamiltonian equations, where the vector $s(t) = -\lambda(t)$ plays the role of the adjoint variables vector:

$$\begin{aligned}&\dot{x}(t) = -(1/\alpha^2)G(x(t), t)G^T(x(t), t)s(t), \\ &\dot{s}_i(t) = x_i(t) - y_i^\delta(t) + \frac{1}{\alpha^2}\langle s(t), \frac{\partial G}{\partial x_i}(x(t), t)G^T(x(t), t)s(t)\rangle, \quad i = 1, \ldots, n.\end{aligned} \tag{7.14}$$

By substituting (7.13) into (7.9), one can obtain boundary conditions, written for system (7.14):

$$x(T) = y^\delta(T), \quad s(T) = -\alpha^2\big(G(y^\delta(T), T)G^T(y^\delta(T), T)\big)^{-1}\dot{y}^\delta(T). \tag{7.15}$$

Thus, we have got the necessary optimality conditions for the AVP (7.1), (7.10), (7.9) in Hamiltonian form (7.14), (7.15).

7.6.3 A Solution of the Reconstruction Problem

Let's introduce the function

$$u^{\delta,\alpha}(\cdot) = -(1/\alpha^2)G^T(x^{\delta,\alpha}(\cdot), \cdot)s^{\delta,\alpha}(\cdot), \tag{7.16}$$

where $x^{\delta,\alpha}(\cdot)$, $s^{\delta,\alpha}(\cdot)$ are the solutions of system (7.14) with boundary conditions (7.15).

We now introduce the cut-off functions

$$\hat{u}_i^{\delta}(t) = \begin{cases} \overline{U}, & u_i^{\delta,\alpha}(t) \geq \overline{U}, \\ u_i^{\delta,\alpha}(t), & |u_i^{\delta,\alpha}(t)| < \overline{U}, \\ -\overline{U}, & u_i^{\delta,\alpha}(t) \leq -\overline{U}. \end{cases} \quad i = 1, \ldots, n. \tag{7.17}$$

We consider the functions $\hat{u}_i^{\delta}(\cdot)$ as the solutions of the inverse problem described in Sect. 7.5. We choose $\alpha = \alpha(\delta)$ in a such way that $\alpha(\delta) \stackrel{\delta \to 0}{\longrightarrow} 0$.

7.6.4 Convergence of the Solution

In this paper a justification for the suggested method is presented for one sub-class of considered differential games (7.1), (7.2). Namely, we consider from now dynamics of form (7.1), where matrixes $G(x, t)$ are diagonal with non-zero elements on the diagonals. The dynamics in such case have the form

$$\dot{x}_i(t) = g_i(x(t), t)u_i(t), \quad i = 1, ..., n,$$

where the functions $g_i(x, t) = g_{ii}(x, t)$, $i = 1, \ldots, n$ are the elements on the diagonal of the matrix $G(x, t)$.

Condition $\underline{\omega}^2 \leq |detG(x, t)| \leq \overline{\omega}^2$ in Hypothesis 7.1 in such case is replaced by equal condition

$$\underline{\omega}^2 \leq g_i^2(x, t) \leq \overline{\omega}^2, \quad i = 1, \ldots, n. \tag{7.18}$$

Necessary optimality conditions (7.14) has now the form

$$\begin{gathered} \dot{x}_i(t) = -s_i(t)\frac{g_i^2(x(t), t)}{\alpha^2}, \\ \dot{s}_i(t) = x_i(t) - y_i^{\delta}(t) + \frac{1}{\alpha^2}\sum_{j=1}^{n}\left[s_j^2(t)\frac{\partial g_j(x(t), t)}{\partial x_i(t)}g_j(x(t), t))\right], \\ i = 1, \ldots, n \end{gathered} \tag{7.19}$$

with boundary conditions

$$x_i(T) = y_i^{\delta}(T), \quad s_i(T) = -\alpha^2 \dot{y}_i^{\delta}(T)/g_i^2(y^{\delta}(T), T), \quad i = 1, \ldots, n. \tag{7.20}$$

The following lemma is true.

Lemma 7.1 *For $\delta \in (0, \delta_0]$ twice continuously differentiable measurement functions $y_i^\delta(\cdot)$, $i = 1, \ldots, n$ satisfying estimates (7.3) and Hypothesis 7.2 fulfill the following relations*

$$\lim_{\delta \to 0} \|y_i^\delta(\cdot) - x_i^*(\cdot)\|_{C[0,T]} = 0, \ \lim_{\delta \to 0} \|\frac{\dot{y}_i^\delta(\cdot)}{g_i(y^\delta(\cdot), \cdot)} - u_i^*(\cdot)\|_{L_{2,[0,T]}} = 0, \ i = 1, ..., n. \tag{7.21}$$

Proof The first relation in (7.21) is true due to (7.3). Let's prove the second one.

Relying upon Luzin's theorem [2] one can find for the piecewise continuous function $u^*(\cdot)$ such constant $\overline{Y}^u$ that for any $\delta \in (0, \delta_0]$ and all $i = 1, \ldots, n$ there exist such twice continuously differentiable functions $\overline{u}_i^\delta(\cdot) : [0, T] \to R$ and such set $\Omega_u^\delta \subset R$ with measure $\mu\Omega_u^\delta = \beta_u^\delta$ that

$$\begin{gathered} |\overline{u}_i^\delta(t)| \le \overline{Y}^u, \ t \in [0, T], \quad |\dot{\overline{u}}_i^\delta(t)| \le \overline{Y}^u, \ t \in [0, T] \setminus \Omega_u^\delta, \\ \beta_u^\delta \max_{t \in \Omega_u^\delta} |\dot{\overline{u}}_i^\delta(t)| \le \overline{Y}^u, \\ \|\overline{u}_i^\delta(\cdot) - u_i^*(\cdot)\|_{L_{2,[0,T]}} \le \delta, \quad i = 1, \ldots, n. \end{gathered} \tag{7.22}$$

Let's estimate the following expression first (hereinafter in the proof $i = 1, \ldots, n$).

$$\|\dot{y}_i^\delta(\cdot) - \overline{u}_i^\delta(\cdot) g_i(y^\delta(t), t)\|^2_{L_{2,[0,T]}} = \int_0^T \big(\dot{y}_i^\delta(t) - \overline{u}_i^\delta(t) g_i(y^\delta(t), t)\big)^2 dt. \tag{7.23}$$

The integral in (7.23) can be calculated by parts.

$$\begin{aligned} &\int_0^T \underbrace{\big(\dot{y}_i^\delta(t) - \overline{u}_i^\delta(t) g_i(y^\delta(t), t)\big)}_{\mathbf{U}} \underbrace{\big(\dot{y}_i^\delta(t) - \overline{u}_i^\delta(t) g_i(y^\delta(t), t)\big) dt}_{\mathbf{dV}} \\ &= \Bigg[\underbrace{\big(\dot{y}_i^\delta(t) - \overline{u}_i^\delta(t) g_i(y^\delta(t), t)\big)}_{\mathbf{U}} \underbrace{\Big(y_i^\delta(t) - x_i^*(0) - \int_0^t \overline{u}_i^\delta(\tau) g_i(y^\delta(\tau), \tau) d\tau\Big)}_{\mathbf{V}} \Bigg]\Bigg|_0^T \\ &\quad - \int_0^T \underbrace{\Big(y_i^\delta(t) - x_i^*(0) - \int_0^t \overline{u}_i^\delta(\tau) g_i(y^\delta(\tau), \tau) d\tau\Big)}_{\mathbf{V}} \\ &\cdot \underbrace{\Big(\ddot{y}_i^\delta(t) - \dot{\overline{u}}_i^\delta(t) g_i(y^\delta(t), t) - \overline{u}_i^\delta(t) \Big(\sum_{j=1}^n [g'_{i,x_j}(y_i^\delta(t), t) \dot{y}_j^\delta] + g'_{i,t}(y_i^\delta(t), t)\Big)\Big) dt}_{\mathbf{dU}} \end{aligned} \tag{7.24}$$

To estimate the whole expression (7.24) we first estimate the difference $\mathbf{V} = y_i^\delta(t) - x_i^*(0) - \int_0^t \overline{u}_i^\delta(\tau) g_i(y^\delta(\tau), \tau) d\tau$. In order to do this, we estimate integral

$$\int_0^t \left(\overline{u}_i^\delta(\tau) - u_i^*(\tau)\right) g_i(y^\delta(\tau), \tau) d\tau = \int_{\Omega^t_{\geq\delta}} \left(\overline{u}_i^\delta(\tau) - u_i^*(\tau)\right) g_i(y^\delta(\tau), \tau) d\tau + \int_{\Omega^t_{<\delta}} \left(\overline{u}_i^\delta(\tau) - u_i^*(\tau)\right) g_i(y^\delta(\tau), \tau) d\tau, \tag{7.25}$$

where set $\Omega^t_{\geq\delta} = \{\tau \in [0, t] : |\overline{u}_i^\delta(\tau) - u_i^*(\tau)| \geq \delta\}$ and set $\Omega^t_{<\delta} = \{\tau \in [0, t] : |\overline{u}_i^\delta(\tau) - u_i^*(\tau)| < \delta\}$.

The first term in (7.25)

$$\left| \int_{\Omega^t_{<\delta}} \left(\overline{u}_i^\delta(\tau) - u_i^*(\tau)\right) g_i(y^\delta(\tau), \tau) d\tau \right| \leq \delta\mu(\Omega^t_{<\delta})\overline{\omega} \leq \delta T \overline{\omega}. \tag{7.26}$$

Remark 7.4 Let's remember that hereinafter when the first argument of functions $g_i(x, t),\ i = 1, \ldots, n$ belongs to compact Ψ from Hypothesis 7.1, relations (7.18) are true.

Using (7.22), the second term in (7.25)

$$\begin{aligned} \left| \int_{\Omega^t_{<\delta}} \left(\overline{u}_i^\delta(\tau) - u_i^*(\tau)\right) g_i(y^\delta(\tau), \tau) d\tau \right| &= \left| \int_{\Omega^t_{<\delta}} \left(\overline{u}_i^\delta(\tau) - u_i^*(\tau)\right)^2 \frac{g_i(y^\delta(\tau), \tau)}{\overline{u}_i^\delta(\tau) - u_i^*(\tau)} d\tau \right| \\ &\leq \max_{\tau \in \Omega^t_{<\delta}} \left| \frac{g_i(y^\delta(\tau), \tau)}{\left(\overline{u}_i^\delta(\tau) - u_i^*(\tau)\right)} \right| \int_{\Omega^t_{<\delta}} \left(\overline{u}_i^\delta(\tau) - u_i^*(\tau)\right)^2 d\tau \\ &\leq \frac{\overline{\omega}}{\delta} \int_0^T \left(\overline{u}_i^\delta(\tau) - u_i^*(\tau)\right)^2 d\tau \leq \delta T \overline{\omega}. \end{aligned} \tag{7.27}$$

From (7.25), (7.26) and (7.27) follows that

$$\left| \int_0^t \left(\overline{u}_i^\delta(\tau) - u_i^*(\tau)\right) g_i(y^\delta(\tau), \tau) dt \right| \leq \delta 2 T \overline{\omega}. \tag{7.28}$$

We can now estimate function $\mathbf{V}$ in (7.24):

$$\Big|y_i^\delta(t) - x_i^*(0) - \int_0^t \overline{u}_i^\delta(\tau) g_i(y^\delta(\tau),\tau)d\tau\Big|$$

$$\leq \Big|y_i^\delta(t) - x_i^*(0) - \int_0^t u_i^*(\tau) g_i(y^\delta(\tau),\tau)d\tau\Big|$$

$$+\Big|\int_0^t \big(\overline{u}_i^\delta(\tau) - u_i^*(\tau)\big) g_i(y^\delta(\tau),\tau)d\tau\Big|$$

$$\leq \Big|y_i^\delta(t) - x_i^*(0) - \int_0^t u_i^*(\tau) g_i(x^*(\tau),\tau)d\tau\Big|$$

$$+\Big|\int_0^t u_i^*(\tau)\big(g_i(y^\delta(\tau),\tau) - g_i(x^*(\tau),\tau)\big)d\tau\Big|$$

$$+\delta 2\overline{\omega}T \leq \Big|y_i^\delta(t) - x_i^*(t)\Big|$$

$$+\Big|\int_0^t \overline{U}\big(n \max_{\theta\in[0,T],\ j=1,\dots,n} |g'_{i,x_j}(y^\delta(\theta),\theta)(x_j^*(\theta) - y_j^\delta(\theta))|\big)d\tau\Big| + \delta 2T\overline{\omega} \leq$$

$$\delta(1 + T\overline{U}n\omega' + 2\overline{\omega}T) \overset{def}{=} \delta R_u. \tag{7.29}$$

Thus, the term $\mathbf{U}\mathbf{V}|_0^T$ in sum (7.24) can be estimated as

$$\Big(\dot{y}_i^\delta(t) - \overline{u}_i^\delta(t) g_i(y^\delta(t),t)\Big)\Big(y_i^\delta(t) - x_i^*(0) - \int_0^t \overline{u}_i^\delta(\tau) g_i(y^\delta(\tau),\tau)d\tau\Big)\Big|_0^T$$

$$\leq 2\delta\big(\overline{Y} + \overline{Y}^u\overline{\omega}\big)R_u. \tag{7.30}$$

Using (7.7), (7.22) and (7.29), the term $\int_0^T \mathbf{V}d\mathbf{U}dt$ in (7.24) can be estimated in the following way

$$\Big|\int_0^T \Big[\Big(y_i^\delta(t) - x_i^*(0) - \int_0^t \overline{u}_i^\delta(\tau) g_i(y^\delta(\tau),\tau)d\tau\Big)$$

$$\cdot\Big(\ddot{y}_i^\delta(t) - \dot{\overline{u}}_i^\delta(t) g_i(y^\delta(t),t) - \overline{u}_i^\delta(t)\big(\sum_{j=1}^n [g'_{i,x_j}(y_i^\delta(t),t)\dot{y}_j^\delta] + g'_{i,t}(y_i^\delta(t),t)\big)\Big)\Big]dt\Big|$$

$$\leq \Big| \int\limits_{[0,T]\setminus\Omega^\delta} \Big(y_i^\delta(t) - x_i^*(0) - \int\limits_0^t \overline{u}_i^\delta(\tau) g_i(y^\delta(\tau),\tau)d\tau\Big)\ddot{y}_i^\delta(t)dt\Big|$$

$$+\Big| \int\limits_{\Omega^\delta} \Big(y_i^\delta(t) - x_i^*(0) - \int\limits_0^t \overline{u}_i^\delta(\tau) g_i(y^\delta(\tau),\tau)d\tau\Big)\ddot{y}_i^\delta(t)dt\Big|$$

$$+\Big| \int\limits_{[0,T]\setminus\Omega_u^\delta} \Big(y_i^\delta(t) - x_i^*(0) - \int\limits_0^t \overline{u}_i^\delta(\tau) g_i(y^\delta(\tau),\tau)d\tau\Big)\dot{\overline{u}}_i^\delta(t) g_i(y^\delta(t),t)dt\Big|$$

$$+\Big| \int\limits_{\Omega_u^\delta} \Big(y_i^\delta(t) - x_i^*(0) - \int\limits_0^t \overline{u}_i^\delta(\tau) g_i(y^\delta(\tau),\tau)d\tau\Big)\dot{\overline{u}}_i^\delta(t) g_i(y^\delta(t),t)dt\Big|$$

$$+\Big| \int\limits_0^T \Big(y_i^\delta(t) - x_i^*(0) - \int\limits_0^t \overline{u}_i^\delta(\tau) g_i(y^\delta(\tau),\tau)d\tau\Big)$$

$$\cdot\Big(\sum_{j=1}^n [g'_{i,x_j}(y_i^\delta(t),t)\dot{y}_j^\delta] + g'_{i,t}(y_i^\delta(t),t)\Big)dt\Big|$$

$$\leq \delta T R_u \overline{Y} + \delta R_u \overline{Y} + \delta T R_u \overline{Y}^u \overline{\omega} + \delta R_u \overline{Y}^u \overline{\omega} + \delta T R_u \omega'(n\overline{Y} + 1) \stackrel{def}{=} \delta \overline{R}_u. \tag{7.31}$$

Combining estimates (7.30) and (7.31), we can now estimate expression (7.24).

$$\int\limits_0^T \big(\dot{y}_i^\delta(t) - \overline{u}_i^\delta(t) g_i(y^\delta(t),t)\big)^2 dt \leq \delta\big(\overline{R}_u + 2(\overline{Y} + \overline{Y}^u\overline{\omega})R_u\big) \tag{7.32}$$

Finally, we can use the first mean value theorem for definite integrals and estimate (7.32) to get

$$\Big|\int\limits_0^T \Big(\frac{\dot{y}^\delta(t)}{g_i(y^\delta(t),t)} - \overline{u}_i^\delta(t)\Big)^2 dt\Big|$$

$$\leq \max_{t\in[0,T]}\left[\frac{1}{g_i^2(y^\delta(t),t)}\right]\int\limits_0^T \big(\dot{y}_i^\delta(t) - \overline{u}_i^\delta(t) g_i(y^\delta(t),t)\big)^2 dt \tag{7.33}$$

$$\leq \delta\frac{2\big(\overline{Y} + \overline{Y}^u\overline{\omega}\big)R_u + \overline{R}_u}{\underline{\omega}^2} \xrightarrow{\delta\to 0} 0.$$

It follows from (7.33) that

$$\lim_{\delta\to 0}\left\|\frac{\dot{y}^\delta(\cdot)}{g_i(y^\delta(\cdot),\cdot)}-\overline{u}_i^\delta(\cdot)\right\|_{L_{2,[0,T]}}=0.$$

Remember that we consider such function $\overline{u}_i^\delta(\cdot)$ that $\lim\limits_{\delta\to 0}\left\|\overline{u}_i^\delta(\cdot)-u_i^*(\cdot)\right\|_{L_{2,[0,T]}}=0$. So, from the triangle inequality $\|f_1(\cdot)+f_2(\cdot)\|_{L_{2,[0,T]}}\le\|f_1(\cdot)\|_{L_{2,[0,T]}}+\|f_2(\cdot)\|_{L_{2,[0,T]}}$ follows that

$$\lim_{\delta\to 0}\left\|\frac{\dot{y}^\delta(\cdot)}{g_i(y^\delta(\cdot),\cdot)}-u_i^*(\cdot)\right\|_{L_{2,[0,T]}}=0,\quad i=1,\dots,n,$$

which was to be proved. □

Theorem 7.1 *For any fixed $\delta\in(0,\delta_0]$ there exists such parameter $\alpha_0^\delta=\alpha_0^\delta(\delta)$ that the solution $x^{\delta,\alpha_0^\delta}(\cdot)$, $s^{\delta,\alpha_0^\delta}(\cdot)$ of system (7.19) with boundary conditions (7.20) is extendable and unique on $t\in[0,T]$.*

Moreover, $\lim\limits_{\delta\to 0}\alpha_0^\delta(\delta)=0$ and

$$\lim_{\delta\to 0}\|x_i^{\delta,\alpha_0^\delta}(\cdot)-x_i^*(\cdot)\|_{C_{[0,T]}}=0,\ \lim_{\delta\to 0}\|u_i^{\delta,\alpha_0^\delta}(\cdot)-u_i^*(\cdot)\|_{L_{2,[0,T]}}=0,\ i=1,...,n,\tag{7.34}$$

where

$$u_i^{\delta,\alpha_0^\delta}(\cdot)=-(1/(\alpha_0^\delta)^2)g_i(x^{\delta,\alpha_0^\delta}(\cdot),\cdot)s_i^{\delta,\alpha_0^\delta}(\cdot),\quad i=1,\dots,n.\tag{7.35}$$

Proof Let's introduce new variables:

$$z_i(t)=x_i(t)-y_i^\delta(t),\quad w_i(t)=s_i(t)+\frac{\alpha^2\dot{y}_i^\delta(t)}{g_i^2(x(t),t)},\quad i=1,\dots,n.\tag{7.36}$$

Their derivatives are

$$\begin{aligned}\dot{z}_i(t)&=\dot{x}_i(t)-\dot{y}_i^\delta(t),\\ \dot{w}_i(t)&=\dot{s}_i(t)+\frac{\alpha^2\ddot{y}_i^\delta(t)}{g_i^2(x(t),t)}-2\frac{\alpha^2\sum\limits_{j=1}^n\left[g_i{}'_{x_j}(x(t),t)\dot{x}_j(t)\right]}{g_i^3(x(t),t)}\\ &=\dot{s}_i(t)+\frac{\alpha^2\ddot{y}_i^\delta(t)}{g_i^2(x(t),t)}+2\frac{\sum\limits_{j=1}^n\left[g_i{}'_{x_j}(x(t),t)s_j(t)g_j^2(x(t),t)\right]}{g_i^3(x(t),t)},\quad i=1,\dots,n.\end{aligned}\tag{7.37}$$

System (7.19) can be rewritten in this variables as

$$\begin{aligned}\dot{z}_i(t) &= -w_i(t)\frac{g_i^2(z(t)+y^\delta(t),t)}{\alpha^2},\\ \dot{w}_i(t) &= z_i(t) + F_i(z(t),w(t),t), \quad i=1,\ldots,n.\end{aligned} \tag{7.38}$$

where

$$\begin{gathered}F_i(z(t),w(t),t) = \alpha^2\Bigg(\frac{\ddot{y}_i^\delta(t)}{g_i^2(z(t)+y^\delta(t),t)} + \\ \sum_{j=1}^{n}\Bigg[2\frac{g_j^2(z(t)+y^\delta(t),t)g_i{}'_{x_j}(z(t)+y^\delta(t),t)\left(\frac{w_j(t)}{\alpha^2}-\frac{\dot{y}_j^\delta(t)}{g_j^2(z(t)+y^\delta(t),t)}\right)}{g_i^3(z(t)+y^\delta(t),t)} \\ +\frac{g_j{}'_{x_i}(z(t)+y^\delta(t),t)(\dot{y}_j^\delta(t))^2}{g_j^3(z(t)+y^\delta(t),t)} + \frac{g_j(z(t)+y^\delta(t),t)g_j{}'_{x_i}(z(t)+y^\delta(t),t)w_j^2(t)}{\alpha^4} \\ -2\frac{g_j{}'_{x_i}(z(t)+y^\delta(t),t)\dot{y}_j^\delta(t)w_j(t)}{\alpha^2 g_j(z(t)+y^\delta(t),t)}\Bigg]\Bigg), \quad i=1,\ldots,n.\end{gathered} \tag{7.39}$$

Boundary conditions (7.20) in new variables take the form

$$z(T) = 0, \quad w(T) = 0. \tag{7.40}$$

As it follows from Hypothesis 7.1, the right hand side of system (7.38) is locally Lipschitz on $\Psi \times [0,T]$—so, by Cauchy theorem there exists such interval $[T_0,T] \subset [0,T]$ that solutions $z^{\delta,\alpha}(\cdot) : [T_0,T] \to R^n$, $w^{\delta,\alpha}(\cdot) : [T_0,T] \to R^n$ of system (7.38) with boundary conditions (7.40) exist and are unique on $t \in [T_0,T]$. Moreover, due to continuity of the solutions and zero boundary conditions (7.40), there exists such interval $[t_1,T] \subset [T_0,T]$ that

$$|z_i^{\delta,\alpha}(t)| \le \alpha\delta R_w, \quad |w_i^{\delta,\alpha}(t)| \le \alpha^2\delta R_w, \quad i=1,\ldots,n, \quad t\in[t_1,T],$$

where the constant R_w is defined in (7.5).

Let's now extend the solution further in reverse time (to the left from t_1 on time axis). As the solution is continuous, we can always extend it up to such moment t_0 that either $z_i^{\delta,\alpha}(t_0) = 2\alpha\delta R_w$, $i \in \{1,\ldots,n\}$ or $w_i^{\delta,\alpha}(t_0) = 2\alpha^2\delta R_w$, $i \in \{1,\ldots,n\}$ or extend it up to $t = 0$. If we are able to extend it up to $t = 0$ without reaching values $2\alpha\delta R_w$, $2\alpha^2\delta R_w$ (the second case), then

$$|z_i^{\delta,\alpha}(t)| \le 2\alpha\delta R_w, \quad |w_i^{\delta,\alpha}(t)| \le 2\alpha^2\delta R_w, \quad i=1,\ldots,n, \quad t\in[0,T].$$

In the first case there exists such moment $t_0 \in [0, T]$ that

$$z_i^{\delta,\alpha}(t) \le 2\alpha\delta R_w, \quad w_i^{\delta,\alpha}(t) \le 2\alpha^2\delta R_w, \quad i = 1, \ldots, n, \quad t \in [t_0, T]. \tag{7.41}$$

Let's consider this case closer.

We introduce a new system of ODEs for functions $\overline{z}_i(\cdot), \overline{w}_i(\cdot),\ i = 1, \ldots, n$

$$\begin{aligned} \dot{\overline{z}}_i(t) &= -\overline{w}_i(t)\frac{g_i^2(z^{\delta,\alpha}(t) + y^\delta(t), t)}{\alpha^2}, \\ \dot{\overline{w}}_i(t) &= \overline{z}_i(t) + F_i(z^{\delta,\alpha}(t), w^{\delta,\alpha}(t), t), \\ & i = 1, \ldots, n, \quad t \in [t_0, T] \end{aligned} \tag{7.42}$$

with boundary conditions

$$\overline{z}(T) = 0, \quad \overline{w}(T) = 0, \tag{7.43}$$

where $z_i^{\delta,\alpha}(t)$, $w_i^{\delta,\alpha}(t)$ are solutions of system (7.38) with boundary conditions (7.40), constrained by (7.41).

System (7.42) is a heterogeneous linear system of ODEs with time-dependent coefficients, continuous on $t \in [t_0, T]$. So, the solution of (7.42), (7.43) exists and is unique on $t \in [t_0, T]$.

Let's now prove that the solutions of (7.42), (7.43) coincide with the solutions of (7.38), (7.40). To do this, we introduce residuals

$$\Delta z(t) = z^{\delta,\alpha}(t) - \overline{z}(t), \quad \Delta w(t) = w^{\delta,\alpha}(t) - \overline{w}(t).$$

Subtracting Eq. (7.42) from (7.38) (with substituted solutions $z^{\delta,\alpha}(t)$, $w^{\delta,\alpha}(t)$), we get

$$\begin{aligned} \Delta z_i(t) &= -w_i^{\delta,\alpha}(t)\frac{g_i^2(z^{\delta,\alpha}(t) + y^\delta(t), t)}{\alpha^2} + \overline{w}_i(t)\frac{g_i^2(z^{\delta,\alpha}(t) + y^\delta(t), t)}{\alpha^2} \\ &= -\Delta w_i(t)\frac{g_i^2(z^{\delta,\alpha}(t) + y^\delta(t), t)}{\alpha^2}, \\ \Delta w_i(t) &= \Delta z_i(t) + F_i(z^{\delta,\alpha}(t), w^{\delta,\alpha}(t), t) - F_i(z^{\delta,\alpha}(t), w^{\delta,\alpha}(t), t) = \Delta z_i(t), \\ & i = 1, \ldots, n \end{aligned} \tag{7.44}$$

with boundary conditions

$$\Delta z(T) = 0, \quad \Delta w(T) = 0. \tag{7.45}$$

As a homogenous system of linear ODEs with continuous time-dependent coefficients, system (7.44) with zero boundary conditions has the only trivial solution

$$\Delta z(t) \equiv 0, \quad \Delta w(t) \equiv 0, \quad t \in [t_0, T]. \tag{7.46}$$

That means that $z^{\delta,\alpha}(t) = \overline{z}(t),\ w^{\delta,\alpha}(t) = \overline{w}(t),\ t \in [t_0, T]$.

Now let's study the properties of the solutions $\overline{z}(t)$, $\overline{w}(t)$ of system (7.42) with boundary conditions (7.43). System (7.42) can be rewritten in vector form

$$\dot{Z}(t) = A(t)Z(t) + F(t), \tag{7.47}$$

where

$$\begin{aligned} Z(\cdot) &= (\overline{z}_1(\cdot), \dots, \overline{z}_n(\cdot), \overline{w}_1(\cdot), \dots, \overline{w}_n(\cdot)), \\ F(\cdot) &= (\underbrace{0, \dots 0}_{n}, F_1(z^{\delta,\alpha}(\cdot), w^{\delta,\alpha}(\cdot), \cdot), \dots, F_n(z^{\delta,\alpha}(\cdot), w^{\delta,\alpha}(\cdot), \cdot)) \end{aligned} \tag{7.48}$$

and the $2n \times 2n$ matrix $A(t)$ can be written in the block form $A(t) = \begin{pmatrix} O & G_A(x,t) \\ I_n & O \end{pmatrix}$, where I_n is an identity matrix, O is an $n \times n$ zero matrix,

$$G_A(x,t) = \begin{pmatrix} -g_1^2(x^{\delta,\alpha}(t), t) & 0 & \dots & 0 \\ 0 & -g_2^2(x^{\delta,\alpha}(t), t) & \dots & 0 \\ \dots & \dots & \dots & \dots \\ 0 & 0 & \dots & -g_n^2(x^{\delta,\alpha}(t), t) \end{pmatrix}.$$

Solutions of system (7.42) can be written in the following form with the help of Cauchy formula for solutions of a heterogenous system of linear ODEs with time-dependent coefficients. One can easily check that for boundary conditions, given at the point $t = T$ (instead of $t = 0$), it has the form

$$Z(t) = \Phi(t)\Phi^{-1}(T)Z(T) - \Phi(t)\int_t^T \Phi^{-1}(\tau)F(z^{\delta,\alpha}(\tau), w^{\delta,\alpha}(\tau), \tau)d\tau, \tag{7.49}$$

were $\Phi(\cdot)$ is an $n \times n$ fundamental matrix of solutions for the homogenous part of system (7.42). This matrix can be chosen as

$$\Phi(t) = \exp\left[-\int_t^T A(\tau)d\tau\right] = \sum_{k=0}^{\infty} \frac{1}{k!}\left(-\int_t^T A(\tau)d\tau\right)^k. \tag{7.50}$$

One can check that after expanding the kth powers in the sum in the latter formula and folding the sum again, using the Taylor series for *sin* and *cos* functions, we can get that $\Phi(t) = \begin{pmatrix} \Phi_1(t) & \Phi_2(t) \\ \Phi_3(t) & \Phi_1(t) \end{pmatrix}$, where $\Phi_1(t)$, $\Phi_2(t)$, $\Phi_3(t)$ are diagonal matrixes with ith elements on diagonals

$$\begin{aligned}\Phi_{1ii}(t) &= \cos\left(\frac{1}{\alpha}\sqrt{(T-t)\int_t^T g_i^2(x^{\delta,\alpha}(\tau),\tau)d\tau}\right),\\ \Phi_{2ii}(t) &= \frac{1}{\alpha}\tilde{\Phi}_i(t)\sin\left(\frac{1}{\alpha}\sqrt{(T-t)\int_t^T g_i^2(x^{\delta,\alpha}(\tau),\tau)d\tau}\right),\\ \Phi_{3ii}(t) &= -\alpha\frac{1}{\tilde{\Phi}_i(t)}\sin\left(\frac{1}{\alpha}\sqrt{(T-t)\int_t^T g_i^2(x^{\delta,\alpha}(\tau),\tau)d\tau}\right),\end{aligned} \tag{7.51}$$

where continuous function

$$\tilde{\Phi}_i(t) = \begin{cases} \dfrac{\sqrt{\int_t^T g_i^2(x^{\delta,\alpha}(\tau),\tau)d\tau}}{\sqrt{T-t}}, & t \in [t_0, T),\\ g_i(x^{\delta,\alpha}(T),T), & t = T,\end{cases} \tag{7.52}$$

$$i = 1, \ldots, n.$$

Using (7.18), one can obtain that

$$\underline{\omega} \le \left|\tilde{\Phi}_i(t)\right| \le \overline{\omega}, \quad i = 1, \ldots, n. \tag{7.53}$$

Due to simple structure of matrix $\Phi(t)$, one can check that inverse matrix $\Phi^{-1}(t) = \begin{pmatrix} \Phi_1(t) & -\Phi_2(t) \\ -\Phi_3(t) & \Phi_1(t) \end{pmatrix}$.

Let's return to (7.49). Here $Z(T) = \overrightarrow{0}$, so vector $Z(t) = -\Phi(t)\int_t^T \Phi^{-1}(\tau) F(\tau)d\tau$ has the following coordinates

$$\begin{aligned} Z_i(t) = \overline{z}_i(t) &= \Phi_{1,ii}(t)\int_t^T \Phi_{2,ii}(\tau)F_i(z^{\delta,\alpha}(\tau), w^{\delta,\alpha}(\tau),\tau)d\tau\\ &\quad -\Phi_{2,ii}(t)\int_t^T \Phi_{1,ii}(\tau)F_i(z^{\delta,\alpha}(\tau), w^{\delta,\alpha}(\tau),\tau)d\tau,\\ Z_{i+1}(t) = \overline{w}_i(t) &= \Phi_{3,ii}(t)\int_t^T \Phi_{2,ii}(\tau)F_i(z^{\delta,\alpha}(\tau), w^{\delta,\alpha}(\tau),\tau)d\tau\\ &\quad -\Phi_{1,ii}(t)\int_t^T \Phi_{1,ii}(\tau)F_i(z^{\delta,\alpha}(\tau), w^{\delta,\alpha}(\tau),\tau)\Big]d\tau,\\ &\quad t \in [t_0, T], \quad i = 1, \ldots, n. \end{aligned} \tag{7.54}$$

To estimate these expressions, we consider the following expression

$$\int_{t_0}^{T} \cos\left(\frac{1}{\alpha}\sqrt{(T-\tau)\int_{\tau}^{T} g_i^2(x^{\delta,\alpha}(\theta),\theta)d\theta}\right) f_i^{\delta}(\tau)d\tau, \quad i=1,\ldots,n \qquad (7.55)$$

where function $f_i^{\delta}(\cdot) = f_i^{\delta}(\cdot,\delta) : [0,T] \to R$ depends on δ and is continuous in the first argument for any $\delta \in (0,\delta_0]$.

Let's introduce functions $\varphi_i(\tau) = \left(\sqrt{(T-\tau)\int_{\tau}^{T} g_i^2(x^{\delta,\alpha}(\theta),\theta)d\theta}\right)$, $i = 1,\ldots,n$, which are continuously differentiable in τ.

Note that all following calculations in the proof are true for all $i \in \{1,\ldots,n\}$.

Using Hypothesis 7.1, we can estimate the derivative

$$\dot{\varphi}_i(\tau) = \frac{-\int_{\tau}^{T} g_i^2(x^{\delta,\alpha}(\theta),\theta)d\theta - (T-\tau)g_i^2(x^{\delta,\alpha}(\tau),\tau)}{2\sqrt{(T-\tau)\int_{\tau}^{T} g_i^2(x^{\delta,\alpha}(\theta),\theta)d\theta}}$$
$$\geq -\frac{2\overline{\omega}^2(T-\tau)}{2\sqrt{(T-\tau)^2\underline{\omega}^2}} = -\frac{\overline{\omega}^2}{\underline{\omega}}. \qquad (7.56)$$
$$\text{Similarly, } \dot{\varphi}_i(\tau) \leq -\frac{\underline{\omega}^2}{\overline{\omega}}, \quad \tau \in [t_0,T].$$

So, $\varphi_i(\tau)$ is a decreasing function with restricted derivative and $\varphi_i(T) = 0$. This means that we can construct a finite increasing sequence $\{\tau_1 < \tau_2 < \ldots < \tau_{n_{\varphi_i}},\ n_{\varphi_i} \in N\}$ that has the following properties:

$$\varphi_i(\tau_{(n_{\varphi_i}-k)}) = \alpha(0.5+k)\pi, \quad k = 0,\ldots,(n_{\varphi_i}-1);$$

$$\alpha\frac{\pi\underline{\omega}}{\overline{\omega}^2} \leq (\tau_{j+1}-\tau_j) \leq \alpha\frac{\pi\overline{\omega}}{\underline{\omega}^2}, \quad n_{\varphi_i} \leq \frac{T\overline{\omega}}{\alpha\underline{\omega}^2}, \qquad (7.57)$$

as the derivative $\dot{\varphi}(t)$ is restricted by (7.56).

Let's add to this sequence elements $\tau_0 = t_0$ and $\tau_{(n_{\varphi_i}+1)} = T$.

Integral (7.55) can be rewritten as

$$\int_{t_0}^{T} \cos\left(\frac{\varphi_i(\tau)}{\alpha}\right) f_i^{\delta}(\tau)d\tau = \sum_{j=0}^{n_{\varphi_i}} \int_{\tau_j}^{\tau_{j+1}} \cos\left(\frac{\varphi_i(\tau)}{\alpha}\right) f_i^{\delta}(\tau)d\tau. \qquad (7.58)$$

Because $\cos\left(\frac{\varphi_i(\tau)}{\alpha}\right)$ is sign-definite on $\tau \in [\tau_j, \tau_{j+1}]$, $j = 0, \ldots, n_{\varphi_i}$ and $f_i^\delta(\tau)$ is continuous, it follows from the first mean value theorem for definite integrals that for each $j = 0, \ldots, n_{\varphi_i}$ there exists such point $\tilde{\tau}_j \in [\tau_j, \tau_{j+1}]$ that $\int\limits_{\tau_j}^{\tau_{j+1}} \cos\left(\frac{\varphi_i(\tau)}{\alpha}\right) f_i^\delta(\tau) d\tau = f_i^\delta(\tilde{\tau}_j) \int\limits_{\tau_j}^{\tau_{j+1}} \cos\left(\frac{\varphi_i(\tau)}{\alpha}\right) d\tau$. Combining the terms of sum (7.58) by pairs $[\tau_j, \tau_{j+1}]$, $[\tau_{j+1}, \tau_{j+2}]$, we get

$$\begin{aligned} &\int\limits_{\tau_j}^{\tau_{j+2}} \cos\left(\frac{\varphi_i(\tau)}{\alpha}\right) f_i^\delta(\tau) d\tau \\ &= f_i^\delta(\tilde{\tau}_j) \int\limits_{\tau_j}^{\tau_{j+1}} \cos\left(\frac{\varphi_i(\tau)}{\alpha}\right) d\tau + f_i^\delta(\tilde{\tau}_{j+1}) \int\limits_{\tau_{j+1}}^{\tau_{j+2}} \cos\left(\frac{\varphi_i(\tau)}{\alpha}\right) d\tau. \end{aligned} \tag{7.59}$$

To estimate expression (7.59), we first make the following estimates:

$$\int\limits_{\tau_j}^{\tau_{j+2}} \cos\left(\frac{\varphi_i(\tau)}{\alpha}\right) d\tau = \int\limits_{\tau_j}^{\tau_{j+2}} \frac{\alpha}{\dot{\varphi}_i(\tau)} \frac{\dot{\varphi}_i(\tau)}{\alpha} \cos\left(\frac{\varphi_i(\tau)}{\alpha}\right) d\tau, \; j = 0, \ldots, n_{\varphi_i} - 3, \tag{7.60}$$

as $(T - \tau) \neq 0$ for $j < n_{\varphi_i}$. We can integrate (7.60) by parts.

$$\begin{aligned} &\int\limits_{\tau_j}^{\tau_{j+2}} \underbrace{\frac{\alpha}{\dot{\varphi}_i(\tau)}}_{\mathbf{U}} \underbrace{\frac{\dot{\varphi}_i(\tau)}{\alpha} \cos\left(\frac{\varphi_i(\tau)}{\alpha}\right) d\tau}_{\mathbf{dV}} = \underbrace{\frac{\alpha}{\dot{\varphi}_i(\tau)}}_{\mathbf{U}} \underbrace{\sin\left(\frac{\varphi_i(\tau)}{\alpha}\right)}_{\mathbf{V}} \Bigg|_{\tau_j}^{\tau_{j+2}} \\ &-\alpha \int\limits_{\tau_j}^{\tau_{j+2}} \underbrace{\sin\left(\frac{\varphi_i(\tau)}{\alpha}\right)}_{\mathbf{V}} \underbrace{\frac{d}{d\tau}\left(\frac{1}{\dot{\varphi}_i(\tau)}\right) d\tau}_{\mathbf{dU}}, \; j = 0, \ldots, n_{\varphi_i} - 3. \end{aligned} \tag{7.61}$$

Here

$$\left| \frac{\alpha}{\dot{\varphi}_i(\tau)} \sin\left(\frac{\varphi_i(\tau)}{\alpha}\right) \Bigg|_{\tau_j}^{\tau_{j+2}} \right| = \left| \frac{\alpha}{\dot{\varphi}_i(\tau)} \Bigg|_{\tau_j}^{\tau_{j+2}} \right| \leq \alpha \sup_{t \in [\tau_j, \tau_{j+2}]} \left| \frac{d}{d\tau} \frac{1}{\dot{\varphi}_i(\tau)} \right| (\tau_{j+2} - \tau_j).$$

One can check that the derivative

$$
\begin{aligned}
&\left|\frac{d}{d\tau}\frac{1}{\dot{\varphi}_i(\tau)}\right| = \Bigg|\frac{1}{\sqrt{(T-\tau)\int\limits_{\tau}^{T} g_i^2(x^{\delta,\alpha}(\theta),\theta)d\theta}} \\
&-\frac{2}{\dot{\varphi}(\tau)}\Bigg(-g_i^2(x^{\delta,\alpha}(\tau),\tau)+(T-\tau)g_i(x^{\delta,\alpha}(\tau),\tau) \\
&\cdot\Bigg(\sum_{i=1}^{n}\Bigg[\frac{\partial g_i(x^{\delta,\alpha}(\tau),\tau)}{\partial x_i}\Bigg(-\frac{w^{\delta,\alpha}(\tau)g_i^2(x^{\delta,\alpha}(\tau),\tau)}{\alpha^2}\Bigg)\Bigg]+\frac{\partial g_i(x^{\delta,\alpha}(\tau),\tau)}{\partial t}\Bigg)\Bigg)\Bigg| \\
&\le \frac{1}{\underline{\omega}(T-\tau_{j+2})}+\frac{2\overline{\omega}}{\underline{\omega}^2}(\overline{\omega}^2+T\overline{\omega}(\delta n\omega'\overline{\omega}^2R_w+\omega')).
\end{aligned}
\tag{7.62}
$$

So, the term $\mathbf{UV}\big|_{\tau_j}^{\tau_{j+1}}$ in (7.61) can be estimated by using (7.57) and (7.62) as

$$
\left|\frac{\alpha}{\dot{\varphi}_i(\tau)}\sin\left(\frac{\varphi_i(\tau)}{\alpha}\right)\Big|_{\tau_j}^{\tau_{j+2}}\right| \le \alpha^2\left(\frac{R_1}{T-\tau_{j+2}}+R_2+\delta R_3R_w\right), \tag{7.63}
$$

where the constants R_1, R_2, R_3 are defined in (7.5). Let's emphasize that these constants don't depend on δ and α.

Now let's estimate the term $\int_{\tau_j}^{\tau_{j+1}}\mathbf{VdU}$ in (7.61).

$$
\begin{aligned}
&\alpha\left|\int\limits_{\tau_j}^{\tau_{j+2}}\sin\left(\frac{\varphi_i(\tau)}{\alpha}\right)\frac{d}{d\tau}\left(\frac{1}{\dot{\varphi}_i(\tau)}\right)d\tau\right| \\
&\le \alpha\sup_{\tau\in[\tau_j,\tau_{j+2}]}\left|\sin\left(\frac{\varphi_i(\tau)}{\alpha}\right)\frac{d}{d\tau}\left(\frac{1}{\dot{\varphi}_i(\tau)}\right)\right|(\tau_{j+2}-\tau_j) \\
&\le \alpha^2\left(\frac{R_1}{T-\tau_{j+2}}+R_2+\delta R_3R_w\right).
\end{aligned}
\tag{7.64}
$$

Applying estimates (7.63) and (7.64) to (7.60)–(7.61), we get

$$
\left|\int\limits_{\tau_j}^{\tau_{j+2}}\cos\left(\frac{\varphi_i(\tau)}{\alpha}\right)d\tau\right| \le 2\alpha^2\left(\frac{R_1}{T-\tau_{j+2}}+R_2+\delta R_3R_w\right). \tag{7.65}
$$

Now let's return to expression (7.59). By splitting the last integral term in (7.59) as $\int\limits_{\tau_{j+1}}^{\tau_{j+2}} = \int\limits_{\tau_j}^{\tau_{j+2}} - \int\limits_{\tau_j}^{\tau_{j+1}}$, we get

$$\begin{aligned} &\left| \int\limits_{\tau_j}^{\tau_{j+2}} \cos\left(\frac{\varphi_i(\tau)}{\alpha}\right) f_i^\delta(\tau)d\tau \right| \\ \leq &\left| \int\limits_{\tau_j}^{\tau_{j+1}} \cos\left(\frac{\varphi_i(\tau)}{\alpha}\right) d\tau \left(f_i^\delta(\tilde{\tau}_j) - f_i^\delta(\tilde{\tau}_{j+1})\right) \right| \\ &+ \left| f_i^\delta(\tilde{\tau}_{j+1}) \int\limits_{\tau_j}^{\tau_{j+2}} \cos\left(\frac{\varphi_i(\tau)}{\alpha}\right) d\tau \right|. \end{aligned} \tag{7.66}$$

By Heine–Cantor theorem, every continuous function defined on a closed interval is uniformly continuous. So, continuous $f_i^\delta(\tau)$ is uniformly continuous on $[t_0, T]$. In other words,

$$\begin{aligned} &\forall \delta > 0 \ \exists \alpha_1^\delta = \alpha_1^\delta(\delta) > 0 : \ \forall \tau_1, \tau_2 \in [\tau_j, \tau_{j+2}] \\ &\left(|\tau_1 - \tau_2| < \alpha_1^\delta 2 \frac{\pi \overline{\omega}}{\underline{\omega}^2}\right) \Rightarrow \left(|f_i^\delta(\tau_1) - f_i^\delta(\tau_2)| < \delta\right). \end{aligned} \tag{7.67}$$

Remark 7.5 As $f_i^\delta(\tau)$ is uniformly continuous on $[t_0, T]$, we are able to choose the same $\alpha_1^\delta = \alpha_1^\delta(\delta)$ in (7.67) for each $j = 0, \ldots, (n_{\varphi_i} + 1)$ as $[\tau_j, \tau_{j+2}] \subset [t_0, T]$.

Combining (7.57), (7.65), (7.66), (7.67), we get

$$\begin{aligned} &\left| \int\limits_{\tau_j}^{\tau_{j+2}} \cos\left(\frac{\varphi_i(\tau)}{\alpha}\right) f_i^\delta(\tau)d\tau \right| \\ &\leq \alpha\delta \frac{\pi \overline{\omega}}{\underline{\omega}^2} + 2\alpha^2 \max_{\tau \in [\tau_j, \tau_{j+2}]} f_i^\delta(\tau) \left(\frac{R_1}{T - \tau_{j+2}} + R_2 + \delta R_3 R_w\right). \end{aligned} \tag{7.68}$$

To be specific, let's assume that the number n_{φ_i} is odd. Then

$$\begin{aligned} &\int\limits_{t_0}^{T} \cos\left(\frac{\varphi_i(\tau)}{\alpha}\right) f_i^\delta(\tau)d\tau \\ &= \int\limits_{t_0}^{\tau_1} \cos\left(\frac{\varphi_i(\tau)}{\alpha}\right) f_i^\delta(\tau)d\tau + \sum_{j=1}^{0.5(n_{\varphi_i}-1)-1} \int\limits_{\tau_{2j-1}}^{\tau_{2j+1}} \cos\left(\frac{\varphi_i(\tau)}{\alpha}\right) f_i^\delta(\tau)d\tau \\ &+ \int\limits_{\tau_{n_{\varphi_i}-2}}^{\tau_{n_{\varphi_i}}} \cos\left(\frac{\varphi_i(\tau)}{\alpha}\right) f_i^\delta(\tau)d\tau + \int\limits_{n_{\varphi_i}}^{T} \cos\left(\frac{\varphi_i(\tau)}{\alpha}\right) f_i^\delta(\tau)d\tau. \end{aligned} \tag{7.69}$$

Using (7.68), let's first estimate the sum

$$\left| \sum_{j=1}^{0.5(n_{\varphi_i}-1)-1} \left[\int_{\tau_{2j-1}}^{\tau_{2j+1}} \cos\left(\frac{\varphi_i(\tau)}{\alpha}\right) f_i^\delta(\tau)d\tau \right] \right| \\ \leq \sum_{j=1}^{0.5(n_{\varphi_i}-1)-1} \left[\alpha\delta \frac{\pi\overline{\omega}}{\underline{\omega}^2} + 2\alpha^2 \overline{f}_i^\delta \left(\frac{R_1}{T-\tau_{2j+1}} + R_2 + \delta R_3 R_w \right) \right], \tag{7.70}$$

where $n_{\varphi_i} \leq \frac{T\overline{\omega}}{\alpha\underline{\omega}^2}$ and $\overline{f}_i^\delta = \max_{\tau\in[t_0,T]} f_i^\delta(\tau)$.

The following sum can be estimated by substituting the denominator in the fraction with it's minimal possible value (7.57) and reversing the order of terms in the sum.

$$\sum_{j=1}^{0.5(n_{\varphi_i}-1)-1} \frac{\alpha}{T-\tau_{2j+1}} \leq \sum_{j=1}^{0.5(n_{\varphi_i}-1)-1} \frac{\alpha}{(\alpha\pi\underline{\omega}/\overline{\omega}^2)j} = \frac{\overline{\omega}^2}{\pi\underline{\omega}} \sum_{j=1}^{0.5(n_{\varphi_i}-1)-1} \frac{1}{j}.$$

The partial sum $\sum_{j=1}^{0.5(n_{\varphi_i}-1)-1} \frac{1}{j}$ of a harmonic series can be estimated by Euler–Mascheroni formula $\sum_{n=1}^{k} \frac{1}{n} \leq (\ln k) + 1$. Thus, continuing estimates (7.70) we get

$$\left| \sum_{j=1}^{0.5(n_{\varphi_i}-1)-1} \left[\int_{\tau_{2j-1}}^{\tau_{2j+1}} \cos\left(\frac{\varphi_i(\tau)}{\alpha}\right) f_i^\delta(\tau)d\tau \right] \right| \\ \leq \delta \frac{T\pi\overline{\omega}^2}{2\underline{\omega}^4} + 2\alpha\overline{f}_i^\delta \left(\frac{\pi\overline{\omega}^2 R_1}{\underline{\omega}} (\ln \frac{0.5T\overline{\omega}}{\alpha\underline{\omega}^2} + 1) + R_2 + \delta R_3 R_w \right). \tag{7.71}$$

We have estimated the second term of sum in the right hand side of (7.69). Using (7.57), one can get the following relations for the first, third and forth terms in (7.69).

$$\left| \int_{t_0}^{\tau_1} \cos\left(\frac{\varphi_i(\tau)}{\alpha}\right) f_i^\delta(\tau)d\tau + \int_{\tau_{n_{\varphi_i}-2}}^{\tau_{n_{\varphi_i}}} \cos\left(\frac{\varphi_i(\tau)}{\alpha}\right) f_i^\delta(\tau)d\tau \right. \\ \left. + \int_{n_{\varphi_i}}^{T} \cos\left(\frac{\varphi_i(\tau)}{\alpha}\right) f_i^\delta(\tau)d\tau \right| \leq 4\alpha \frac{\pi\overline{\omega}^2}{\underline{\omega}} \overline{f}_i^\delta. \tag{7.72}$$

Remark 7.6 We assumed that the number n_{φ_i} is odd. In the case of even n_{φ_i} the calculations are similar, because the only difference is in formula (7.69), where the lower limit of the integral $\int\limits_{\tau_{n_{\varphi_i}-2}}^{\tau_{n_{\varphi_i}}} \cos\left(\frac{\varphi_i(\tau)}{\alpha}\right) f_i^\delta(\tau)d\tau$ is exchanged for $\tau_{n_{\varphi_i}-1}$.

Finally, applying (7.71) and (7.72) to (7.69), we get

$$\left|\int\limits_{t_0}^{T} \cos\left(\frac{\varphi_i(\tau)}{\alpha}\right) f_i^\delta(\tau)d\tau\right| \leq \delta\frac{T\pi\overline{\omega}^2}{2\underline{\omega}^4}$$

$$+\alpha\overline{f}_i^\delta\left(4\frac{\pi\overline{\omega}^2}{\underline{\omega}} + 2\frac{R_1\pi\overline{\omega}^2}{\underline{\omega}}\left(\ln\frac{T\overline{\omega}}{2\underline{\omega}^2} + 1\right) + R_2\right) + \alpha|\ln\alpha|\overline{f}_i^\delta + 2\alpha\delta\overline{f}_i^\delta R_3 R_w.$$

For any given $\delta \in (0, \delta_0]$ there exists a constant $\overline{f}_i^\delta = \overline{f}_i^\delta(\delta)$ (as $f_i^\delta(\tau)$ is continuous for $\delta \in (0, \delta_0]$). We can always find such parameter $\alpha_2^\delta = \alpha_2^\delta(\delta)$ that

$$\alpha_2^\delta(\delta)|\ln\alpha_2^\delta(\delta)|\overline{f}_i^\delta(\delta) \leq \delta. \tag{7.73}$$

This is possible because $\lim\limits_{\alpha\to 0}\alpha|\ln\alpha| = 0$. Thus, for any

$$\alpha \leq \alpha_0 = \min\{\alpha_1^\delta, \alpha_2^\delta, 1\}, \quad \text{(where } \alpha_1^\delta \text{ is from (7.67), } \alpha_2^\delta \text{ is from (7.73)),} \tag{7.74}$$

we have

$$\left|\int\limits_{t_0}^{T} \cos\left(\frac{\varphi_i(\tau)}{\alpha}\right) f_i^\delta(\tau)d\tau\right| \leq \delta R_4 + 2\delta^2 R_3 R_w, \tag{7.75}$$

where the constants $R_3,\ R_4$ are defined in (7.5).

We can apply this result to expressions (7.54). First, let's estimate expression

$$\alpha^2\Phi_{2,ii}(t)\int\limits_{t_0}^{T}\Phi_{1,ii}(\tau)\frac{F_i(z^{\delta,\alpha}(\tau), w^{\delta,\alpha}(\tau), \tau)}{\alpha^2}d\tau, \tag{7.76}$$

for which $f_i^\delta(\tau) = F_i(z^{\delta,\alpha}(\tau), w^{\delta,\alpha}(\tau), \tau)/\alpha^2 \overset{not}{=} f_{i,1}^\delta(\tau)$ in the sense of (7.55). It follows from (7.39), (7.41) and Hypotheses 7.2, 7.1 that

$$\overline{f}_i^\delta = \overline{f}_{i,1}^\delta = \max_{\tau\in[t_0,T]}\left|F_i(z^{\delta,\alpha}(\tau), w^{\delta,\alpha}(\tau), \tau)/\alpha^2\right|$$

$$\leq \left(\frac{\max\limits_{\tau\in[t_0,T]}\ddot{y}^\delta(\tau)}{\underline{\omega}^2} + n\frac{\overline{\omega}^2\omega'\overline{Y} + \omega'\overline{Y}^2}{\underline{\omega}^3}\right) + \delta n R_w\left(\frac{\overline{\omega}^2\omega'}{\underline{\omega}^3} + 2\frac{\omega'\overline{Y}}{\underline{\omega}}\right) + \delta^2 R_w^2\overline{\omega}\omega'. \tag{7.77}$$

For $\alpha \leq \alpha_0^1$, where α_0^1 is defined in the same way as α_0 in (7.67), (7.73), (7.74), but assuming $f_i^\delta(\tau) = f_{i,1}^\delta(\tau)$ and $\overline{f}_i^\delta(\tau) = \overline{f}_{i,1}^\delta(\tau)$, estimates (7.75) and (7.53) give us

$$\left| \Phi_{2,ii}(t) \int_{t_0}^{T} \Phi_{1,ii}(\tau) F_i(z^{\delta,\alpha}(\tau), w^{\delta,\alpha}(\tau), \tau) d\tau \right| \\ \leq \alpha\overline{\omega}(\delta R_4 + 2\delta^2 R_3 R_w), \ t \in [t_0, T]. \tag{7.78}$$

Let's introduce α_0^2 that is defined in the same way as α_0 in (7.67), (7.73), (7.74), but assuming

$$f_i^\delta(\tau) = \tilde{\Phi}_i(\tau) F_i(z^{\delta,\alpha}(\tau), w^{\delta,\alpha}(\tau), \tau)/\alpha^2 \overset{not}{=} f_{i,2}^\delta(\tau), \quad \overline{f}_{i,2}^\delta = \overline{\omega}\overline{f}_{i,1}^\delta.$$

One can use the scheme of proof (7.55)–(7.78) and (7.51)–(7.53) to obtain that for $\alpha \leq \min\{\alpha_0^1, \alpha_0^2\}$ the following estimates are true as well

$$\left| \Phi_{1,ii}(t) \int_{t_0}^{T} \Phi_{2,ii}(\tau) F_i(z^{\delta,\alpha}(\tau), w^{\delta,\alpha}(\tau), \tau) d\tau \right| \\ \leq \alpha\overline{\omega}(\delta R_4 + 2\delta^2 R_3 R_w), \quad t \in [t_0, T]; \tag{7.79}$$

$$\left| \Phi_{3,ii}(t) \int_{t}^{T} \Phi_{2,ii}(\tau) F_i(z^{\delta,\alpha}(\tau), w^{\delta,\alpha}(\tau), \tau) d\tau \right| \\ \leq \alpha^2 \frac{1}{\underline{\omega}}(\delta R_4 + 2\delta^2 R_3 R_w), \quad t \in [t_0, T]; \tag{7.80}$$

$$\left| \Phi_{1,ii}(t) \int_{t}^{T} \Phi_{1,ii}(\tau) F_i(z^{\delta,\alpha}(\tau), w^{\delta,\alpha}(\tau), \tau) \Big] d\tau \right| \\ \leq \alpha^2(\delta R_4 + 2\delta^2 R_3 R_w), \quad t \in [t_0, T]. \tag{7.81}$$

Remark 7.7 Estimates (7.78)–(7.81) are true under combined condition

$$\alpha \leq \alpha_0^\delta \overset{def}{=} \min\{\alpha_0^1, \alpha_0^2\}. \tag{7.82}$$

Combining (7.54) and (7.78)–(7.81), we get

$$|\overline{z}_i(t)| \leq \alpha\delta(1+\overline{\omega})(R_4 + 2\delta R_3 R_w), \\ |\overline{w}_i(t)| \leq \alpha^2\delta\frac{1+\underline{\omega}}{\underline{\omega}}(R_4 + 2\delta R_3 R_w), \quad t \in [t_0, T], \quad i = 1, \ldots, n.$$

For $\delta : 0 < \delta \le \delta_0 \le \frac{1}{R_w}$, $\alpha \in (0, \alpha_0^\delta]$, as far as $z^{\delta,\alpha}(t) = \overline{z}(t)$, $w^{\delta,\alpha}(t) = \overline{w}(t)$, $t \in [t_0, T]$,

$$\begin{gathered} |z_i^{\delta,\alpha}(t)| = |\overline{z}_i(t)| \le \alpha\delta(1+\overline{\omega})(R_4 + 2R_3) \le \alpha\delta R_w, \\ |w_i^{\delta,\alpha}(t)| = |\overline{w}_i(t)| \le \alpha^2\delta\tfrac{1+\omega}{\omega}(R_4 + 2R_3) \le \alpha^2\delta R_w, \\ t \in [t_0, T], \quad i = 1, \ldots, n. \end{gathered} \tag{7.83}$$

Remark 7.8 Estimates (7.83) are true for $t_0 \in [0, T)$ as long as solutions $z^{\delta,\alpha}(\cdot)$, $w^{\delta,\alpha}(\cdot)$ of system (7.38) with boundary conditions (7.40) exist and are unique on $t \in [t_0, T]$ and (7.41) is true.

But (7.83) means that for $\delta \in (0, \delta_0]$, $\alpha \in (0, \alpha_0^\delta]$ at $t = t_0$ (in particular)

$$|z_i^{\delta,\alpha}(t_0)| \le \alpha\delta R_w, \quad |w_i^{\delta,\alpha}(t_0)| \le \alpha^2\delta R_w, \quad i = 1, \ldots, n,$$

which is contrary to the assumption that either $z_i^{\delta,\alpha}(t_0) = 2\alpha\delta R_w,\ i \in \{1, \ldots, n\}$ or $w_i^{\delta,\alpha}(t_0) = 2\alpha^2\delta R_w,\ i \in \{1, \ldots, n\}$. That means that such moment t_0 does not exist.

In other words, we have proved that we can extend the solutions $z^{\delta,\alpha}(\cdot)$, $w^{\delta,\alpha}(\cdot)$ up to $t = 0$ and

$$\begin{gathered} |z_i^{\delta,\alpha}(t)| \le \alpha\delta 2R_w, \\ |w_i^{\delta,\alpha}(t)| \le \alpha^2\delta 2R_w, \quad t \in [0, T], \quad i = 1, \ldots, n \end{gathered} \tag{7.84}$$

for $\delta \in (0, \delta_0]$ and $\alpha \in (0, \alpha_0^\delta]$.

As far as we can extend solutions $z^{\delta,\alpha}(\cdot)$, $w^{\delta,\alpha}(\cdot)$ on $t \in [0, T]$, we can return to variables (7.36)

$$x_i^{\delta,\alpha}(t) = z_i^{\delta,\alpha}(t) + y_i^\delta(t), \quad u_i^{\delta,\alpha}(t) = -\frac{g_i(x^{\delta,\alpha}(t), t)}{\alpha^2} w_i^{\delta,\alpha}(t) + \frac{\dot{y}^\delta(t)}{g_i(x^{\delta,\alpha}(t), t)}.$$

Applying the result (7.84) (see Remark 7.8), we get that

$$\begin{gathered} |x_i^{\delta,\alpha}(t) - y_i^\delta(t)| \le \alpha\delta 2R_w, \quad |u_i^{\delta,\alpha}(t) - \frac{\dot{y}_i^\delta(t)}{g_i(x^{\delta,\alpha}(t), t)}| \le \delta 2R_w, \\ i = 1, ..., n, \quad t \in [0, T] \end{gathered} \tag{7.85}$$

for $\delta \le (0, \delta_0]$ and $\alpha \in (0, \alpha_0^\delta]$.

It follow from (7.85) and Hypothesis 7.2 that

$$\begin{aligned} |x_i^{\delta,\alpha_0^\delta}(t) - x_i^*(t)| &\le |x_i^{\delta,\alpha_0^\delta}(t) - y_i^\delta(t)| + |x_i^*(t) - y_i^\delta(t)| \\ &\le \alpha\delta 2R_w + \delta \xrightarrow{\delta \to 0} 0, \quad i = 1, ..., n, \quad t \in [0, T], \end{aligned}$$

which means that

$$\lim_{\delta\to 0} \|x_i^{\delta,\alpha_0^\delta}(\cdot) - x_i^*(\cdot)\|_{C[0,T]} = 0 \tag{7.86}$$

Let's now make the following calculations.

$$\begin{aligned} &\left| \frac{\dot{y}_i^\delta(t)}{g_i(x^{\delta,\alpha}(t),t)} - \frac{\dot{y}_i^\delta(t)}{g_i(y^\delta(t),t)} \right| = \left| \frac{\dot{y}_i^\delta(t)\big(g_i(y^\delta(t),t) - g_i(x^{\delta,\alpha}(t),t)\big)}{g_i(x^{\delta,\alpha}(t),t)g_i(y^\delta(t),t)} \right| \\ &\quad \le \frac{\overline{Y}n\omega' 2\alpha\delta R_w}{\underline{\omega}^2}, \quad i = 1,...,n, \quad t \in [0,T] \end{aligned} \tag{7.87}$$

for $\delta \le (0, \delta_0]$ and $\alpha \in (0, \alpha_0^\delta]$.

It follows from (7.87) and (7.85) that

$$\begin{aligned} |u_i^{\delta,\alpha_0^\delta}(t) - \frac{\dot{y}_i^\delta(t)}{g_i(y^\delta(t),t)}| &\le |u_i^{\delta,\alpha_0^\delta}(t) - \frac{\dot{y}_i^\delta(t)}{g_i(x^{\delta,\alpha_0^\delta}(t),t)}| \\ &\quad + \left| \frac{\dot{y}_i^\delta(t)}{g_i(x^{\delta,\alpha_0^\delta}(t),t)} - \frac{\dot{y}_i^\delta(t)}{g_i(y^\delta(t),t)} \right| \\ \le \delta 2R_w + \frac{\overline{Y}n\omega' 2\alpha_0^\delta\delta R_w}{\underline{\omega}^2}, &\quad i = 1,...,n, \quad t \in [0,T]. \end{aligned} \tag{7.88}$$

Relation (7.88) and Lemma 1 imply that

$$\begin{aligned} &\|u_i^{\delta,\alpha_0^\delta}(\cdot) - u_i^*(t)\|^2_{L_{2,[0,T]}} = \int_0^T (u_i^{\delta,\alpha_0^\delta}(t) - u_i^*(t))^2 dt = \int_0^T \Bigg[\left(u_i^{\delta,\alpha_0^\delta}(t) - \frac{\dot{y}_i^\delta(t)}{g_i(y^\delta(t),t)} \right)^2 \\ &+2\left(u_i^{\delta,\alpha_0^\delta}(t) - \frac{\dot{y}_i^\delta(t)}{g_i(y^\delta(t),t)} \right)\left(\frac{\dot{y}_i^\delta(t)}{g_i(y^\delta(t),t)} - u_i^*(t) \right) + \left(u_i^*(t) - \frac{\dot{y}_i^\delta(t)}{g_i(y^\delta(t),t)} \right)^2 \Bigg] dt \\ &\quad \le T\left(\delta 2R_w + \frac{\overline{Y}n\omega' 2\alpha_0^\delta\delta R_w}{\underline{\omega}^2} \right)^2 + T\left(\delta 2R_w + \frac{\overline{Y}n\omega' 2\alpha_0^\delta\delta R_w}{\underline{\omega}^2} \right)\left(\frac{\overline{Y}}{\underline{\omega}} + \overline{U} \right) \\ &\qquad + \left\| \frac{\dot{y}_i^\delta(t)}{g_i(y^\delta(t),t)} - u_i^*(t) \right\|^2_{L_{2,[0,T]}} \overset{\delta\to 0}{\longrightarrow} 0, \end{aligned}$$

which was to be proved. □

Let's now consider for a fixed $\delta \in (0, \delta_0]$ cut-off functions

$$\hat{u}_i^\delta(t) = \begin{cases} \overline{U}, & u_i^{\delta,\alpha_0^\delta}(t) \ge \overline{U}, \\ u_i^{\delta,\alpha_0^\delta}(t), & |u_i^{\delta,\alpha_0^\delta}(t)| < \overline{U}, \\ -\overline{U}, & u_i^{\delta,\alpha_0^\delta}(t) \le -\overline{U}, \end{cases}, \quad i = 1,\ldots,n, \tag{7.89}$$

where the functions $u_i^{\delta,\alpha_0^\delta}(\cdot),\ i=1,\dots,n$ are defined in (7.35) and α_0^δ is introduced in Theorem 7.1 in (7.82).

It follows from Theorem 7.1 that

$$\begin{aligned}\|u_i^{\delta,\alpha_0^\delta}(\cdot)-u_i^*(\cdot)\|^2_{L_{2,[0,T]}} &= \|(u_i^{\delta,\alpha_0^\delta}(\cdot)-\hat{u}_i^\delta(\cdot))+(\hat{u}_i^\delta(\cdot)-u_i^*(\cdot))\|^2_{L_{2,[0,T]}}\\ &= \|u_i^{\delta,\alpha_0^\delta}(\cdot)-\hat{u}_i^\delta(\cdot)\|^2_{L_{2,[0,T]}}+\|\hat{u}_i^\delta(\cdot)-u_i^*(\cdot)\|^2_{L_{2,[0,T]}}\\ &+2\int_0^T\big(u_i^{\delta,\alpha_0^\delta}(t)-\hat{u}_i^\delta(t)\big)\big(\hat{u}_i^\delta(t)-u_i^*(t)\big)dt \xrightarrow{\delta\to 0} 0.\end{aligned} \tag{7.90}$$

Combining (7.89) and constraints (7.2), we get

$$\big(u_i^{\delta,\alpha_0^\delta}(t)-\hat{u}_i^\delta(t)\big)\big(\hat{u}_i^\delta(t)-u_i^*(t)\big)\ge 0,\quad t\in[0,T].$$

Since all terms in the last expression in (7.90) are non-negative, we obtain

$$\|u_i^{\delta,\alpha_0^\delta}(\cdot)-\hat{u}_i^\delta(\cdot)\|^2_{L_{2,[0,T]}}\xrightarrow{\delta\to 0} 0, \tag{7.91}$$

$$\|\hat{u}_i^\delta(\cdot)-u_i^*(\cdot)\|^2_{L_{2,[0,T]}}\xrightarrow{\delta\to 0} 0. \tag{7.92}$$

Now let's prove the following lemma

Lemma 7.2 *The system of differential equations*

$$\dot{x}_i(t)=g_i(x(t),t)\hat{u}_i^\delta(t),\quad x_i(T)=y_i^\delta(T),\quad i=1,\dots,n,\quad t\in[0,T], \tag{7.93}$$

where $\hat{u}_i^\delta(\cdot)$ *is defined in (7.89) for a fixed* $\delta\le(0,\delta_0]$, *have a unique solution* $x(\cdot)\overset{not}{=}\hat{x}^\delta(\cdot):[0,T]\to R^n$. *Moreover,*

$$\lim_{\delta\to 0}\|x_i^*(\cdot)-\hat{x}_i^\delta(\cdot)\|_{C_{[0,T]}}=0,\quad i=1,\dots,n.$$

Proof Let's introduce new variables

$$\Delta x_i(t)=x_i(t)-x_i^{\delta,\alpha_0^\delta}(t),\quad i=1,\dots,n,$$

where $x^{\delta,\alpha_0^\delta}(t)$ is the solution of system (7.19) with boundary conditions (7.20).

System (7.93) in this variables has the form

$$\begin{gathered}\dot{\Delta x}_i(t)=g_i\big(\Delta x(t)+x^{\delta,\alpha_0^\delta}(t),t\big)\hat{u}_i^\delta(t)-g_i\big(x^{\delta,\alpha_0^\delta}(t),t\big)u_i^{\delta,\alpha_0^\delta}(t),\\ \Delta x_i(T)=0,\quad i=1,\dots,n.\end{gathered} \tag{7.94}$$

The right-hand sides of this equations

$$
\begin{aligned}
&\Big|g_i\big(\Delta x(t)+x^{\delta,\alpha_0^\delta}(t),t\big)\hat{u}_i^\delta(t)-g_i\big(x^{\delta,\alpha_0^\delta}(t),t\big)u_i^{\delta,\alpha_0^\delta}(t)\pm g_i\big(x^{\delta,\alpha_0^\delta}(t),t\big)\hat{u}_i^\delta(t)\Big|\\
&\quad=\Big|\hat{u}_i^\delta(t)\Big(g_i\big(\Delta x(t)+x^{\delta,\alpha_0^\delta}(t),t\big)-g_i\big(x^{\delta,\alpha_0^\delta}(t),t\big)\Big)\\
&\qquad+g_i\big(x^{\delta,\alpha_0^\delta}(t),t\big)\big(\hat{u}_i^\delta(t)-u_i^{\delta,\alpha_0^\delta}(t)\big)\Big|\\
&\quad\le\overline{U}\sum_{j=1}^{n}[\omega'|\Delta x_j(t)|]+\overline{\omega}|\hat{u}_i^\delta(t)-u_i^{\delta,\alpha_0^\delta}(t)|\\
&\quad\le\overline{U}\omega' n\|\Delta x(t)\|+\overline{\omega}|\hat{u}_i^\delta(t)-u_i^{\delta,\alpha_0^\delta}(t)|.
\end{aligned}
\tag{7.95}
$$

Since estimates (7.95) are true and the function $|\hat{u}_i^\delta(\cdot)-u_i^{\delta,\alpha_0^\delta}(\cdot)|$ is continuous, the solution of system (7.94) is unique and can be extended on $[0,T]$ [13]. Thus, the solutions $\hat{x}_i^\delta(t)=\Delta x_i(t)-x_i^{\delta,\alpha_0^\delta}(t),\ i=1,\dots,n$ of system (7.93) can be extended on $t\in[0,T]$ as well.

From (7.95) it follows that

$$
\begin{aligned}
\Big|\|\Delta x(t)\|_t'\Big| &= \left|\frac{\sum_{i=1}^{n}[\Delta x_i(t)\dot{\Delta x}_i(t)]}{\|\Delta x(t)\|}\right| \le \frac{\sum_{i=1}^{n}[\|\Delta x(t)\|\cdot|\dot{\Delta x}_i(t)|]}{\|\Delta x(t)\|}\\
&\le n(\overline{U}\omega' n\|\Delta x(t)\|+\overline{\omega}|\hat{u}_i^\delta(t)-u_i^{\delta,\alpha_0^\delta}(t)|).
\end{aligned}
$$

Hence,

$$
\begin{aligned}
\|\Delta x(t)\| &\le \|\Delta x(T)\|+\int_t^T n\big(\overline{U}\omega' n\|\Delta x(\tau)\|+\overline{\omega}|\hat{u}_i^\delta(\tau)-u_i^{\delta,\alpha_0^\delta}(\tau)|\big)d\tau\\
&\le \|\Delta x(T)\|+n\overline{\omega}\int_t^T|\hat{u}_i^\delta(\tau)-u_i^{\delta,\alpha_0^\delta}(\tau)|d\tau+n^2\overline{U}\omega'\int_t^T\|\Delta x(\tau)\|d\tau.
\end{aligned}
$$

Applying the Grönwall–Bellman inequality, we get

$$
\|\Delta x(t)\|\le\Big(\|\Delta x(T)\|+n\overline{\omega}\int_t^T|\hat{u}_i^\delta(\tau)-u_i^{\delta,\alpha_0^\delta}(\tau)|d\tau\Big)\exp(n^2\overline{U}\omega' T).
$$

Here $\|\Delta x(T)\| \leq \sqrt{n}\delta \overset{\delta\to 0}{\longrightarrow} 0$. Since (7.91), $\int\limits_t^T |\hat{u}_i^\delta(\tau) - u_i^{\delta,\alpha_0^\delta}(\tau)|d\tau \overset{\delta\to 0}{\longrightarrow} 0,\ t \in [0, T]$. So, $\|\Delta x(t)\| \overset{\delta\to 0}{\longrightarrow} 0,\ t \in [0, T]$. In other words,

$$\lim_{\delta\to 0} \|x_i^{\delta,\alpha_0^\delta}(\cdot) - \hat{x}_i^\delta(\cdot)\|_{C[0,T]} = 0, \quad i = 1, ..., n.$$

Combining this result with result of Theorem 7.1 (7.34), we get

$$\lim_{\delta\to 0} \|\hat{x}_i^\delta(\cdot) - x_i^*(\cdot)\|_{C[0,T]} = 0, \quad i = 1, ..., n,$$

which was to be proved.

Lemma 7.2, definition (7.89) and formula (7.92) mean that functions (7.89) can be considered as solution of the inverse problem described in Sect. 7.5.

7.7 Remarks on the Suggested Method

Note that Hypotheses 7.2 and 7.1, Theorem 7.1 and Lemmas 7.1 and 7.2 provide that in case of diagonal matrix $G(x, t)$ the solution for the inverse problem described in Sect. 7.5 can be found as

$$\hat{u}_i^\delta(t) = \begin{cases} \overline{U}, & u_i^\delta(t) \geq \overline{U}, \\ u_i^\delta(t), & |u_i^\delta(t)| < \overline{U}, \\ -\overline{U}, & u_i^\delta(t) \leq -\overline{U}. \end{cases} \text{ where } u_i^\delta(\cdot) = \frac{\dot{y}_i^\delta(\cdot)}{g_i(y^\delta(\cdot), \cdot)}, \quad i = 1, \ldots, n.$$

The case of non diagonal non degenerate matrix $G(x, t)$ is more interesting. In this case the solution can still be found by inversing the matrix $G(y^\delta(t), t)$

$$u^\delta(\cdot) = G^{-1}(y^\delta(\cdot), \cdot)\dot{y}^\delta(\cdot), \tag{7.96}$$

but it involves finding the inverse matrix $G^{-1}(y^\delta(t), t)$ for each $t \in [0, T]$.

One can modify the algorithm suggested in Sect. 7.6 to solve the inverse problem for the case of non-diagonal matrix $G(y^\delta(t), t)$ as well. The justification uses the same scheme of proof, but is more complex due to more complicated form of system (7.19). It will be published in later works.

Comparing the direct approach (7.96) and the approach suggested in this paper, one can see that the second one reduces the task of inversing non-constant $n \times n$ matrix $G(y^\delta(t), t)$ to the task of solving systems of non-linear ODEs. In some applications numerical integration of ODE systems may be more preferable than matrix inversing. Accurate comparing of this approaches (including numerical computations issues) is the matter of the upcoming studies and also will be published in later works.

7.8 Example

To illustrate the work of the suggested method let's consider a model of a macroeconomic process, which can be described by a differential game with the dynamics

$$\begin{aligned} \frac{dx_1(t)}{dt} &= \frac{\partial G(x_1(t), x_2(t))}{\partial x_1} u_1(t), \\ \frac{dx_2(t)}{dt} &= \frac{\partial G(x_1(t), x_2(t))}{\partial x_2} u_2(t). \end{aligned} \tag{7.97}$$

Here $t \in [0, T]$, x_1 is the product, x_2 is the production cost. $G(x_1, x_2)$ is the profit, which is described as

$$G(x_1, x_2) = x_1 x_2 (a_0 + a_1 x_1 + a_2 x_2), \tag{7.98}$$

where $a_0 = 0.008$, $a_1 = 0.00019$, $a_2 = -0.00046$ are parameters of the macroeconomic model [1]. The functions $u_1(t)$, $u_2(t)$ are bounded piecewise continuous controls

$$|u_1| \le \overline{U}, \quad |u_2| \le \overline{U}, \quad \overline{U} = 200, \quad t \in [0, T]. \tag{7.99}$$

The control u_1 has the meaning of the scaled coefficient of the production increase speed and u_2 has the meaning of the scaled coefficient of the speed of the production cost changing.

This model has been suggested by Albrecht [1].

We assume that some base trajectories $x_1^*(t)$, $x_2^*(t)$ of system (7.97) have been realized on the time interval $t \in [0, T]$ (time is measured in years). This trajectory is supposed to be generated by some admissible controls $u_1^*(\cdot)$, $u_2^*(\cdot)$. We also assume that we know inaccurate measurements of $x_1^*(t)$, $x_2^*(t)$—twice continuously differentiable functions $y_1^\delta(t)$, $y_2^\delta(t)$ that fulfill Hypothesis 7.2.

Remark 7.9 To model measurement functions $y_1^\delta(t)$ and $y_2^\delta(t)$ real statistics on Ural region's industry during 1970–1985 [1] have been used. They satisfy Hypothesis 7.2.

We consider the inverse problem described in Sect. 7.5 for dynamics (7.97)–(7.99) and functions $x_1^*(t)$, $x_2^*(t)$, $u_1^*(\cdot)$, $u_2^*(\cdot)$ and $y_1^\delta(t)$, $y_2^\delta(t)$. We assume in our example that we don't know the base trajectory and controls, but know the inaccurate measurements $y_1^\delta(t)$, $y_2^\delta(t)$.

The trajectories $x_1^{\alpha,\delta}(t)$, $x_2^{\alpha,\delta}(t)$ and controls $\hat{u}_1^{\alpha,\delta}(t)$, $\hat{u}_2^{\alpha,\delta}(t)$, generating them, were obtained numerically. The results are presented on Figs. 7.1, 7.2, and 7.3. On Figs. 7.1 and 7.2 time interval is reduced for better scaling.

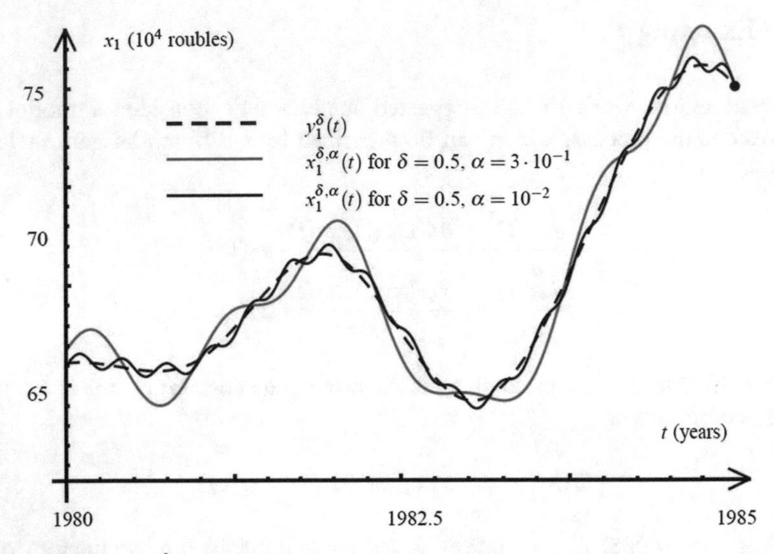

Fig. 7.1 Graphics of $x_1^{\delta,\alpha}(t)$, $t \in [1980, 1985]$ for various values of approximation parameters

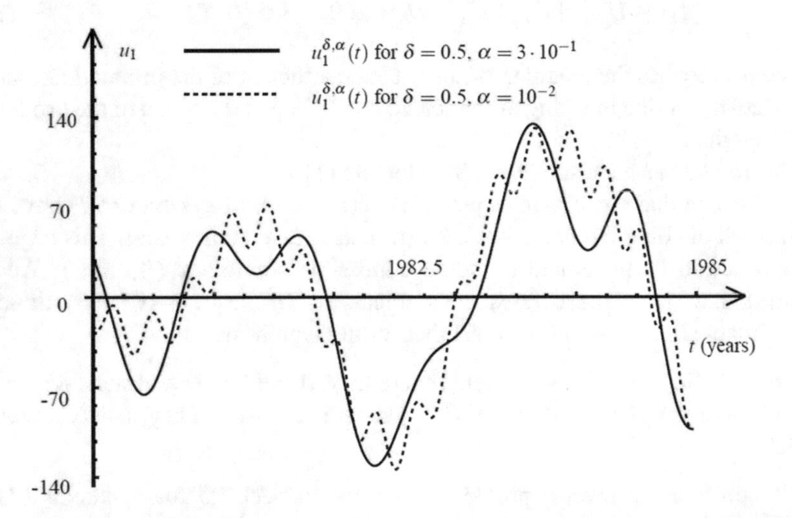

Fig. 7.2 Graphics of $u_1^{\delta,\alpha}(t)$, $t \in [1980, 1985]$ for various values of approximation parameters

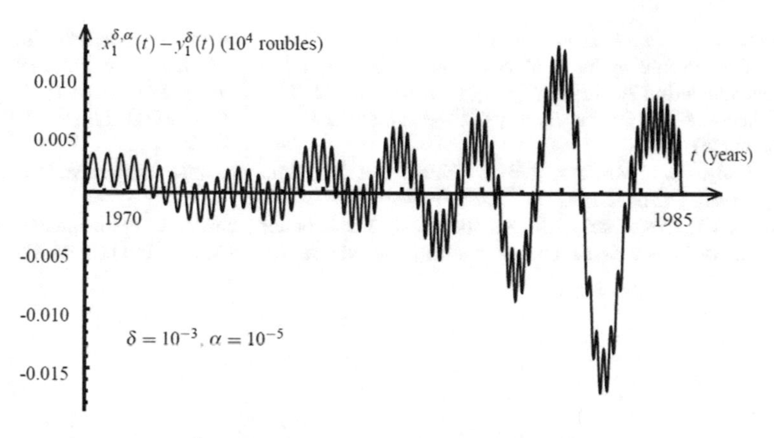

Fig. 7.3 Graphic of error $x_1^{\delta,\alpha}(t) - y_1^{\delta}(t)$ for $\alpha = 10^{-5}$, $t \in [1970, 1985]$

Acknowledgements This work was supported by the Russian Foundation for Basic Research (project no. 17-01-00074) and by the Ural Branch of the Russian Academy of Sciences (project no. 18-1-1-10).

References

1. Albrecht, E.G.: Metodika postroenija i identifikacii matematicheskih modelej makrojekonomicheskih processov (Methodics of construction and identification of mathematic models of macroeconomic processes). Elektronnyj zhurnal Issledovano v Rossii **5**, 54–86 (2002)
2. Kolmogorov, A.N., Fomin, S.V.: Elementy teorii funktsii i funktsionalnogo analiza. [Elements of the theory of functions and functional analysis] With a supplement, Banach algebras, by V. M. Tikhomirov, 6th edn. Nauka, Moscow, 624 pp. (1989) (in Russian)
3. Krasovskii, N.N.: Theory of Motion Control. Nauka, Moscow (1968) (in Russian)
4. Krasovskii, N.N., Subbotin, A.I.: Positional Differential Games. Nauka, Moscow (1974) (in Russian)
5. Krupennikov, E.A.: Validation of a solution method for the problem of reconstructing the dynamics of a macroeconomic system. Trudy Inst. Mat. Mekh. UrO RAN **21**(2), 102–114 (2015)
6. Kryazhimskii, A.V., Osipov, Y.S.: Modelling of a control in a dynamic system. Eng. Cybern. **21**(2), 38–47 (1984)
7. Osipov, Y.S., Kryazhimskii, A.V.: Inverse Problems for Ordinary Differential Equations: Dynamical Solutions. Gordon and Breach, London (1995)
8. Subbotina, N.N., Krupennikov, E.A.: Dynamic programming to identification problems. World J. Eng. Technol. **4**(3B), 228–234 (2016)
9. Subbotina, N.N., Kolpakova, E.A., Tokmantsev, T.B., Shagalova, L.G.: The Method of Characteristics for Hamilton–Jacobi–Bellman Equations. Izd. UrO RAN, Yekaterinburg (2013) (in Russian)
10. Subbotina, N.N., Tokmantsev, T.B., Krupennikov, E.A.: On the Solution of Inverse Problems of Dynamics of Linearly Controlled Systems by the Negative Discrepancy Method. In: Optimal control, Collected Papers. In commemoration of the 105th anniversary of Academician Lev Semenovich Pontryagin, Tr. Mat. Inst. Steklova, vol. 291, MAIK Nauka/Interperiodica, Moscow, pp. 266–275 (2015)

11. Subbotina, N.N., Tokmantsev, T.B., Krupennikov, E.A.: Dynamic Programming to Reconstruction Problems for a Macroeconomic Model. In: IFIP Advances in Information and Communication Technology. [S.l.], vol. 494, pp. 472–481. Springer, Berlin (2017)
12. Tikhonov, A.N.: On the stability of inverse problems. Dokl. Acad. Sci. URSS (N.S.) **39**, 195–198 (1943)
13. Tikhonov, A.N., Vasileva, A.B., Sveshnikov, A.G.: Differentsialnye uravneniya [Differential equations]. Fizmalit, Moscow (2005) (in Russian)
14. Vanko, V.I., Ermoshina, O.V., Kuvirkin G.N.: Variacionnoe ischislenie i optimalnoe upravlenie [Variational calculus and optimal control], 2nd edn. MGU, Moscow (2001) (in Russian)

Chapter 8
Evolution of Risk-Statuses in One Model of Tax Control

Suriya Kumacheva, Elena Gubar, Ekaterina Zhitkova, and Galina Tomilina

Abstract Nowadays information is an important part of social life and economic environment. One of the principal elements of economics is the system of taxation and therefore tax audit. However total audit is expensive, hence fiscal system should choose new instruments to force the tax collections. In current study we consider an impact of information spreading about future tax audits in a population of taxpayers. It is supposed that all taxpayers pay taxes in accordance with their income and individual risk-status. Moreover we assume that each taxpayer selects the best method of behavior, which depends on the behavior of her social neighbors. Thus if any agent receives information from her contacts that the probability of audit is high, then she might react according to her risk-status and true income. Such behavior forms a group of informed agents which propagate information further then the structure of population is changed. We formulate an evolutionary model with network structure which describes the changes in the population of taxpayers under the impact of information about future tax audit. The series of numerical simulation shows the initial and final preferences of taxpayers depends on the received information.

8.1 Introduction

The tax system is one of the most important mechanisms of state regulation. A significant part of this system is tax control, which provides receiving taxes and fees in the state budget. For a wide class of models, such as [4, 6, 16, 25], which describe tax control with static game-theoretical attitude, "the threshold rule" was formulated. This rule defines the value of auditing probability which is critical for

S. Kumacheva (✉) · E. Gubar · E. Zhitkova · G. Tomilina
Saint Petersburg State University, Saint Petersburg, Russia
e-mail: s.kumacheva@spbu.ru; e.gubar@spbu.ru; e.zhitkova@spbu.ru; g.tomilina@yandex.ru

L. A. Petrosyan et al. (eds.), *Frontiers of Dynamic Games*,
Static & Dynamic Game Theory: Foundations & Applications,
https://doi.org/10.1007/978-3-319-92988-0_8

the decision of taxpayers to evade taxation or not. However, in real life it is difficult to implement tax inspections with the threshold probability because this process requires large investments from the tax authority, while it has substantially limited budget. Hence, the tax authority needs to find a way to stimulate the population to pay taxes in accordance with their true level of income.

Previous studies [18, 22] have shown that information dissemination has a significant impact on the behavior of agents in various environments, such as the urban population, the social network, labor teams, etc. Taking into account previous research [1–3], the current paper studies the propagation of information about upcoming tax inspections as a way to stimulate the population to pay taxes honestly. This approach allows tax authority to optimize the collection of taxes within the strong limitation of budget.

Let's suppose that the population of taxpayers is heterogeneous in its perception of such information. Additionally to previous research [9, 10] susceptibility of each agent depends on its risk-status, due to her natural propensity to risk. Economic environment of each individual also impacts on the perceiving of incoming information. In contrast to many different works, where information spreads during random matches of agents, here we consider only structured population and we suppose that information can be transferred only between connected agents. Social connections of each taxpayer can be described mathematically by using networks of various modifications. We also assume that tax authority injects information about future audits to the population and thereby the share of Informed agents is formed. Agents from the Informed group can spread it over their network of contacts and thereby the structure of population is changed. According to all these reasons we formulate an evolutionary model on the network which describes the variation of taxpayers' behavior. We estimate the initial and final distribution of taxpayers which prefer to evade taxation in series of numerical simulations. Numerical experiments include two different approaches that characterize the evolutionary process: special imitation rule for evolutionary game on the network and Markov's chain which define random process on the network.

The paper is organized as follows. Section 8.2 presents the mathematical model of tax audit in classical formulation. Section 8.3 introduces an idea of risk propensity of taxpayers. Section 8.4 shows the dynamic model of tax control, which includes the knowledge about additional information and presents two different approaches to find a solution. Numerical examples are presented in Sect. 8.5.

8.2 Static Model of Tax Audit

As a basic model we consider a game-theoretical static model of tax control, in which the players are the tax authority and n taxpayers as a basis for the following study. Every taxpayer has true income ξ and declares income η after each tax period, $\eta \leq \xi$.

As it was studied in the classical works, such as [6, 25], to simplify the model, we suppose, that the total set of taxpayers is divided into the groups of low level income agents and high level income agents. Obviously, the number of partitions can be increased, but it does not effect on the following arguments and conclusions. In other words, taxpayers' incomes can take only two values: $\xi \in \{L,\ H\}$, where L is the low level and H is the high level of income $(0 \leq L < H)$. Declared income η also can take values from the mentioned binary set $\eta \in \{L,\ H\}$.

Thus, in this model there are three different groups of taxpayers, depending on the relation $\eta(\xi)$ between true and declared incomes:

1. $\eta(\xi) = L(L)$;
2. $\eta(\xi) = H(H)$;
3. $\eta(\xi) = L(H)$.

Obviously that the taxpayers from the first and the second groups declare their income correspondingly to its true level and they do not try to evade. The third group is the group of tax evaders. In other words, this group is of interest of the tax authority.

The tax authority audits those taxpayers, who declared $\eta = L$, with the probability P_L every tax period. Let's suppose that tax audit is absolutely effective, i.e. it reveals the existing evasion. The proportional case of penalty is considered: when the tax evasion is revealed, the evader must pay $(\theta+\pi)(\xi-\eta)$, where constants θ and π are the tax and the penalty rates correspondingly. For the agents from the studied groups the payoffs are:

$$u\,(L(L)) = (1-\theta)\cdot L; \tag{8.1}$$

$$u\,(H(H)) = (1-\theta)\cdot H; \tag{8.2}$$

$$u\,(L(H)) = H - \theta L - P_L(\theta+\pi)(H-L). \tag{8.3}$$

The fraction of audited taxpayers is P_L. It's obvious that either the agents from the first group (who actually have true income $\xi = L$) or the evaders from the third group are both in this fraction of audited taxpayers.

The total set of the taxpayers is divided into the next groups: wealthy taxpayers, who pay taxes honestly ($\eta(\xi) = H(H)$), insolvent taxpayers ($\eta(\xi) = L(L)$) and wealthy evaders ($\eta(\xi) = L(H)$). The diagram, presented in Fig. 8.1a, illustrates this distribution.

In Fig. 8.1, cases: b, c, d, the little circle, inscribed in the diagram, corresponds to the fraction of audited taxpayers. Let's call it an "audited circle". Only those, who declared $\eta = L$, are in the interest for auditing. Therefore, the mentioned circle is inscribed into the sectors, which correspond to the situations $\eta(\xi) = L(H)$ and $\eta(\xi) = L(L)$.

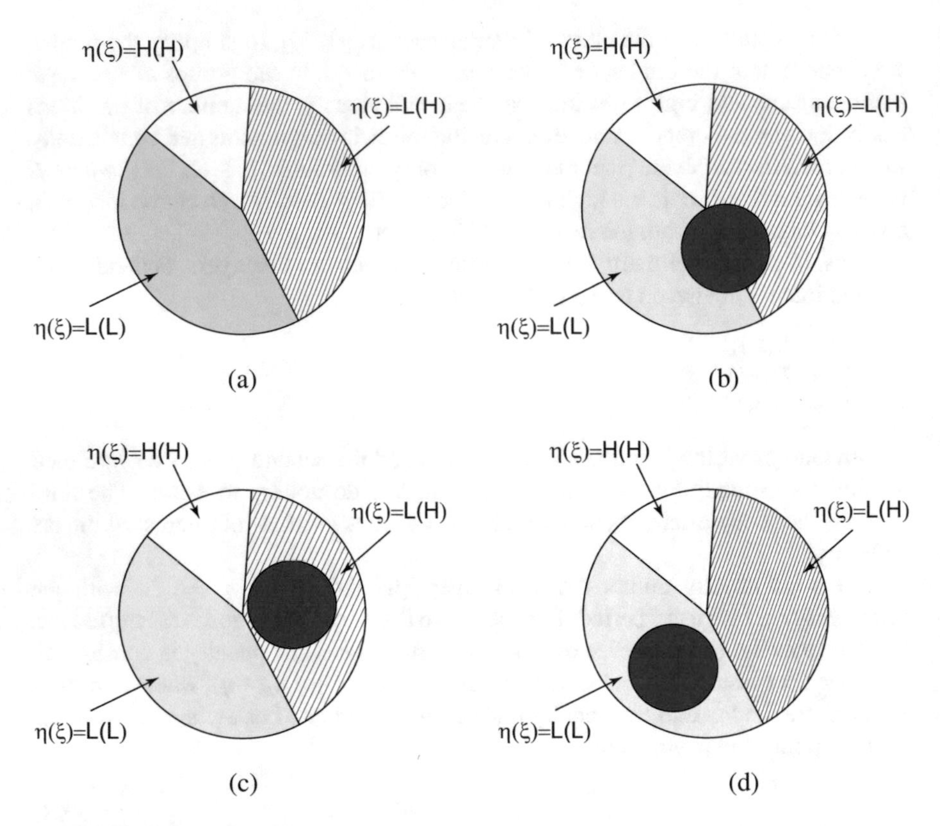

Fig. 8.1 Auditing of the different groups of the taxpayers

On one hand in Fig. 8.1b the case, when either evaders or simply insolvent taxpayers are audited, is considered. In Fig. 8.1c the "audited circle" is contained in the area, which satisfies the condition $\eta(\xi) = L(H)$. This is the optimistic situation, when every audit reveals the existing tax evasion. On the other hand, the Fig. 8.1d illustrates the pessimistic case, when none of the audits reveals the evasion, because only insolvent taxpayers, declared their true income $\eta(\xi) = L(L)$, are audited. Certainly, all presented diagrams illustrate only some boundary ideal cases, however the tax authority's aim is obviously to lead the real situation closer to the illustration in Fig. 8.1c.

Hence, the following arguments, related to the searching of possible tax evasions, apply to the third group of the agents, declared $\eta(\xi) = L(H)$. Thereby, precisely this group is expedient to be considered as the studied population, speaking in the terms of the evolutionary games.

Risk neutral taxpayers' behaviour supposed to be absolutely rational: their tax evasion is impossible only if the risk of punishment is so high that the tax evader's profit is less or equal to his expected post-audit payments (in the case when his evasion is revealed):

$$P_L(\theta + \pi)(H - L) \geq \theta(H - L).$$

Therefore, the critical value of audit probability P_L (due to the taxpayer's decision to evade or not) is

$$P_L^* = \frac{\theta}{\theta + \pi}. \tag{8.4}$$

For this type of models the optimal solution is usually presented in the form of the "threshold rule" in various modifications (see, for example, [6] or [25]). In [4] this rule is formulated so that the optimal value P_L^* of the auditing probability is defined from (8.4), and for the risk neutral taxpayer the optimal strategy is

$$\eta^*(\xi) = \begin{cases} H, \ P_L \geq P_L^*; \\ L, \ P_L < P_L^*. \end{cases} \tag{8.5}$$

Nevertheless there are some problems which should be fixed to make the static model described above close to real-life process. The first problem is that the players are supposed to be risk neutral. However in real life there are also risk averse and risk loving economic agents. Another problem is that we consider the game with complete information. It is assumed that the taxpayers know (or can estimate) the value of the auditing probability, but, by considering the static model, we do not take into account the method of receiving information. The third problem is that the auditing with optimal probability (8.4) is excessively expensive and the tax authority usually has strongly limited budget, thus, the actual value of P_L should be substantially less than P_L^* in real life. By taking into consideration all mentioned reasons, we formulate an extended model of tax audit which includes an information component and an evolutionary process of adaptation of taxpayers to the changes in the economic environment.

8.3 Model with Different Risk-Statuses

Now let's consider the homogeneous population of n taxpayers, where agents possess one of three risk-statuses: risk averse, risk neutral and risk loving agents. Let's restrict the considered population by the subpopulation of taxpayers with high level H of income. This restriction is natural because there is no reason and ability to evade for the taxpayers with low level L of income, independently on their risk-status.

The size of this subpopulation is n_H ($n_L + n_H = n$, where n_L is a number of the taxpayers with income level L). Now let ν_a be the share of risk averse agents with income H ($\nu_a = \frac{n_a}{n_H}$), ν_n be the share of risk neutral agents with high-level income

($\nu_n = \frac{n_n}{n_H}$) and ν_l be the share of risk loving agents with income H ($\nu_l = \frac{n_l}{n_H}$) respectively. Naturally, $\nu_a+\nu_n+\nu_l = 1$ (or, equivalent equation, $n_a+n_n+n_l = n_H$).

We also assume that each risk-status has its "threshold of sensitivity". This term means that each taxpayer with income H compares the real and critical values of the auditing probability P_L before to make a decision to evade or to pay taxes honestly. Based on the results obtained for the static model 8.2 it is obvious that for the risk neutral agent this threshold value is P_L^* from the Eq. (8.4). Let $\underline{P_L}$ and $\overline{P_L}$ be the sensitivity thresholds for the risk averse and risk loving agents correspondingly. These values satisfy the inequality [17]:

$$0 \leq \underline{P_L} < P_L{}^* < \overline{P_L} \leq 1.$$

It is natural to suppose that the agents' behavior depends on their statuses. Here, risk-status defines the relation between obtained information about future auditing and agent's own sensitivity threshold. If the information of tax audit is absent then we assume that initially the population is sure that the probability of future audits takes its values from the interval $(\underline{P_L}, P_L{}^*)$.

This value is less than the threshold for the risk neutral agents, therefore, they evade of taxation, moreover, risk loving taxpayers, which are sure in small possibility of auditing, also do not pay. The only payers are agents with risk-averse status form the considered subpopulation. In this situation the total tax revenue is

$$TTR_1 = n_L\theta L + n_H\,(\nu_a\theta H + (1-\nu_a)\,(\theta\, L + P_L\,(\theta+\pi)(H-L))) - n\,P_L\,c, \tag{8.6}$$

where c is the cost of one audit.

8.4 The Evolutionary Model on the Network

In Sect. 8.2 we have discussed that it is extremely expensive to audit taxpayers with the optimal probability (8.4). Thus the tax authority needs to find additional ways to stimulate taxpayers' fees. One of these ways is the injection of information about future auditing (which possibly can be false) into the population of taxpayers. Following [9–11], in current study we discuss that information contains a message "$P_L \geq P_L^*$". We suppose that the dissemination of such information over the population will impact on the behavior of taxpayers and their risk-statuses. The cost of unit of information is c_{inf}. We assume that c_{inf} is significantly less than the auditing casts ($c_{inf} \ll c$). Every taxpayer who receives the information can use it by choosing the strategy to pay or not to pay taxes due to her true income level. Additionally, in contrast to the standard approach of evolutionary games [21, 26], in which it is assumed that meetings between agents occur randomly in the population, here we will consider only the connected agents. For example, any taxpayer has a social environment such as family, relatives, friends, neighbors.

Communicating with them, agents choose opponents at random to transmit information, but taking into account the existing connections. Thus, to describe possible interactions between agents, the population can be represented by a network where the nodes are taxpayers transmitting information to each other during the process of communication and links are the connections between them. Earlier such approach was considered in [9].

Therefore, the taxpayer's decision about her risk-status (risk averse, risk neutral or risk loving) depends on two important factors: her own (natural) risk propensity and the behavior of her neighbors in the population (those with whom she communicates). As a result of the dissemination of information, the entire population of agents is divided into two subgroups: those who received and used the information (the share n_{inf}), and those who do not intend to use the information (the share n_{noinf}), or, we can say, those who have a propensity to perceive or not to perceive the received information. Thus, at the initial time moment this population can be presented as a sum of the mentioned shares:

$$n = n_{inf}(t_0) + n_{noinf}(t_0),$$

but at each following moment the ratio of the fractions $n_{inf}(t)$ and $n_{noinf}(t)$ will differ from the previous one.

If one taxpayer from a subgroup of those who use information meets another one from the same subgroup, they will get the payoffs (U_{inf}, U_{inf}). In this case both of them know the same information and pay, hence their payoff is defined from the Eq. (8.2). Similarly, if the taxpayer who does not perceive the information (and therefore wants to evade) meets the same taxpayer, they will get the payoffs (U_{ev}, U_{ev}), which are defined from the Eq. (8.3).

We denote the taxpayer's propensity to perceive the information as δ, $0 \leq \delta \leq 1$, and consider a case when the uninformed taxpayer meets the informed taxpayer. As a result of such meeting, uninformed taxpayer obtains information and should pay the payoff (8.2) with probability δ if she believes in this information, or the payoff (8.3) if she does not believe.

In the current study we use evolutionary game approach to describe the dynamic nature of such economic process. Thus, we have a well-mixed population of economic agents (taxpayers), where the instant communications between taxpayers can be defined by two-players bimatrix game [21]. For the cases, when taxpayers of different types meet each other, the matrix of payoffs can be written in the form:

	A	B
A	(U_{inf}, U_{inf})	$(U_{inf}, \delta U_{inf} + (1-\delta)U_{ev})$
B	$(\delta U_{inf} + (1-\delta)U_{ev}, U_{inf})$	$(\delta U_{inf} + (1-\delta)U_{ev}, \delta U_{inf} + (1-\delta)U_{ev})$

where A is the strategy of taxpayer if she is informed (she perceives the information) and B is the strategy of taxpayer if she is uninformed.

Suppose that at a finite time moment T the system reached its stationary state. Then let's denote by ν_{inf} the share of those who used the received information and paid taxes according to their true income level ($\nu_{inf} = \nu_{inf}(T) = \frac{n_{inf}(T)}{n}$), while the share of those who evades taxation despite information received is denoted by ν_{ev} ($\nu_{ev} = \nu_{ev}(T) = \frac{n_{ev}(T)}{n}$).

Hence the total income received from taxation of the entire population is

$$TTR_2 = n_L\theta L + n_H\left(\nu_{inf}\theta H + \nu_{ev}\left(\theta L + P_L(\theta+\pi)(H-L)\right)\right) - \\ -n(P_L c + \nu_{inf}^0 c_{inf}), \qquad (8.7)$$

where $\nu_{inf}^0 = \nu_{inf}(t_0)$.

The papers [12, 14, 15] have studied the game of a large number of agents and obtained the results which can be used to give precise quantitative predictions and proper stability analysis of equilibria.

In this paper we present a comparative analysis of two evolutionary approaches applied to the model of tax control. The first approach is to define propagation information as a random process on the network and use the model by De Groot [7] as a mathematical tool. The second approach is built on the special stochastic imitation rule for evolutionary dynamics on the network [20, 21]. Now let's examine how these ideas can be applied to the modeling of dynamic processes on networks.

8.4.1 The Model Based on the Markov Process on the Network

Now let's refuse the assumption that agents can estimate the choice of each strategy absolutely correctly. This refusion allows us to consider a model of random dissemination of information about the probability of future auditing. One of the first models describing this problem is the model by De Groot [7], then similar attitude was also studied in [8] and [5].

Based on the previous research we consider a direct network $G = (N, P)$, where N is the set of economic agents (for the present study $N = \{1, \dots, n\}$), and P is a stochastic matrix of connections between agents: p_{ij} is an element of the matrix P which characterizes the connection between agents i and j. $p_{ij} > 0$ in the case when there exists a social connection between taxpayers i and j, $(i, j \in N)$. The value of this parameter is close to 1 if the ith agent has a reason to assume that the agent j has an expert knowledge about the probability of auditing, and otherwise it is close to 0.

Let's assume that at the initial time moment each agent has a certain belief, f_i^0 about the value of the auditing probability. Moreover, we suppose that she decides to evade taxation, comparing f_i^0 with the threshold of sensitivity $P_L{}^*$, in this case if $f_i^0 < P_L{}^*$ then the ith agent evades paying taxes, else if the threshold rises, then she prefers not to risk.

Interaction of agents leads to the updating of their knowledge on the auditing probability at each iteration:

$$f_i^k = \sum_{j=1}^{n} p_{ij} f_j^{k-1}.$$

The interaction continues infinitely or until the moment when for some k the condition $f_i^k \approx f_i^{k-1}$ holds for every i.

Now, we assume that information centers can be artificially introduced into the natural population of agents, which is presented as a network. Here, as information centers (further principals) we set the agents who seek to convince the other agents that value of the auditing probability is equal to some determined value.

The role of such information center can be performed by any of the agents j. Let S be the set of agents for which the condition $p_{ij} > 0, i \neq j$, (it is clear because this inequality is formulated for pairs of different agents). We assign the parameter α_j, which expresses the degree of confidence of the value f_j^0, with this agent j. Then the updated elements of the jth row of the matrix P will have the following form:

$$p_{ij} = \begin{cases} \alpha_j, & i = j \\ \frac{1-\alpha_j}{|S|}, & i \in S \\ 0, & i \notin S, i \neq j \end{cases} \tag{8.8}$$

Described model can be used, for example, to present a goal of the tax authority to overstate the value of the auditing probability. In this case, the parameters of the information center have the following form: $f_j^0 = 1, \alpha_j \approx 1$.

8.4.2 *The Model Based on the Proportional Imitation Rule*

In the current paragraph we present a different approach to describe the sequence of changes in the population of taxpayers. Let $G = (N, L)$ denote an indirect network, where N is a set of economic agents ($N = \{1, \ldots, n\}$ as in Sect. 8.4.1) and $L \subset N \times N$ is an edge set. Each edge in L represents two-player symmetric game between connected taxpayers. The taxpayers choose strategies from a binary set $X = \{A, B\}$ and receive payoffs according to the matrix of payoffs in Sect. 8.4. Each instant time moment agents use a single strategy against all opponents and thus the games occurs simultaneously. We define the strategy state by $x(T) = (x_1(t), \ldots, x_n(t))^T$, $x_i(t) \in X$. Here $x_i(t) \in X$ is a strategy of taxpayer i, $i = \overline{1, n}$, at time moment t. Aggregated payoff of agent i is defined as in [20]:

$$u_i = \omega_i \sum_{j \in M_i} a_{x_i(t), x_j(t)}, \tag{8.9}$$

where $a_{x_i(t),x_i(t)}$ is a component of payoff matrix, $M_i := \{j \in L : \{i, j\} \in L\}$ is a set of neighbors for taxpayer i, weighted coefficient $\omega_i = 1$ for cumulative payoffs and $\omega_i = \frac{1}{|M_i|}$ for averaged payoffs. Vector of payoffs of the total population is $u(t) = (u_1(t), \ldots, u_n(t))^T$.

The state of population is changed according to the rule, which is a function of the strategies and payoffs of neighboring agents:

$$x_i(t+1) = f(\{x_j(t), u_j(t) : j \in N_i \cup \{i\}\}). \quad (8.10)$$

Here we suppose that taxpayer changes her behavior if at least one neighbor has better payoff. As the example of such dynamics we can use the proportional imitation rule [21, 26], in which each agent chooses a neighbor randomly and if this neighbor received a higher payoff by using a different strategy, then the agent will switch with a probability proportional to the payoff difference. The proportional imitation rule can be presented as:

$$p\left(x_i(t+1) = x_j(t)\right) := \left[\frac{\lambda}{|M_i|}(u_j(t) - u_i(t))\right]_0^1 \quad (8.11)$$

for each agent $i \in L$ where $j \in M_i$ is a uniformly randomly chosen neighbor, $\lambda > 0$ is an arbitrary rate constant, and the notation $[z]_0^1$ indicates $\max(0, \min(1, z))$.

Below we present two cases of the changing rule [10, 11]:

- **Case 1.** *Initial distribution of agents is nonuniform.* When agent i receives an opportunity to revise her strategy then she considers her neighbors as one homogeneous player with aggregated payoff function. This payoff function is equal to the mean value of payoffs of players who form a homogeneous player. It is assumed that the agent meets any neighbor with uniform probability, then mixed strategy of such homogeneous player is a distribution vector of pure strategies of included players. If payoff function of homogeneous player is better, then player i changes her strategy to the most popular strategy of her neighbors.
- **Case 2.** *Initial distribution of agents is uniform.* In this case agent i keeps her own strategy.

8.5 Numerical Simulations

In this section we present numerical examples to support the approaches described in Sects. 8.4.1 and 8.4.2. Based on these simulation we analyze the following factors of influence on the population of taxpayers:

1. the structure of the network: we use grid and random structures of graphs;
2. the way of information dissemination: we consider the processes based on the Markov processes and the proportional imitation rule;

Table 8.1 The distribution of income among taxpayers

Group	Income interval (rub. per month)	Share of population (%)
1	Less 7500	1.8
2	7500.1–10,600	6.1
4	10,600.1–17,000	15.2
6	17,000.1–25,000	19.9
6	25,000.1–50,000	36.1
6	50,000.1–10,0000	16.4
7	100,000.1–250,000	4.0
8	More 250,000	0.5

Table 8.2 Two modeled groups and average income

Group	Income interval (rub. per month)	Average income	Share of population (%)
L	Less 25,000	$L = 12{,}500$	43
H	More 25,000	$H = 50{,}000$	57

3. the initial distribution of risk-status in the population, i.e. what part of the population has a certain propensity to risk;
4. the value of the information injection, i.e. the portion of Informed agents at the initial time moment.

In all experiments we use the distribution of the income among the population of Russian Federation in 2017 [23] (see Table 8.1).

According to the model we suppose that only two level of income is accessible for each taxpayer: low and high (L and H). After the unification of groups with different levels of income according to the economic reasons, we calculate the average levels of income L and H (the mathematical expectations of the uniform and Pareto distributions—see 8.6) and receive the corresponding shares of the population (see Table 8.2).

For all experiments we fix the following values of parameters:

- share of risk-averse taxpayers in population is $\nu_a = 17\%$ due to the psychological research [19];
- tax and penalty rates are $\theta = 0.13$ due to the income tax rate in Russia [24], $\pi = 0.065$ (for bigger values of π, we obtain even bigger values of optimal audit probability $P_L{}^*$);
- optimal value of the probability of audit is $P_L{}^* = 0.67$;
- actual value of the probability of audit is $P_L = 0.1$;
- unit cost of auditing is $c = 7455$ (minimum wage in St. Petersburg [23]);
- unit cost of information injection is $c_{inf} = 10\%{}^*c = 745.5$;
- under the implementation of the approximate equality $f_i^k \approx f_i^{k-1}$ from the Sect. 8.4.1 we have that the inequality $|f_i^k - f_i^{k-1}| < 10^{-3}$ holds.

As we described in Sect. 8.4, we use the network G to define the structure of population. The number of considered taxpayers is defined from the Table 8.2. If, for example, the size of the total population is $n = 30$, the size of subpopulation with income level H is $n_H = 0.57 \cdot n = 17.10$, when $n = 25$ we obtain that $n_H = 14.25$ and so on. For the network we use the relation $\frac{\lambda}{|M_i|} = 1$ and vary values of other parameters in different examples.

Let the number of nodes in the population be $n = 30$. For the initial model, which does not include the process of information dissemination the value of total tax revenue (8.6) is $TTR_1 = 50{,}935.26$. For the network of $n = 25$ nodes $TTR_1 = 42{,}446.05$.

For the model which takes into account dissemination of information, we apply two algorithms: the first is based on the Markov process on the network (see Sect. 8.4.1), the second is based on the proportional imitation rule (Sect. 8.4.2).

Several results of numerical modeling of the Markov process in the network are presented in Figs. 8.2, 8.3, 8.4, and 8.5. Blue dots correspond to evaders and yellow

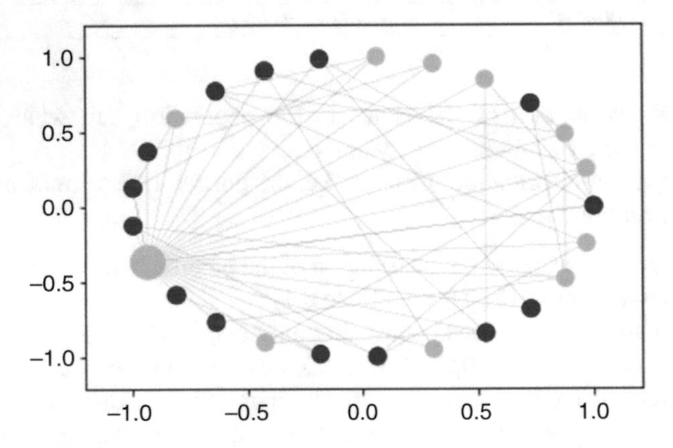

Fig. 8.2 Markov process. Agents are considered in pairs. Direct link from one to the other is formed with a probability of 1/20, $n = 25$. Initial state is $(\nu_{inf}, \nu_{ev}) = (1, 24)$

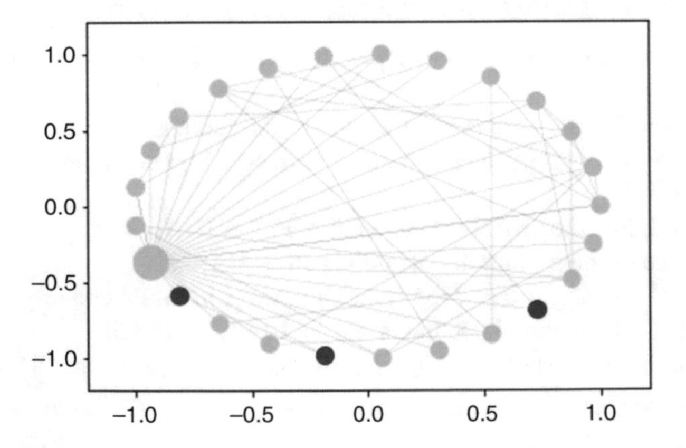

Fig. 8.3 Markov process. Stationary state is $(\nu_{inf}, \nu_{ev}) = (22, 3)$. Total tax revenue $TTR_2 = 83{,}624.94$

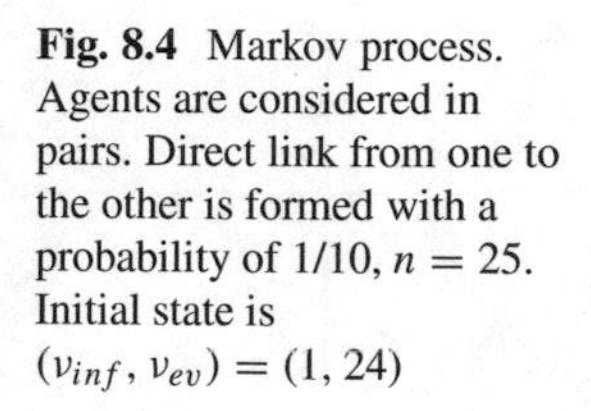

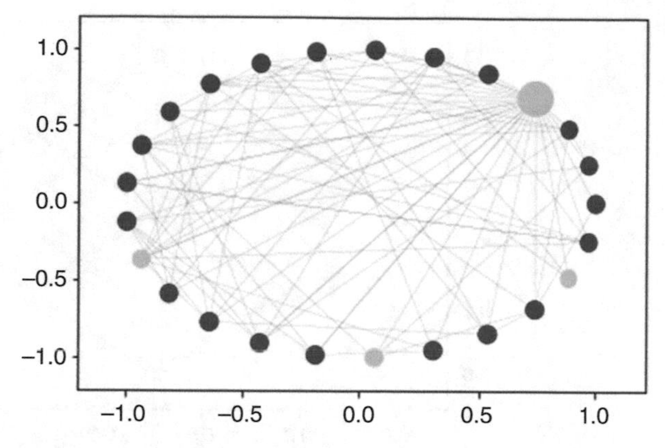

Fig. 8.4 Markov process. Agents are considered in pairs. Direct link from one to the other is formed with a probability of 1/10, $n = 25$. Initial state is $(\nu_{inf}, \nu_{ev}) = (1, 24)$

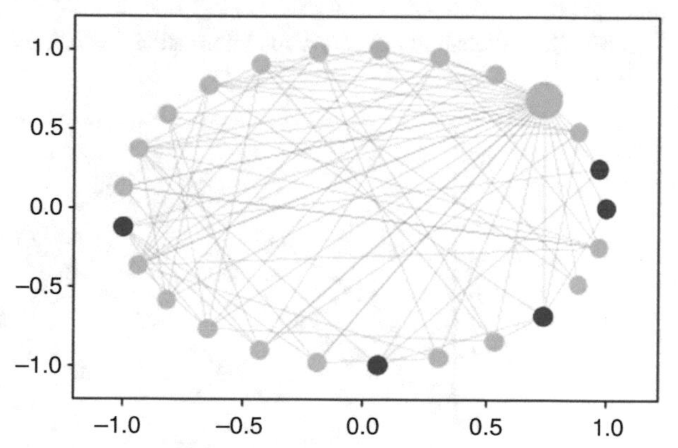

Fig. 8.5 Markov process. Stationary state is $(\nu_{inf}, \nu_{ev}) = (20, 5)$. Total tax revenue $TTR_2 = 78{,}901.06$

dots correspond to those who pay honestly, respectively. The value of TTR_2 exceeds TTR_1 by almost two times, thus the existence of an information center is actual because the number of evaders has decreased.

For the second algorithm we compute the payoff functions of taxpayers $U_{inf} = 43{,}500$, $U_{ev} = 47{,}643.75$.

The results of numerical simulation obtained from the proportional imitation rule are shown in Figs. 8.6, 8.7, 8.8, 8.9, 8.10, and 8.11. Despite the fact that the probability of information perception is high ($\delta = 0.9$), the network structure is such that most agents become evaders, and the total tax revenue is significantly reduced. From the experiments it follows that the value of TTR_2 is significantly lower than the value of TTR_1.

From the series of numerical experiments by using two alternative algorithms for the model of tax control which takes into account the risk statuses of economical agents we can summarize the following.

Firstly, the algorithm based on proportional imitation rule (see Sect. 8.4.2) is not very effective for the considered model. In contrast to the previous study (see. [10] or [11]), where we considered the average income of each agent, in the current paper the structure of the network influences on the population of taxpayers such

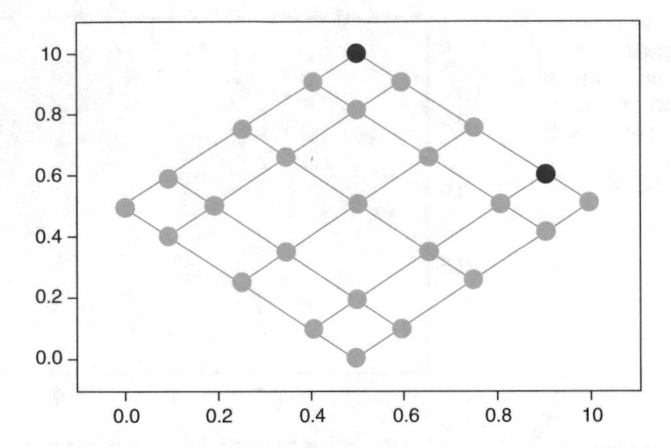

Fig. 8.6 Grid. Probability of perception information $\delta = 0.9$, $n = 25$. Initial state is $(\nu_{inf}, \nu_{ev}) = (23, 2)$

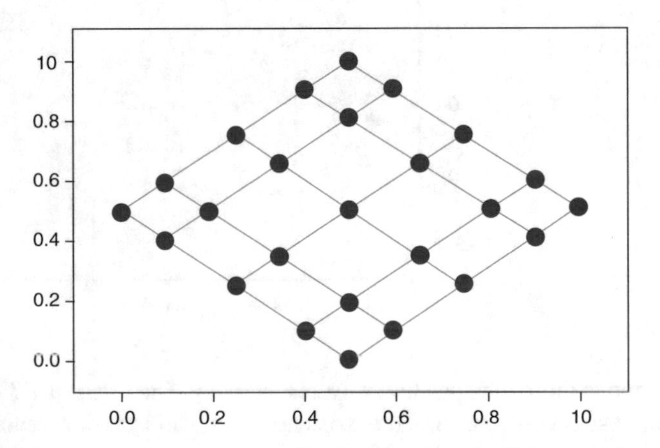

Fig. 8.7 Grid. Probability of perception information $\delta = 0.9$, $n = 25$. Stationary state is $(\nu_{inf}, \nu_{ev}) = (0, 25)$. Total tax revenue $TTR_2 = 15{,}261.31$

as the imitation of behavior of nearest neighbor increases the share of evaders, even with a high probability of information perception and a relatively small number of the evaders at the initial moment. Therefore, the total revenue of the system is significantly reduced. Hence we can conclude that if the income of taxpayers is differentiated than we need to apply an alternative approach to estimate the effectiveness of the propagated information.

In contrast, the new approach considers the algorithm which is based on the Markov process (see Sect. 8.4.1). And the second conclusion is that the mentioned algorithm is effective. In the framework of this attitude only one agent injects information and hence she can be considered as an information center in the network. This approach significantly minimizes the costs of spreading information over the population of taxpayers. Thus the total revenue of the system is increased

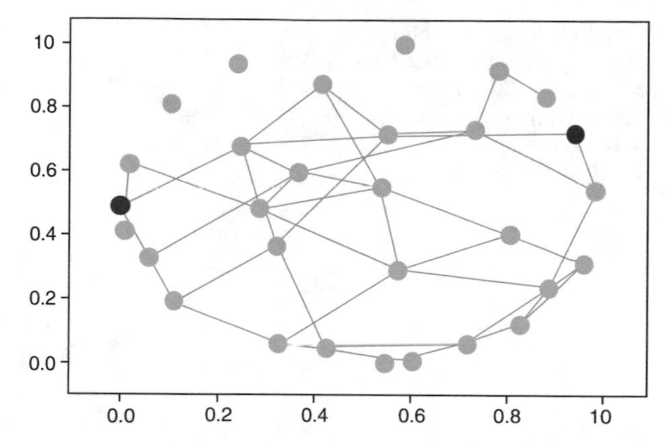

Fig. 8.8 Random. Probability of perception information $\delta = 0.9$, $n = 30$. Initial state is $(\nu_{inf}, \nu_{ev}) = (28, 2)$

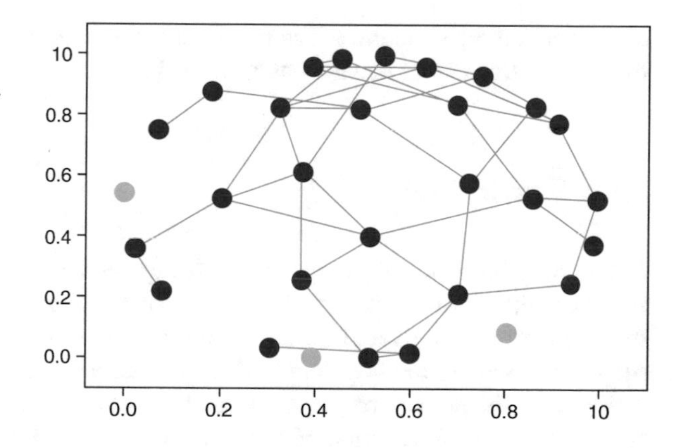

Fig. 8.9 Random. Probability of perception information $\delta = 0.9$, $n = 30$. Stationary state is $(\nu_{inf}, \nu_{ev}) = (3, 27)$. Total tax revenue $TTR_2 = 25{,}101.19$

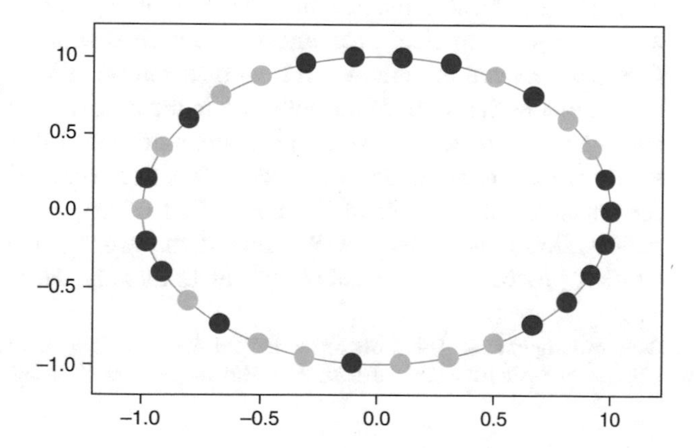

Fig. 8.10 Ring. Probability of perception information $\delta = 0.9$, $n = 30$. Initial state is $(\nu_{inf}, \nu_{ev}) = (13, 17)$

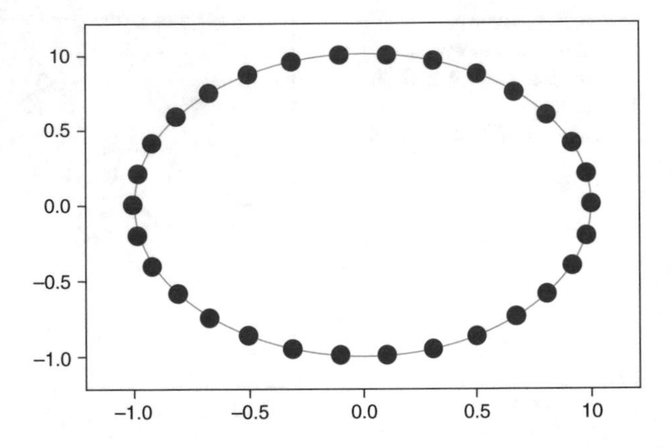

Fig. 8.11 Ring. Probability of perception information $\delta = 0.9$, $n = 30$. Stationary state is $(\nu_{inf}, \nu_{ev}) = (0, 30)$. Total tax revenue $TTR_2 = 29{,}197.88$

depending on the structure of the population and a number of connections between the information center and the outer network.

8.6 Conclusion

In the study presented above we have investigated the problem of tax control taking into account two important factors: the difference between propensity to risk of the economical agents and the propagation of the information about possible tax inspections among the population of taxpayers. We used two different approaches to illustrate the process of propagation of information on network such as Markov process and stochastic imitation rule for evolutionary dynamics. For the mentioned models we presented mathematical formulations, analysis of the agents' behavior and series of numerical experiments. Numerical simulation demonstrates the differences between the initial and final distribution of honest taxpayers and evaders and estimates the profit of tax authority in the case of using information as a tool to stimulate tax collection. We also can summarize that if agents behave accordingly to the proportional imitation rule then numerical simulation has demonstrated that the injected information in the population of taxpayers and the process of it's propagation is not effective. Whereas if the propagation process is described by Markov process with one authorized center then the described model is valid.

Acknowledgements This work are supported the research grant "Optimal Behavior in Conflict-Controlled Systems" (17-11-01079) from Russian Science Foundation.

Appendix

In Appendix we present additional information about probability distribution, used in this paper. Let's recall that the density $f(x)$ and function $F(x)$ of the uniform distribution of the value X on the interval $(b-a, b+a)$ are defined by the next way [13]:

$$f(x)=\begin{cases}\dfrac{1}{2a}, & \text{if } |x-b|\le a,\\ 0, & \text{if } |x-b|>a,\end{cases}$$

$$F(x)=\begin{cases}0, & \text{if } x<b-a,\\ \dfrac{1}{2a}(x-b+a), & \text{if } |x-b|\le a,\\ 1, & \text{if } x>b+a,\end{cases}$$

The mathematical expectation MX of the uniform distribution is $MX=b$.

The Pareto distribution [13], which is often used in the modeling and prediction of an income, has the next density

$$f(x)=\begin{cases}\dfrac{ab^a}{x^{a+1}}, & \text{if } x\ge b,\\ 0, & \text{if } x<b,\end{cases}$$

function

$$F(x)=\begin{cases}1-\left(\dfrac{b}{x}\right)^a, & \text{if } x\ge b,\\ 0, & \text{if } x<b,\end{cases}$$

and the mathematical expectation $MX=\dfrac{a}{(a-1)}\cdot b$.

The scatter of income levels in the group with the highest income may be extremely wide. Therefore, as a value of parameter of the distribution we consider $a=2$: higher or lower values significantly postpone or approximate average value to the lower limit of income.

References

1. Antoci, A., Russu, P., Zarri, L.: Tax evasion in a behaviorally heterogeneous society: an evolutionary analysis. Econ. Model. **42**, 106–115 (2014)

2. Antunes, L., Balsa, J., Urbano, P., Moniz, L., Roseta-Palma, C.: Tax compliance in a simulated heterogeneous multi-agent society. Lect. Notes Comput. Sci. **3891**, 147–161 (2006)
3. Bloomquist, K.M.: A comparison of agent-based models of income tax evasion. Soc. Sci. Comput. Rev. **24**(4), 411–425 (2006)
4. Boure, V., Kumacheva, S.: A game theory model of tax auditing using statistical information about taxpayers. St. Petersburg, Vestnik SPbGU, series 10, **4**, 16–24 (2010) (in Russian)
5. Bure, V.M., Parilina, E.M., Sedakov, A.A.: Consensus in a social network with two principals. Autom. Remote Control **78**(8), 1489–1499 (2017)
6. Chander, P., Wilde, L.L.: A general characterization of optimal income tax enforcement. Rev. Econ. Stud. **65**, 165–183 (1998)
7. DeGroot, M.H.: Reaching a consensus. J. Am. Stat. Assoc. **69**(345), 118–121 (1974)
8. Gubanov, D.A., Novikov, D.A., Chkhartishvili, A.G.: Social Networks: Models of Informational Influence, Control and Confrontation. Fizmatlit, Moscow (2010) (in Russian)
9. Gubar, E.A., Kumacheva, S.S., Zhitkova, E.M., Porokhnyavaya, O.Y.: Propagation of information over the network of taxpayers in the model of tax auditing. In: International Conference on Stability and Control Processes in Memory of V.I. Zubov, SCP 2015 – Proceedings, IEEE Conference Publications. INSPEC Accession Number: 15637330, pp. 244–247 (2015)
10. Gubar, E., Kumacheva, S., Zhitkova, E., Kurnosykh, Z.: Evolutionary Behavior of Taxpayers in the Model of Information Dissemination. In: Constructive Nonsmooth Analysis and Related Topics (Dedicated to the Memory of V.F. Demyanov), CNSA 2017 - Proceedings, IEEE Conference Publications, pp. 1–4 (2017)
11. Gubar, E., Kumacheva, S., Zhitkova, E., Kurnosykh, Z., Skovorodina, T.: Modelling of information spreading in the population of taxpayers: evolutionary approach. Contributions Game Theory Manag. **10**, 100–128 (2017)
12. Katsikas, S., Kolokoltsov, V., Yang, W.: Evolutionary inspection and corruption games. Games **7**(4), 31 (2016). https://doi.org/10.3390/g7040031. http://www.mdpi.com/2073-4336/7/4/31/html
13. Kendall, M.G., Stuart, A.: Distribution Theory. Nauka, Moscow (1966) (in Russian)
14. Kolokoltsov, V.: The evolutionary game of pressure (or interference), resistance and collaboration. Math. Oper. Res. **42**(4), 915944 (2017). http://arxiv.org/abs/1412.1269
15. Kolokoltsov, V., Passi, H., Yang, W.: Inspection and crime prevention: an evolutionary perspective. http://arxiv.org/abs/1306.4219 (2013)
16. Kumacheva, S.S.: Tax Auditing Using Statistical Information about Taxpayers. Contributions to Game Theory and Management, vol. 5. Graduate School of Management, SPbU, Saint Petersburg, pp. 156–167 (2012)
17. Kumacheva, S.S., Gubar, E.A.: Evolutionary model of tax auditing. Contributions Game Theory Manag. **8**,164–175 (2015)
18. Nekovee A.M., Moreno, Y., Bianconi G., Marsili, M.: Theory of rumor spreading in complex social networks. Phys. A **374**, 457–470 (2007)
19. Niazashvili, A.: Individual Differences in Risk Propensity in Different Social Situations of Personal Development. Moscow University for the Humanities, Moscow (2007)
20. Riehl, J.R., Cao, M.: Control of stochastic evolutionary games on networks. In: 5th IFAC Workshop on Distributed Estimation and Control in Networked Systems, Philadelphia, pp. 458–462 (2015)
21. Sandholm, W.H.: Population Games and Evolutionary Dynamics, 616 pp. MIT Press, Cambridge (2010)
22. Tembine, H., Altman, E., Azouzi, R., Hayel, Y.: Evolutionary games in wireless networks. IEEE Trans. Syst. Man Cybern. B Cybern. **40**(3), 634–646 (2010)
23. The web-site of the Russian Federation State Statistics Service. http://www.gks.ru/
24. The web-site of the Russian Federal Tax Service. https://www.nalog.ru/
25. Vasin, A., Morozov, V.: The Game Theory and Models of Mathematical Economics. MAKS Press, Moscow (2005) (in Russian)
26. Weibull, J.: Evolutionary Game Theory, 265 pp. MIT Press, Cambridge (1995)

Chapter 9
Stationary Nash Equilibria for Average Stochastic Positional Games

Dmitrii Lozovanu

Abstract An *average stochastic positional game* is a stochastic game with average payoffs in which the set of states is divided into several disjoint subsets such that each subset represents the position set for one of the player and each player controls the Markov process only in his position set. In such a game each player chooses actions in his position set in order to maximize his average reward per transition. We show that an arbitrary average stochastic positional game possesses a stationary Nash equilibrium. Based on this result we propose an approach for determining the optimal stationary strategies of the players.

9.1 Introduction

The problem of the existence and determination of stationary Nash equilibria in average stochastic games is a relevant problem extensively studied in game theory. However the existence of Nash equilibria in stationary strategies actually is shown only for special classes of average stochastic games. In [12] the existence of stationary Nash equilibria has been proven for the games where the probability transition matrices induced by any stationary strategies of the players are unichain. Important results concerned with the existence of Nash equilibria have been obtained for two-player stochastic games in [14]. In general case, for an average stochastic game with m players ($m \geq 3$) a stationary Nash equilibrium may not exist. This fact has been shown in [5] where an example of 3-player average stochastic game that has no stationary Nash equilibrium is presented.

D. Lozovanu (✉)
Institute of Mathematics and Computer Science of Moldova Academy of Sciences, Chisinau, Moldova
e-mail: dmitrii.lozovanu@math.md

L. A. Petrosyan et al. (eds.), *Frontiers of Dynamic Games*,
Static & Dynamic Game Theory: Foundations & Applications,
https://doi.org/10.1007/978-3-319-92988-0_9

In this paper we prove the existence of stationary Nash equilibria for a class of average stochastic games that we call *average stochastic positional games*. This class of games generalizes the deterministic positional games from [3, 6–9]. We formulate and study the considered class of games by applying the concept of positional games to average Markov decision processes with finite state and action spaces. We assume that a Markov decision process is controlled by m players where the set of states is divided into m disjoint subsets such that each subset represents the position set for one of the players and each player controls the Markov process only in his position set. In the control process each player chooses actions in his position set in order to maximize his average payoff.

Note that some special classes of average stochastic positional games has been considered in [8, 9]. In [8] the existence of pure Nash equilibria for the average stochastic positional games with unichain property is proven and in [9] the existence of pure Nash equilibria for two-player zero-sum average stochastic games is shown. In general, a pure Nash equilibrium for a non-zero average positional game with m players may not exist even for the deterministic case. This fact has been shown in [6] where an example of a non-zero two-player cyclic game that has no pure Nash equilibrium is constructed.

In this paper we show that for an arbitrary m-player average stochastic positional game there exists a Nash equilibrium in mixed stationary strategies. Based on constructive proof of this result we suggest an approach for determining the optimal stationary strategies of the players.

The paper is organized as follows. In Sect. 9.2 we formulate the average stochastic positional game and specify the formulation of the game when players use pure and mixed stationary strategies. In Sect. 9.3 we present some necessary preliminary results from [2, 4] concerned with the existence of Nash equilibria in m-player noncooperative games with quasi-concave and graph-continuous payoffs. In Sect. 9.4 we show that an average Markov decision problem can be formulated in the terms of stationary strategies as a nonlinear optimization problem where the object function is quasi-monotonic (i.e. it is quasi-concave and quasi-convex). In Sect. 9.5, based on results from Sects. 9.3 and 9.4, we present the proof of the main result, i.e. we prove the existence of Nash equilibria in mixed stationary strategies for an average stochastic positional game.

9.2 Formulation of Average Stochastic Positional Games

We first present the framework of a m-person stochastic positional game and then specify the formulation of stochastic positional games with average payoffs when the players use pure and mixed stationary strategies.

9.2.1 *The Framework of an Average Stochastic Positional Game*

A stochastic positional game with m players consists of the following elements:

- a state space X (which we assume to be finite);
- a partition $X = X_1 \cup X_2 \cup \cdots \cup X_m$ where X_i represents the position set of player $i \in \{1, 2, \ldots, m\}$;
- a finite set $A(x)$ of actions in each state $x \in X$;
- a step reward $f^i(x, a)$ with respect to each player $i \in \{1, 2, \ldots, m\}$ in each state $x \in X$ and for an arbitrary action $a \in A(x)$;
- a transition probability function $p : X \times \prod_{x \in X} A(x) \times X \rightarrow [0, 1]$ that gives the probability transitions $p^a_{x,y}$ from an arbitrary $x \in X$ to an arbitrary $y \in X$ for a fixed action $a \in A(x)$, where $\sum_{y \in X} p^a_{x,y} = 1, \ \forall x \in X, \ a \in A(x)$;
- a starting state $x_0 \in X$.

The game starts at the moment of time $t = 0$ in the state x_0 where the player $i \in \{1, 2, \ldots, m\}$ who is the owner of the state position x_0 ($x_0 \in X_i$) chooses an action $a_0 \in A(x_0)$ and determines the rewards $f^1(x_0, a_0), f^2(x_0, a_0), \ldots, f^m(x_0, a_0)$ for the corresponding players $1, 2, \ldots, m$. After that the game passes to a state $y = x_1 \in X$ according to probability distribution $\{p^{a_0}_{x_0,y}\}$. At the moment of time $t = 1$ the player $k \in \{1, 2, \ldots, m\}$ who is the owner of the state position x_1 ($x_1 \in X_k$) chooses an action $a_1 \in A(x_1)$ and players $1, 2, \ldots, m$ receive the corresponding rewards $f^1(x_1, a_1), f^2(x_1, a_1), \ldots, f^m(x_1, a_1)$. Then the game passes to a state $y = x_2 \in X$ according to probability distribution $\{p^{a_1}_{x_1,y}\}$ and so on indefinitely. Such a play of the game produces a sequence of states and actions $x_0, a_0, x_1, a_1, \ldots, x_t, a_t, \ldots$ that defines a stream of stage rewards $f^1(x_t, a_t), f^2(x_t, a_t), \ldots, f^m(x_t, a_t), \quad t = 0, 1, 2, \ldots$. *The average stochastic positional game* is the game with payoffs of the players

$$\omega^i_{x_0} = \liminf_{t \to \infty} \frac{1}{t} \sum_{\tau=0}^{t-1} \mathsf{E}(f^i(x_\tau, a_\tau)), \ \ i = 1, 2, \ldots, m$$

where E is the expectation operator with respect to the probability measure in the Markov process induced by actions chosen by players in their position sets and given starting state x_0. Each player in this game has the aim to maximize his average reward per transition. In the case $m = 1$ this game becomes the average Markov decision problem with given action sets $A(x)$ for $x \in X$, a transition probability function $p : X \times \prod_{x \in X} A(x) \times X \rightarrow [0, 1]$ and step rewards $f(x, a) = f^1(x, a)$ for $x \in X$ and $a \in A(x)$.

In the paper we will study the average stochastic positional game when players use pure and mixed stationary strategies of choosing the actions in the states.

9.2.2 Pure and Mixed Stationary Strategies of the Players

A strategy of player $i \in \{1, 2, \ldots, m\}$ in a stochastic positional game is a mapping s^i that provides for every state $x_t \in X_i$ a probability distribution over the set of actions $A(x_t)$. If these probabilities take only values 0 and 1, then s^i is called *a pure strategy*, otherwise s^i is called *a mixed strategy*. If these probabilities depend only on the state $x_t = x \in X_i$ (i.e. s^i do not depend on t), then s^i is called *a stationary strategy*, otherwise s^i is called a non-stationary *strategy*.

In the following we can identify a pure stationary strategy $s^i(x)$ of player i with the set of boolean variables $s^i_{x,a} \in \{0, 1\}$, where for a given $x \in X_i$ $s^i_{x,a} = 1$ if and only if player i fixes the action $a \in A(x)$. So, we can represent the set of pure stationary strategies S^i of player i as the set of solutions of the following system:

$$\begin{cases} \sum\limits_{a \in A(x)} s^i_{x,a} = 1, & \forall x \in X_i; \\ s^i_{x,a} \in \{0, 1\}, & \forall x \in X_i, \ \forall a \in A(x). \end{cases} \tag{9.1}$$

Obviously the sets of pure strategies $S^1, S^2, \ldots, S^m$ of players are finite sets. If in system (9.1) we change the restrictions $s^i_{x,a} \in \{0, 1\}$ for $x \in X_i, \ a \in A(x)$ by the conditions $0 \le s^i_{x,a} \le 1$ then we obtain the set of stationary strategies in the sense of Shapley [13], where $s^i_{x,a}$ is treated as the probability of the choices of the action a by player i every time when the state x is reached by any route in the dynamic stochastic game. Thus, we can identify the set of mixed stationary strategies $\mathbf{S}^i$ of player i with the set of solutions of the system

$$\begin{cases} \sum\limits_{a \in A(x)} s^i_{x,a} = 1, & \forall x \in X_i; \\ s^i_{x,a} \ge 0, & \forall x \in X_i, \ \forall a \in A(x) \end{cases} \tag{9.2}$$

and for a given profile $\mathbf{s} = (s^1, s^2, \ldots, s^m) \in \mathbf{S} = \mathbf{S}^1 \times \mathbf{S}^2 \times \cdots \times \mathbf{S}^m$ of mixed strategies $s^1, s^2, \ldots, s^m$ of the players the probability transition matrix $P^{\mathbf{s}} = (p^{\mathbf{s}}_{x,y})$ induced by $\mathbf{s}$ can be calculated as follows

$$p^{\mathbf{s}}_{x,y} = \sum_{a \in A(x)} s^i_{x,a} p^a_{x,y} \quad \text{for} \quad x \in X_i, \ \ i = 1, 2, \ldots, m. \tag{9.3}$$

In the sequel we will distinguish stochastic games in pure and mixed stationary strategies.

9.2.3 Average Stochastic Games in Pure and Mixed Stationary Strategies

Let $\mathbf{s} = (s^1, s^2, \ldots, s^m)$ be a profile of stationary strategies (pure or mixed strategies) of the players. Then the elements of probability transition matrix $P^{\mathbf{s}} = (p^{\mathbf{s}}_{x,y})$ in the Markov process induced by $\mathbf{s}$ can be calculated according to (9.3). Therefore if $Q^{\mathbf{s}} = (q^{\mathbf{s}}_{x,y})$ is the limiting probability matrix of $P^{\mathbf{s}}$ then the average payoffs per transition $\omega^1_{x_0}(\mathbf{s}),\ \omega^2_{x_0}(\mathbf{s}), \ldots, \omega^m_{x_0}(\mathbf{s})$ for the players are determined as follows

$$\omega^i_{x_0}(\mathbf{s}) = \sum_{k=1}^{m} \sum_{y \in X_k} q^{\mathbf{s}}_{x_0,y} f^i(y, s^k), \quad i = 1, 2, \ldots, m, \tag{9.4}$$

where

$$f^i(y, s^k) = \sum_{a \in A(y)} s^k_{y,a} f^i(y, a), \quad \text{for } y \in X_k,\ k \in \{1, 2, \ldots, m\} \tag{9.5}$$

expresses the average reward (step reward) of player i in the state $y \in X_k$ when player k uses the strategy s^k.

The functions $\omega^1_{x_0}(\mathbf{s}),\ \omega^2_{x_0}(\mathbf{s}),\ \ldots,\ \omega^m_{x_0}(\mathbf{s})$ on $\mathbf{S} = \mathbf{S}^1 \times \mathbf{S}^2 \times \cdots \times \mathbf{S}^m$, defined according to (9.4), (9.5), determine a game in normal form that we denote $\langle \{\mathbf{S}^i\}_{i=\overline{1,m}}, \{\omega^i_{x_0}(\mathbf{s})\}_{i=\overline{1,m}} \rangle$. This game corresponds to the *average stochastic positional game in mixed stationary strategies* that in extended form is determined by the tuple $(\{X_i\}_{i=\overline{1,m}},\ \{A(x)\}_{x \in X},\ \{f^i(x,a)\}_{i=\overline{1,m}},\ p,\ x_0)$. The functions $\omega^1_{x_0}(\mathbf{s}),\ \omega^2_{x_0}(\mathbf{s}), \ldots, \omega^m_{x_0}(\mathbf{s})$ on $S = S^1 \times S^2 \times \cdots \times S^m$, determine the game $\langle \{S^i\}_{i=\overline{1,m}}, \{\omega^i_{x_0}(\mathbf{s})\}_{i=\overline{1,m}} \rangle$ that corresponds to the *average stochastic positional game in pure strategies*. In the extended form this game also is determined by the tuple $(\{X_i\}_{i=\overline{1,m}},\ \{A(x)\}_{x \in X},\ \{f^i(x,a)\}_{i=\overline{1,m}},\ p,\ x_0)$.

9.2.4 Average Stochastic Positional Games with Random Starting State

In the paper we will consider also stochastic positional games in which the starting state is chosen randomly according to a given distribution $\{\theta_x\}$ on X. So, for a given stochastic positional game we will assume that the play starts in the state $x \in X$ with probability $\theta_x > 0$ where $\sum_{x \in X} \theta_x = 1$. If the players use mixed stationary strategies then the payoff functions

$$\psi_\theta^i(\mathbf{s}) = \sum_{x \in X} \theta_x \omega_x^i(\mathbf{s}), \quad i = 1, 2, \ldots, m$$

on $\mathbf{S}$ define a game in normal form $\langle \{\mathbf{S}^i\}_{i=\overline{1,m}}, \; \{\psi_\theta^i(\mathbf{s})\}_{i=\overline{1,m}} \rangle$ that in extended form is determined by $(\{X_i\}_{i=\overline{1,m}}, \{A(x)\}_{x \in X}, \{f^i(x,a)\}_{i=\overline{1,m}}, p, \{\theta_x\}_{x \in X})$. In the case $\theta_x = 0, \forall x \in X \setminus \{x_0\}, \; \theta_{x_o} = 1$ the considered game becomes a stochastic positional game with fixed starting state x_0.

9.3 Preliminaries

In the paper we shall use some results concerned with the existence of Nash equilibria in noncooperative games with quasi-concave and quasi-monotonic payoffs.

A function $f : \mathbf{S} \to \mathbf{R}^1$ on convex set $\mathbf{S} \subseteq \mathbf{R}^n$ is *quasi-concave* [1] if $\forall \mathbf{s}', \mathbf{s}'' \in \mathbf{S}$ and $\forall \lambda \in [0,1]$ holds $f(\lambda \mathbf{s}' + (1-\lambda)\mathbf{s}'') \geq \min\{f(\mathbf{s}'), \; f(\mathbf{s}'')\}$. If $\forall \mathbf{s}', \mathbf{s}'' \in \mathbf{S}$ and $\forall \lambda \in [0,1]$ holds $f(\lambda \mathbf{s}' + (1-\lambda)\mathbf{s}'') \leq \max\{f(\mathbf{s}'), \; f(\mathbf{s}'')\}$ then the function $f : \mathbf{S} \to \mathbf{R}^1$ is called *quasi-convex*. A function $f : \mathbf{S} \to \mathbf{R}^1, \; \mathbf{S} \subseteq \mathbf{R}^n$, which is quasi-concave and quasi-convex is called *quasi-monotonic*. A detailed characterization of quasi-convex, quasi-concave and quasi-monotonic functions with an application to linear-fractional programming problems can be found in [1].

Let $\langle \mathbf{S}^i{}_{i=\overline{1,m}}, \; f^i(\mathbf{s})_{i=\overline{1,m}} \rangle$ be an m-player game in normal form, where $\mathbf{S}^i \subseteq \mathbf{R}^{n_i}, \; i = 1, 2, \ldots, m$, represent the corresponding sets of strategies of the players $1, 2, \ldots, m$, and $f^i : \prod_{j=1}^{m} \mathbf{S}^j \to \mathbf{R}^1, \; i = 1, 2, \ldots, m$, represent the corresponding payoffs of these players. Let $\mathbf{s} = (s^1, s^2, \ldots, s^m)$ be a profile of strategies of the players, $\mathbf{s} \in \mathbf{S} = \prod_{j=1}^{m} \mathbf{S}^j$, and define $\mathbf{s}^{-i} = (s^1, s^2, \ldots, s^{i-1}, s^{i+1}, \ldots, s^m), \; \mathbf{S}^{-i} = \prod_{j=1 (j \neq i)}^{m} \mathbf{S}^j$ where $\mathbf{s}^{-i} \in \mathbf{S}^{-i}$. Thus, for an arbitrary $\mathbf{s} \in \mathbf{S}$ we can write $\mathbf{s} = (s^i, \mathbf{s}^{-i})$.

Fan [4] extended the well-known equilibrium result of Nash [10] to the games with quasi-concave payoffs. He proved the following theorem:

Theorem 9.1 *Let $\mathbf{S}^i \subseteq \mathbf{R}^{n_i}, \; i = 1, 2, \ldots, m$, be non-empty, convex and compact sets. If each payoff $f^i : \mathbf{S} \to \mathbf{R}^1, \; i \in \{1, 2, \ldots, m\}$, is continuous on $\mathbf{S} = \prod_{j=1}^{m} \mathbf{S}^j$ and quasi-concave with respect to s^i on $\mathbf{S}^i$, then the game $\langle \mathbf{S}^i{}_{i=\overline{1,m}}, \; f^i(\mathbf{s})_{i=\overline{1,m}} \rangle$ possesses a Nash equilibrium.*

Dasgupta and Maskin [2] considered a class of games with upper semi-continuous, quasi-concave and graph-continuous payoffs.

Definition 9.1 The payoff $f^i : \prod_{j=1}^{m} \mathbf{S}^j \to \mathbf{R}^1$ of the game $\langle \mathbf{S}^i{}_{i=\overline{1,m}}\, f^i(\mathbf{s})_{i=\overline{1,m}} \rangle$ is *upper semi-continuous* if for any sequence $\{\mathbf{s}_k\} \subseteq \mathbf{S} = \prod_{j=1}^{m} \mathbf{S}^j$ such that $\{\mathbf{s}_k\} \to \mathbf{s}$ it holds $\lim \sup_{k\to\infty} f^i(\mathbf{s}_k) \le f^i(\mathbf{s})$.

Definition 9.2 The payoff $f^i : \prod_{j=1}^{m} \mathbf{S}^j \to \mathbf{R}^1$ of the game $\langle \mathbf{S}^i{}_{i=\overline{1,m}} f^i(\mathbf{s})_{i=\overline{1,m}} \rangle$ is *graph-continuous* if for all $\overline{\mathbf{s}} = (\overline{s}^i, \overline{\mathbf{s}}^{-i}) \in \mathbf{S} = \prod_{j=1}^{m} \mathbf{S}^j$ there exists a function $F^i : \mathbf{S}^{-i} \to \mathbf{S}^i$ with $F^i(\overline{\mathbf{s}}^{-i}) = \overline{s}^i$ such that $f^i(F^i(\mathbf{s}^{-i}), \mathbf{s}^{-i})$ is continuous at $\mathbf{s}^{-i} = \overline{\mathbf{s}}^{-i}$.

Dasgupta and Maskin [2] proved the following theorem.

Theorem 9.2 *Let $\mathbf{S}^i \subseteq \mathbf{R}^{n_i}$, $i = 1, 2, \dots, m$, be non-empty, convex and compact sets. If each payoff $f^i : \prod_{j=1}^{m} \mathbf{S}^j \to \mathbf{R}^1$, $i \in \{1, 2, \dots, m\}$, is quasi-concave with respect to s^i on $\mathbf{S}^i$, upper semi-continuous with respect to $\mathbf{s}$ on $\mathbf{S} = \prod_{j=1}^{m} \mathbf{S}^j$ and graph-continuous, then the game $\langle \{\mathbf{S}^i\}_{i=\overline{1,m}}, \{f^i(\mathbf{s})\}_{i=\overline{1,m}} \rangle$ possesses a Nash equilibrium.*

In the following we shall use this theorem for the case when each payoff $f^i(s^i, \mathbf{s}^{-i})$, $i \in \{1, 2, \dots, m\}$ is quasi-monotonic with respect to s^i on $\mathbf{S}^i$ and graph-continuous. In this case the reaction correspondence of player

$$\phi^i(\mathbf{s}^{-i}) = \{\hat{s}^i \in \mathbf{S}^i \,|\, f^i(\hat{s}^i, \mathbf{s}^{-i}) = \max_{s^i \in \mathbf{S}^i} f^i(s^i, \mathbf{s}^{-i})\}, \quad i = 1, 2, \dots, m$$

are compact and convex valued and therefore the upper semi-continuous condition for the functions $f^i(\mathbf{s})$, $i = 1, 2, \dots, m$ in Theorem 9.2 can be released. So, in this case the theorem can be formulated as follows.

Theorem 9.3 *Let $\mathbf{S}^i \subseteq \mathbf{R}^{n_i}$, $i = \overline{1, m}$ be non-empty, convex and compact sets. If each payoff $f^i : \prod_{j=1}^{m} \mathbf{S}^j \to \mathbf{R}^1, i \in \{1, 2, \dots, n\}$, is quasi-monotonic with respect to s^i on $\mathbf{S}^i$ and graph-continuous, then the game $\langle \{\mathbf{S}^i\}_{i=\overline{1,m}}, \{f^i(\mathbf{s})\}_{i=\overline{1,m}} \rangle$ possesses a Nash equilibrium.*

9.4 Some Auxiliary Results

To prove the main result we need to formulate and study the average Markov decision problem in the terms of stationary strategies. We present such a formulation for the average Markov decision problem and prove some properties of its optimal solutions that we shall use in the following for the average stochastic positional game in mixed stationary strategies

9.4.1 *A Linear Programming Approach for an Average Markov Decision Problem*

It is well-known [11] that an optimal stationary strategy for the infinite horizon average Markov decision problem with finite state and action spaces can be found by using the following linear programming model:
Maximize

$$\varphi(\alpha, \beta) = \sum_{x\in X} \sum_{a\in A(x)} f(x,a)\alpha_{x,a} \tag{9.6}$$

subject to

$$\begin{cases} \sum\limits_{a\in A(y)} \alpha_{y,a} - \sum\limits_{x\in X} \sum\limits_{a\in A(x)} p^a_{x,y}\,\alpha_{x,a} = 0, \ \ \forall y \in X; \\ \sum\limits_{a\in A(y)} \alpha_{y,a} + \sum\limits_{a\in A(y)} \beta_{y,a} - \sum\limits_{x\in X} \sum\limits_{a\in A(x)} p^a_{x,y}\beta_{x,a} = \theta_y, \ \ \forall y \in X; \\ \qquad \alpha_{x,a} \geq 0, \quad \beta_{y,a} \geq 0, \ \ \forall x \in X, \ a \in A(x), \end{cases} \tag{9.7}$$

where θ_y for $y \in X$ represent arbitrary positive values that satisfy the condition $\sum\limits_{y\in X} \theta_y = 1$, where θ_y for $y \in X$ are treated as the probabilities of choosing the starting state $y \in X$. In the case $\theta_y = 1$ for $y = x_0$ and $\theta_y = 0$ for $y \in X \setminus \{x_0\}$ we obtain the linear programming model for an average Markov decision problem with fixed starting state x_0.

This linear programming model corresponds to the multichain case of an average Markov decision problem. If each stationary policy in the decision problem induces an ergodic Markov chain then restrictions (9.7) can be replaced by restrictions

$$\begin{cases} \sum\limits_{a\in A(y)} \alpha_{y,a} - \sum\limits_{x\in X} \sum\limits_{a\in A(x)} p^a_{x,y}\,\alpha_{x,a} = 0, \ \ \forall y \in X; \\ \sum\limits_{y\in X} \sum\limits_{a\in A(y)} \alpha_{y,a} = 1; \\ \qquad \alpha_{y,a} \geq 0, \quad \forall y \in X, \quad a \in A(y). \end{cases} \tag{9.8}$$

In the linear programming model (9.6), (9.7) the restrictions

$$\sum_{a\in A(y)} \alpha_{y,a} + \sum_{a\in A(y)} \beta_{y,a} - \sum_{x\in X}\sum_{a\in A(x)} p^a_{x,y}\beta_{x,a} = \theta_y,\ \forall y \in X$$

with the condition $\sum_{y\in X} \theta_y = 1$ generalize the constraint $\sum_{x\in X}\sum_{a\in A(y)} \alpha_{y,a} = 1$ in the linear programming model (9.6), (9.8) for the ergodic case.

The relationship between feasible solutions of problem (9.6), (9.7) and stationary strategies in the average Markov decision problem is the following:

Let (α, β) be a feasible solution of the linear programming problem (9.6), (9.7) and denote by $X_\alpha = \{x \in X | \sum_{a\in X} \alpha_{x,a} > 0\}$. Then (α, β) possesses the properties that $\sum_{a\in A(x)} \beta_{x,a} > 0$ for $x \in X \setminus X_\alpha$ and a stationary strategy $s_{x,a}$ that corresponds to (α, β) is determined as

$$s_{x,a} = \begin{cases} \dfrac{\alpha_{x,a}}{\sum\limits_{a\in A(x)} \alpha_{x,a}} & \text{if } x \in X_\alpha,\ a \in A(x); \\ \dfrac{\beta_{x,a}}{\sum\limits_{a\in A(x)} \beta_{x,a}} & \text{if } x \in X \setminus X_\alpha,\ a \in A(x), \end{cases} \tag{9.9}$$

where $s_{x,a}$ expresses the probability of choosing the actions $a \in A(x)$ in the states $x \in X$. In [11] it is shown that the set of feasible solutions of problem (9.6), (9.7) generate through (9.17) the set of stationary strategies **S** that corresponds to the set of solutions of the following system

$$\begin{cases} \sum\limits_{a\in A(x)} s_{x,a} = 1, & \forall x \in X; \\ s_{x,a} \geq 0, & \forall x \in X,\ \forall a \in A(x). \end{cases}$$

Remark 9.1 Problem (9.6), (9.7) can be considered also for the case when $\theta_x = 0$ for some $x \in X$. In particular, if $\theta_x = 0,\ \forall x \in X \setminus \{x_0\}$ and $\theta_{x_0} = 1$ then this problem is transformed into the model with fixed starting state x_0. In this case for a feasible solution (α, β) the subset $X \setminus X_\alpha$ may contain states for which $\sum_{a\in A(x)} \beta_{x,a} = 0$. In such states (9.9) cannot be used for determining $s_{x,a}$. Formula (9.9) can be used for determining the strategies $s_{x,a}$ in the states $x \in X$ for which either $\sum_{a\in A(x)} \alpha_{x,a} > 0$ or $\sum_{a\in A(x)} \beta_{x,a} > 0$ and these strategies determine the value of the objective function in the decision problem. In the state $x \in X_0$, where

$$X_0 = \{x \in X | \sum_{a\in A(x)} \alpha_{x,a} = 0,\ \sum_{a\in A(x)} \beta_{x,a} = 0\},$$

the strategies of a selection the actions may be arbitrary because they do not affect the value of the objective function.

9.4.2 Average Markov Decision Problem in the Terms of Stationary Strategies

We show that an average Markov decision problem in the terms of stationary strategies can be formulated as follows:

Maximize

$$\psi(\mathbf{s}, \mathbf{q}, \mathbf{w}) = \sum_{x \in X} \sum_{a \in A(x)} f(x, a) s_{x,a} q_x \tag{9.10}$$

subject to

$$\begin{cases} q_y - \sum\limits_{x \in X} \sum\limits_{a \in A(x)} p_{x,y}^a \, s_{x,a} q_x = 0, & \forall y \in X; \\ q_y + w_y - \sum\limits_{x \subset X} \sum\limits_{a \subset A(x)} p_{x,y}^a s_{x,a} w_x = \theta_y, & \forall y \in X; \\ \sum\limits_{a \in A(y)} s_{y,a} = 1, & \forall y \in X; \\ s_{x,a} \geq 0, \ \forall x \in X, \ \forall a \in A(x); \quad w_x \geq 0, \ \forall x \in X, \end{cases} \tag{9.11}$$

where θ_y are the same values as in problem (9.6), (9.7) and $s_{x,a}$, q_x, w_x for $x \in X$, $a \in A(x)$ represent the variables that must be found.

Theorem 9.4 *Optimization problem (9.10), (9.11) determines the optimal stationary strategies of the multichain average Markov decision problem.*

Proof Indeed, if we assume that each action set $A(x), x \in X$ contains a single action a' then system (9.7) is transformed into the following system of equations

$$\begin{cases} q_y - \sum\limits_{x \in X} p_{x,y} q_x = 0, & \forall y \in X; \\ q_y + w_y - \sum\limits_{x \in X} p_{x,y} w_x = \theta_y, & \forall y \in X \end{cases}$$

with conditions $q_y, w_y \geq 0$ for $y \in X$, where $q_y = \alpha_{y,a'}$, $w_y = \beta_{y,a'}$, $\forall y \in X$ and $p_{x,y} = p_{x,y}^{a'}$, $\forall x, y \in X$. This system uniquely determines q_x for $x \in X$ and determines w_x for $x \in X$ up to an additive constant in each recurrent class of $P = (p_{x,y})$ (see [11]). Here q_x represents the limiting probability in the state x when the system starts in the state $y \in X$ with probabilities θ_y and therefore the condition $q_x \geq 0$ for $x \in X$ can be released. Note that w_x for some states may be negative, however always the additive constants in the corresponding recurrent classes can be chosen so that w_x became nonnegative. In general, we can observe that in (9.11) the condition $w_x \geq 0$ for $x \in X$ can be released and this does not influence the value of objective function of the problem. In the case $|A(x)| = 1, \ \forall x \in X$ the average cost is determined as $\psi = \sum_{x\in X} f(x)q_x$, where $f(x) = f(x,a), \forall x \in X$.

If the action sets $A(x), \ x \in X$ may contain more than one action then for a given stationary strategy $\mathbf{s} \in \mathbf{S}$ of selection of the actions in the states we can find the average cost $\psi(s)$ in a similar way as above by considering the probability matrix $P^{\mathbf{s}} = (p^{\mathbf{s}}_{x,y})$, where

$$p^{\mathbf{s}}_{x,y} = \sum_{a\in A(x)} p^a_{x,y} s_{x,a} \tag{9.12}$$

expresses the probability transition from a state $x \in X$ to a state $y \in X$ when the strategy $\mathbf{s}$ of selections of the actions in the states is applied. This means that we have to solve the following system of equations

$$\begin{cases} q_y - \sum_{x\in X} p^s_{x,y} q_x = 0, & \forall y \in X; \\ q_y + w_y - \sum_{x\in X} p^s_{x,y} w_x = \theta_y, & \forall y \in X. \end{cases}$$

If in this system we take into account (9.12) then this system can be written as follows

$$\begin{cases} q_y - \sum_{x\in X} \sum_{a\in A(x)} p^a_{x,y} s_{x,a} q_x = 0, & \forall y \in X; \\ q_y + w_y - \sum_{x\in X} \sum_{a\in A(x)} p^a_{x,y} s_{x,a} w_x = \theta_y, & \forall y \in X. \end{cases} \tag{9.13}$$

An arbitrary solution $(\mathbf{q}, \mathbf{w})$ of the system of equations (9.13) uniquely determines q_y for $y \in X$ that allows us to determine the average cost per transition

$$\psi(\mathbf{s}) = \sum_{x\in X} \sum_{a\in X} f(x,a) s_{x,a} q_x \tag{9.14}$$

when the stationary strategy $\mathbf{s}$ is applied. If we are seeking for an optimal stationary strategy then we should add to (9.13) the conditions

$$\begin{cases} \sum_{a\in A(x)} s_{x,a} = 1, & \forall x \in X; \\ s_{x,a} \geq 0, & \forall x \in X, \ \forall a \in A(x) \end{cases} \tag{9.15}$$

and to maximize (9.14) under the constraints (9.13), (9.15). In such a way we obtain problem (9.10), (9.11) without conditions $w_x \geq 0$ for $x \in X$. As we have noted the conditions $w_x \geq 0$ for $x \in X$ do not influence the values of the objective function (9.10) and therefore we can preserve such conditions that show the relationship of the problem (9.10), (9.11) with problem (9.6), (9.7). □

The relationship between feasible solutions of problem (9.6), (9.7) and feasible solutions of problem (9.10), (9.11) can be established on the basis of the following lemma.

Lemma 9.1 *Let* $(\mathbf{s}, \mathbf{q}, \mathbf{w})$ *be a feasible solution of problem (9.10), (9.11). Then*

$$\alpha_{x,a} = s_{x,a} q_x, \quad \beta_{x,a} = s_{x,a} w_x, \quad \forall x \in X, a \in A(x) \tag{9.16}$$

represent a feasible solution (α, β) *of problem (9.6), (9.7) and* $\psi(\mathbf{s}, \mathbf{q}, \mathbf{w}) = \varphi(\alpha, \beta)$. *If* (α, β) *is a feasible solution of problem (9.6), (9.7) then a feasible solution* $(\mathbf{s}, \mathbf{q}, \mathbf{w})$ *of problem (9.10), (9.11) can be determined as follows:*

$$s_{x,a} = \begin{cases} \dfrac{\alpha_{x,a}}{\sum\limits_{a \in A(x)} \alpha_{x,a}} & \text{for } x \in X_\alpha, \ a \in A(x); \\ \dfrac{\beta_{x,a}}{\sum\limits_{a \in A(x)} \beta_{x,a}} & \text{for } x \in X \setminus X_\alpha, \ a \in A(x); \end{cases} \tag{9.17}$$

$$q_x = \sum_{a \in A(x)} \alpha_{x,a}, \quad w_x = \sum_{a \in A(x)} \beta_{x,a} \ \text{ for } x \in X.$$

Proof Assume that $(\mathbf{s}, \mathbf{q}, \mathbf{w})$ is a feasible solution of problem (9.10), (9.11) and (α, β) is determined according to (9.16). Then by introducing (9.16) in (9.6), (9.7) we can observe that (9.7) is transformed in (9.11) and $\psi(\mathbf{s}, \mathbf{q}, \mathbf{w}) = \varphi(\alpha, \beta)$, i.e. (α, β) is a feasible solution of problem (9.6), (9.7). The second part of lemma follows directly from the properties of feasible solutions of problems (9.6), (9.7) and (9.10), (9.11). □

Note that a pure stationary strategy s of problem (9.10), (9.11) corresponds to a basic solution (α, β) of problem (9.6), (9.7) for which (9.17) holds, however system (9.7) may contain basic solutions for which stationary strategies determined through (9.17) do not correspond to pure stationary strategies. Moreover, two different feasible solutions of problem (9.6), (9.7) may generate through (9.17) the same stationary strategy. Such solutions of system (9.7) are considered *equivalent solutions* for the decision problem.

Corollary 9.1 *If* $(\alpha^i, \beta^i), \ i = \overline{1, k}$, *represent the basic solutions of system (9.7) then the set of solutions*

$$M = \left\{ (\alpha, \beta) | \ (\alpha, \beta) = \sum_{i=1}^{k} \lambda^i (\alpha^i, \beta^i), \ \sum_{i=1}^{k} \lambda^i = 1, \ \lambda^i > 0, \ i = \overline{1, k} \right\}$$

determines all feasible stationary strategies of problem (9.10), (9.11) through (9.17).

An arbitrary solution (α, β) *of system* (9.7) *can be represented as follows:* $\alpha = \sum_{i=1}^{k} \lambda^i \alpha^i$, *where* $\sum_{i=1}^{k} \lambda^i = 1;\ \lambda^i \geq 0,\ i = \overline{1,k}$, *and* β *represents a solution of the system*

$$\begin{cases} \sum\limits_{a \in A(y)} \beta_{x,a} - \sum\limits_{z \in X} \sum\limits_{a \in A(z)} p_{z,x}^{a} \beta_{z,a} = \theta_x - \sum\limits_{a \in A(x)} \alpha_{x,a}, \ \forall x \in X; \\ \beta_{y,a} \geq 0, \ \forall x \in X,\ a \in A(x). \end{cases}$$

If (α, β) *is a feasible solution of problem* (9.6), (9.7) *and* $(\alpha, \beta) \notin M$ *then there exists a solution* $(\alpha', \beta') \in M$ *that is equivalent to* (α, β) *and* $\varphi(\alpha, \beta) = \varphi(\alpha', \beta')$.

9.4.3 A Quasi-Monotonic Programming Model in Stationary Strategies for an Average Markov Decision Problem

Based on results from previous section we show now that an average Markov decision problem in the terms of stationary strategies can be represented as a quasi-monotonic programming problem.

Theorem 9.5 *Let an average Markov decision problem be given and consider the function*

$$\psi(\mathbf{s}) = \sum_{x \in X} \sum_{a \in A(x)} f(x,a) s_{x,a} q_x, \tag{9.18}$$

where q_x *for* $x \in X$ *satisfy the condition*

$$\begin{cases} q_y - \sum\limits_{x \in X} \sum\limits_{a \in A(x)} p_{x,y}^{a}\, s_{x,a} q_x = 0, & \forall y \in X; \\ q_y + w_y - \sum\limits_{x \in X} \sum\limits_{a \in A(x)} p_{x,y}^{a} s_{x,a} w_x = \theta_y, & \forall y \in X. \end{cases} \tag{9.19}$$

Then on the set $\mathbf{S}$ *of solutions of the system*

$$\begin{cases} \sum\limits_{a \in A(x)} s_{x,a} = 1, \quad \forall x \in X; \\ s_{x,a} \geq 0, \quad \forall x \in X,\ a \in A(x) \end{cases} \tag{9.20}$$

the function $\psi(\mathbf{s})$ *depends only on* $s_{x,a}$ *for* $x \in X$, $a \in A(x)$ *and* $\psi(\mathbf{s})$ *is quasi-monotonic on* $\mathbf{S}$ *(i.e.* $\psi(\mathbf{s})$ *is quasi-convex and quasi-concave on* $\mathbf{S}$*).*

Proof For an arbitrary $\mathbf{s} \in \mathbf{S}$ system (9.19) uniquely determines q_x for $x \in X$ and determines w_x for $x \in X$ up to a constant in each recurrent class of $P^{\mathbf{s}} = (p^{\mathbf{s}}_{x,y})$, where $p^{\mathbf{s}}_{x,y} = \sum_{a \in A(x)} p^a_{x,y} s_{x,a}$, $\forall x, y \in X$. This means that $\psi(\mathbf{s})$ is determined uniquely for an arbitrary $\mathbf{s} \in \mathbf{S}$, i.e. the first part of the theorem holds.

Now let us prove the second part of the theorem.

Assume that $\theta_x > 0, \forall x \in X$ where $\sum_{x \in X} \theta_x = 1$ and consider arbitrary two strategies $\mathbf{s}', \mathbf{s}'' \in \mathbf{S}$ for which $\mathbf{s}' \neq \mathbf{s}''$. Then according to Lemma 9.1 there exist feasible solutions (α', β') and (α'', β'') of linear programming problem (9.6), (9.7) for which

$$\psi(\mathbf{s}') = \varphi(\alpha', \beta'), \quad \psi(\mathbf{s}'') = \varphi(\alpha'', \beta''), \tag{9.21}$$

where

$$\alpha'_{x,a} = s'_{x,a} q'_x, \quad \alpha''_{x,y} = s''_{x,a} q''_x, \quad \forall x \in X,\ a \in A(x);$$

$$\beta'_{x,a} = s'_{x,a} w'_x, \quad \beta''_{x,y} = s''_{x,a} q''_x, \quad \forall x \in X,\ a \in A(x);$$

$$q'_x = \sum_{a \in A(x)} \alpha'_{x,a} \quad w'_{x,a} = \sum_{a \in A(x)} \beta'_{x,a}, \quad \forall x \in X;$$

$$q''_x = \sum_{a \in A(x)} \alpha''_{x,a} \quad w''_{x,a} = \sum_{a \in A(x)} \beta''_{x,a}, \quad \forall x \in X.$$

The function $\varphi(\alpha, \beta)$ is linear and therefore for an arbitrary feasible solution $(\overline{\alpha}, \overline{\beta})$ of problem (9.6), (9.7) holds

$$\varphi(\overline{\alpha}, \overline{\beta}) = t\varphi(\alpha', \beta') + (1-t)\varphi(\alpha'', \beta'') \tag{9.22}$$

if $0 \le t \le 1$ and $(\overline{\alpha}, \overline{\beta}) = t(\alpha', \beta') + (1-t)(\alpha'', \beta'')$.

Note that $(\overline{\alpha}, \overline{\beta})$ corresponds to a stationary strategy $\overline{\mathbf{s}}$ for which

$$\psi(\overline{\mathbf{s}}) = \varphi(\overline{\alpha}, \overline{\beta}), \tag{9.23}$$

where

$$\overline{s}_{x,a} = \begin{cases} \dfrac{\overline{\alpha}_{x,a}}{\overline{q}_x} & \text{if } x \in X_{\overline{\alpha}}; \\ \dfrac{\overline{\beta}_{x,a}}{\overline{w}_x} & \text{if } x \in X \setminus X_{\overline{\alpha}}. \end{cases} \tag{9.24}$$

Here $X_{\overline{\alpha}} = \{x \in X | \sum_{a \in A(x)} \overline{\alpha}_{x,a} > 0\}$ is the set of recurrent states induced by $P^{\overline{s}} = (p^{\overline{s}}_{x,y})$, where $p^{\overline{s}}_{x,y}$ are calculated according to (9.12) for $s = \overline{s}$ and

$$\overline{q}_x = tq'_x + (1-t)q'', \quad \overline{w}_x = tw'_x + (1-t)w''_x, \quad \forall x \in X.$$

We can see that $X_{\overline{\alpha}} = X_{\alpha'} \cup X_{\alpha''}$, where $X_{\alpha'} = \{x \in X | \sum_{a \in A(x)} \alpha'_{x,a} > 0\}$ and $X_{\alpha''} = \{x \in X | \sum_{a \in A(x)} \alpha''_{x,a} > 0\}$. The value

$$\psi(\overline{\mathbf{s}}) = \sum_{x \in X} \sum_{a \in A(x)} f(x,a)\overline{s}_{x,a}\overline{q}_x$$

is determined by $f(x,a)$, $\overline{s}_{x,a}$ and $\overline{q}_x$ in recurrent states $x \in X_{\overline{\alpha}}$ and it is equal to $\varphi(\overline{\alpha}, \overline{\beta})$. If we use (9.24) then for $x \in X_{\overline{\alpha}}$ and $a \in A(x)$ we have

$$\overline{s}_{x,a} = \frac{t\alpha'_{x,a} + (1-t)\alpha''_{x,a}}{tq'_x + (1-t)q''_x} = \frac{ts'_{x,a}q'_x + (1-t)s''_{x,a}q''_x}{tq'_x + (1-t)q''_x} =$$

$$= \frac{tq'_x}{tq'_x + (1-t)q''_x} s'_{x,a} + \frac{(1-t)q''_x}{tq'_x + (1-t)q''_x} s''_{x,a}$$

and for $x \in X \setminus X_{\overline{\alpha}}$ and $a \in A(x)$ we have

$$\overline{s}_{x,a} = \frac{t\beta'_{x,a} + (1-t)\beta''_{x,a}}{tw'_x + (1-t)w''_x} = \frac{ts'_{x,a}w'_x + (1-t)s''_{x,a}w''_x}{tw'_x + (1-t)w''_x} =$$

$$= \frac{tw'_x}{tw'_x + (1-t)w''_x} s'_{x,a} + \frac{(1-t)w''_x}{tw'_x + (1-t)w''_x} s''_{x,a}.$$

So, we obtain

$$\overline{s}_{x,a} = t_x s'_{x,a} + (1-t_x)s''_{x,a}, \quad \forall a \in A(x), \tag{9.25}$$

where

$$t_x = \begin{cases} \dfrac{tq'_x}{tq'_x + (1-t)q''_x} & \text{if } x \in X_{\overline{\alpha}}; \\[2ex] \dfrac{tw'_x}{tw'_x + (1-t)w''_x} & \text{if } x \in X \setminus X_{\overline{\alpha}}. \end{cases} \tag{9.26}$$

and from (9.21)–(9.23) we have

$$\psi(\overline{\mathbf{s}}) = t\psi(\mathbf{s}') + (1-t)\psi(\mathbf{s}''). \tag{9.27}$$

This means that if we consider the set of strategies

$$S(\mathbf{s}', \mathbf{s}'') = \{\overline{\mathbf{s}} |\ \overline{\mathbf{s}}_{x,a} = t_x s'_{x,a} + (1-t_x)s''_{x,a},\ \ \forall x \in X, a \in A(x)\}$$

then for an arbitrary $\overline{\mathbf{s}} \in S(\mathbf{s}', \mathbf{s}'')$ it holds

$$\min\{\psi(\mathbf{s}'), \psi(\mathbf{s}'')\} \le \psi(\overline{\mathbf{s}}) \le \max\{\psi(\mathbf{s}'), \psi(\mathbf{s}'')\}, \tag{9.28}$$

i.e $\psi(\mathbf{s})$ is monotone on $S(\mathbf{s}', \mathbf{s}'')$. Moreover, using (9.25)–(9.28) we obtain that $\overline{\mathbf{s}}$ possesses the properties

$$\lim_{t\to 1} \overline{s}_{x,a} = s'_{x,a},\ \ \forall x \in X,\ a \in A(x);\quad \lim_{t\to 0} \overline{s}_{x,a} = s''_{x,a},\ \ \forall x \in X,\ a \in A(x) \tag{9.29}$$

and respectively

$$\lim_{t\to 1} \psi(\overline{\mathbf{s}}) = \psi(s');\qquad \lim_{t\to 0} \psi(\overline{\mathbf{s}}) = \psi(\mathbf{s}'').$$

In the following we show that the function $\psi(\mathbf{s})$ is quasi-monotonic on $\mathbf{S}$. To prove this it is sufficient to show that for an arbitrary $c \in \mathbf{R}^1$ the *sublevel set*

$$L_c^-(\psi) = \{\mathbf{s} \subset \mathbf{S} |\ \psi(\mathbf{s}) \le c\}$$

and the *superlevel set*

$$L_c^+(\psi) = \{\mathbf{s} \in \mathbf{S} |\ \psi(\mathbf{s}) \ge c\}$$

of function $\psi(\mathbf{s})$ are convex. These sets can be obtained respectively from the *sublevel set*

$$L_c^-(\varphi) = \{(\alpha, \beta) |\ \varphi(\alpha, \beta) \le c\}$$

and the *superlevel set*

$$L_c^+(\varphi) = \{(\alpha, \beta) |\ \varphi(\alpha, \beta) \ge c\}$$

of function $\varphi(\alpha, \beta)$ for linear programming problem (9.6), (9.7) using (9.17).

Denote by (α^i, β^i), $i = \overline{1,k}$ the basic solutions of system (9.7). According to Corollary 9.1 all feasible strategies of problem (9.6), (9.7) can be obtained trough (9.17) using the basic solutions (α^i, β^i), $i = \overline{1,k}$. Each (α^i, β^i), $i = \overline{1,k}$, determines a stationary strategy

$$s^i_{x,a} = \begin{cases} \dfrac{\alpha^i_{x,a}}{q^i_x}, & \text{for } x \in X_{\alpha^i},\ a \in A(x); \\ \dfrac{\beta^i_{x,a}}{w^i_x}, & \text{for } x \in X \setminus X_{\alpha^i},\ a \in A(x) \end{cases} \tag{9.30}$$

for which $\psi(s^i) = \varphi(\alpha^i, \beta^i)$ where

$$X_{\alpha^i} = \{x \in X \mid \sum_{a \in A(x)} \alpha^i_{x,a} > 0\},\ q^i_x = \sum_{a \in A(x)} \alpha^i_{x,a},\ w^i_x = \sum_{a \in A(x)} \beta^i_{x,a},\ \forall x \in X. \tag{9.31}$$

An arbitrary feasible solution (α, β) of system (9.7) determines a stationary strategy

$$s_{x,a} = \begin{cases} \dfrac{\alpha_{x,a}}{q_x}, & \text{for } x \in X_{\alpha},\ a \in A(x); \\ \dfrac{\beta_{x,a}}{w_x}, & \text{for } x \in X \setminus X_{\alpha},\ a \in A(x), \end{cases} \tag{9.32}$$

for which $\psi(\mathbf{s}) = \varphi(\alpha, \beta)$ where

$$X_\alpha = \{x \in X \mid \sum_{a \in A(x)} \alpha_{x,a} > 0\}, \quad q_x = \sum_{a \in A(x)} \alpha_{x,a},\ w_x = \sum_{a \in A(x)} \beta_{x,a},\ \forall x \in X.$$

Taking into account that (α, β) can be represented as

$$(\alpha, \beta) = \sum_{i=1}^{k} \lambda^i (\alpha^i, \beta^i), \text{ where } \sum_{i=1}^{k} \lambda^i = 1,\ \lambda^i \ge 0,\ i = \overline{1,k} \tag{9.33}$$

we have $\varphi(\alpha, \beta) = \sum_{i=1}^{k} \varphi(\alpha^i, \beta^i)\lambda^i$ and we can consider

$$X_\alpha = \bigcup_{i=1}^{k} X_{\alpha^i}; \quad \alpha = \sum_{i=1}^{k} \lambda^i \alpha^i; \quad q = \sum_{i=1}^{k} \lambda^i q^i; \quad w = \sum_{i=1}^{k} \lambda^i w^i. \tag{9.34}$$

Using (9.30)–(9.34) we obtain:

$$s_{x,a} = \frac{\alpha_{x,a}}{q_x} = \frac{\sum_{i=1}^{k} \lambda^i \alpha_{x,a}^k}{q_x} = \frac{\sum_{i=1}^{k} \lambda^i s_{x,a}^i q_x^i}{q_x} = \sum_{i=1}^{k} \frac{\lambda^i q_x^i}{q_x} s_{x,a}^i, \ \forall x \in X_\alpha, \ a \in A(x);$$

$$s_{x,a} = \frac{\beta_{x,a}}{w_x} = \frac{\sum_{i=1}^{k} \lambda^i \beta_{x,a}^k}{w_x} = \frac{\sum_{i=1}^{k} \lambda^i s_{x,a}^i w_x^i}{w_x} = \sum_{i=1}^{k} \frac{\lambda^i w_x^i}{w_x} s_{x,a}^i, \ \forall x \in X \setminus X_\alpha, a \in A(x)$$

and

$$q_x = \sum_{i=1}^{k} \lambda^i q_x^i, \qquad w_x = \sum_{i=1}^{k} \lambda^i w_x^i \ \text{ for } \ x \in X. \tag{9.35}$$

So,

$$s_{x,a} = \begin{cases} \sum_{i=1}^{k} \frac{\lambda^i q_x^i}{q_x} s_{x,a}^i & \text{if } q_x > 0; \\ \sum_{i=1}^{k} \frac{\lambda^i w_x^i}{w_x} s_{x,a}^i & \text{if } q_x = 0, \end{cases} \tag{9.36}$$

where q_x and w_x are determined according to (9.35).

We can see that if $\lambda^i, s^i, q^i, \ i = \overline{1,k}$ are given then the strategy $\mathbf{s}$ defined by (9.36) is a feasible strategy because $s_{x,a} \geq 0, \forall x \in X, a \in A(x)$ and $\sum_{a \in A(x)} s_{x,a} = 1, \ \forall x \in X$. Moreover, we can observe that $q_x = \sum_{i=1}^{k} \lambda^i q_x^i, \ \ w_x = \sum_{i=1}^{k} \lambda^i w_x^i$ for $x \in X$ represent a solution of system (9.19) for the strategy s defined by (9.36). This can be verified by plugging (9.35) and (9.36) into (9.19); after such a substitution all equations from (9.19) are transformed into identities. For $\psi(\mathbf{s})$ we have

$$\psi(\mathbf{s}) = \sum_{x \in X} \sum_{a \in A(x)} f(x,a) s_{x,a} q_x = \sum_{x \in X_\alpha} \sum_{a \in A(x)} f(x,a) \sum_{i=1}^{k} \left(\frac{\lambda^i q_x^i}{q_x} s_{x,a}^i \right) q_x$$

$$= \sum_{i=1}^{k} \Big(\sum_{x \in X_{\alpha^i}} \sum_{a \in A(x)} f(x,a) s_{x,a}^i q_x^i \Big) \lambda^i = \sum_{i=1}^{k} \psi(s^i) \lambda^i,$$

i.e.

$$\psi(\mathbf{s}) = \sum_{i=1}^{k} \psi(s^i)\lambda^i, \tag{9.37}$$

where $\mathbf{s}$ is the strategy that corresponds to (α, β).

Thus, assuming that the strategies $\mathbf{s}^1, \mathbf{s}^2, \ldots, \mathbf{s}^k$ correspond to basic solutions (α^1, β^1), $(\alpha^2, \beta^2), \ldots, (\alpha^k, \beta^k)$ of problem (9.6), (9.7) and $\mathbf{s} \in \mathbf{S}$ corresponds to an arbitrary solution (α, β) of this problem that can be expressed as convex combination of basic solutions of problem (9.6), (9.7) with the corresponding coefficients $\lambda^1, \lambda^2, \ldots, \lambda^k$, we can express the strategy $\mathbf{s}$ and the corresponding value $\psi(\mathbf{s})$ by (9.35)–(9.37). In general the representation (9.35)–(9.37) of strategy $\mathbf{s}$ and of the value $\psi(\mathbf{s})$ is valid for an arbitrary finite set of strategies from $\mathbf{S}$ if (α, β) can be represented as convex combination of the finite number of feasible solutions $(\alpha^1, \beta^1), (\alpha^2, \beta^2), \ldots, (\alpha^k, \beta^k)$ that correspond to $\mathbf{s}^1, \mathbf{s}^2, \ldots, \mathbf{s}^k$; in the case $k = 2$ from (9.35)–(9.37) we obtain (9.25)–(9.27). It is evident that for a feasible strategy $\mathbf{s} \in \mathbf{S}$ the representation (9.35), (9.36) may be not unique, i.e. two different vectors $\overline{\Lambda} = (\overline{\lambda}^1, \overline{\lambda}^2, \ldots, \overline{\lambda}^k)$ and $\overline{\overline{\Lambda}} = \overline{\overline{\lambda}}^1, \overline{\overline{\lambda}}^2, \ldots, \overline{\overline{\lambda}}^k$ may determine the same strategy $\mathbf{s}$ via (9.35), (9.36). In the following we will assume that $\mathbf{s}^1, \mathbf{s}^2, \ldots, \mathbf{s}^k$ represent the system of linear independent basic solutions of system (9.20), i.e. each $\mathbf{s}^i \in \mathbf{S}$ corresponds to a pure stationary strategy.

Thus, an arbitrary strategy $\mathbf{s} \in \mathbf{S}$ is determined according to (9.35), (9.36) where $\lambda^1, \lambda^2, \ldots, \lambda^k$ correspond to a solution of the following system

$$\sum_{i=1}^{k} \lambda^i = 1; \quad \lambda^i \geq 0, \quad i = \overline{1, k}.$$

Consequently, the sublevel set $L_c^-(\psi)$ of function $\psi(s)$ represents the set of strategies s determined by (9.35), (9.36), where $\lambda^1, \lambda^2, \ldots, \lambda^k$ satisfy the condition

$$\begin{cases} \sum_{i=1}^{k} \psi(s^i)\lambda^i \leq c; \\ \sum_{i=1}^{k} \lambda^i = 1; \quad \lambda^i \geq 0, \quad i = \overline{1, k} \end{cases} \tag{9.38}$$

and the superlevel set $L_c^+(\psi)$ of $\psi(s)$ represents the set of strategies s determined by (9.35), (9.36), where $\lambda^1, \lambda^2, \ldots, \lambda^k$ satisfy the condition

$$\begin{cases} \sum_{i=1}^{k} \psi(s^i)\lambda^i \geq c; \\ \sum_{i=1}^{k} \lambda^i = 1; \quad \lambda^i \geq 0, \quad i = \overline{1, k}. \end{cases} \tag{9.39}$$

Respectively the level set $L_c(\psi) = \{s \in \overline{S} | \ \psi(s) = c\}$ of function $\psi(s)$ represents the set of strategies s determined by (9.35), (9.36), where $\lambda^1, \lambda^2, \ldots, \lambda^k$ satisfy the condition

$$\begin{cases} \sum_{i=1}^{k} \psi(s^i)\lambda^i = c; \\ \sum_{i=1}^{k} \lambda^i = 1; \quad \lambda^i \geq 0, \ i = \overline{1,k}. \end{cases} \tag{9.40}$$

Let us show that $L_c^-(\psi), \ L_c^+(\psi), \ L_c(\psi)$ are convex sets. We present the proof of convexity of sublevel set $L_c^-(\psi)$. The proof of convexity of $L_c^+(\psi)$ and $L_c(\psi)$ is similar to the proof of convexity of $L_c^-(\psi)$.

Denote by Λ the set of solutions $(\lambda^1, \lambda^2, \ldots, \lambda^k)$ of system (9.38). Then from (9.35), (9.36), (9.38) we have

$$L_c^-(\psi) = \prod_{x \in X} \hat{S}_x$$

where $\hat{S}_x$ represents the set of strategies

$$s_{x,a} = \begin{cases} \dfrac{\sum_{i=1}^k \lambda^i q_x^i s_{x,a}^i}{\sum_{i=1}^k \lambda^i q_x^i} & \text{if } \sum_{i=1}^k \lambda^i q_x^i > 0, \\ \dfrac{\sum_{i=1}^k \lambda^i w_x^i s_{x,a}^i}{\sum_{i=1}^k \lambda^i w_x^i} & \text{if } \sum_{i=1}^k \lambda^i q_x^i = 0, \end{cases} \qquad a \in A(x)$$

in the state $x \in X$ determined by $(\lambda^1, \lambda^2, \ldots, \lambda^k) \in \Lambda$.

For an arbitrary $x \in X$ the set Λ can be represented as follows $\Lambda = \Lambda_x^+ \cup \Lambda_x^0$, where

$$\Lambda_x^+ = \{(\lambda^1, \lambda^2, \ldots, \lambda^k) \in \Lambda | \ \sum_{i=1}^{k} \lambda^i q_x^i > 0\},$$

$$\Lambda_x^0 = \{(\lambda^1, \lambda^2, \ldots, \lambda^k) \in \Lambda | \ \sum_{i=1}^{k} \lambda^i q_x^i = 0\}$$

and $\sum_{i=1}^k \lambda^i w_x^i > 0$ if $\sum_{i=1}^k \lambda^i q_x^i = 0$. Therefore $\hat{S}_x$ can be expressed as follows $\hat{S}_x = \hat{S}_x^+ \cup \hat{S}_x^0$, where $\hat{S}_x^+$ represents the set of strategies

$$s_{x,a} = \frac{\sum_{i=1}^k \lambda^i q_x^i s_{x,a}^i}{\sum_{i=1}^k \lambda^i q_x^i}, \ \text{for } a \in A(x) \tag{9.41}$$

in the state $x \in X$ determined by $(\lambda^1, \lambda^2, \ldots, \lambda^k) \in \Lambda_x^+$ and $\hat{S}_x^0$ represents the set of strategies

$$s_{x,a} = \frac{\sum_{i=1}^{k} \lambda^i w_x^i s_{x,a}^i}{\sum_{i=1}^{k} \lambda^i w_x^i}, \quad \text{for } a \in A(x) \tag{9.42}$$

in the state $x \in X$ determined by $(\lambda^1, \lambda^2, \ldots, \lambda^k) \in \Lambda_x^0$.

Thus, if we analyze (9.41) then observe that $s_{x,a}$ for a given $x \in X$ represents a linear-fractional function with respect to $\lambda^1, \lambda^2, \ldots, \lambda^k$ defined on convex set Λ_x^+ and $\hat{S}_x^+$ is the image of $s_{x,a}$ on Λ_x^+. Therefore $\hat{S}_x^+$ is a convex set. If we analyze (9.42) then observe that $s_{x,a}$ for given $x \in X$ represents a linear-fractional function with respect to $\lambda^1, \lambda^2, \ldots, \lambda^k$ on the convex set Λ_x^0 and $\hat{S}_x^0$ is the image of $s_{x,a}$ on Λ_x^0. Therefore $\hat{S}_x^0$ is a convex set (see [1]). Additionally, we can observe that $\Lambda_x^+ \cap \Lambda_x^0 = \emptyset$ and in the case $\Lambda_x^+, \Lambda_x^0, \neq \emptyset$ the set Λ_x^0 represents the limit inferior of Λ_x^+. Using this property and taking into account (9.29) we can conclude that each strategy $s_x \in \hat{S}_x^0$ can be regarded as the limit of a sequence of strategies $\{s_x^t\}$ from $\hat{S}_x^+$. Therefore we obtain that $\hat{S}_x = \hat{S}_x^+ \cup \hat{S}_x^0$ is a convex set. This involves the convexity of the sublevel set $L_c^-(\psi)$. In an analogues way using (9.39) and (9.40) we can show that the superlevel set $L_c^+(\psi)$ and the level set $L_c(\psi)$ are convex sets. This means that the function $\psi(\mathbf{s})$ is quasi-monotonic on $\mathbf{S}$. So, if $\theta_x > 0, \forall x \in X$ and $\sum_{x \in X} \theta_x = 1$ then the theorem holds.

If $\theta_x = 0$ for some $x \in X$ then the set $X \setminus X_\alpha$ may contain states for which $\sum_{a \in A(x)} \alpha_{x,a} = 0$ and $\sum_{a \in A(x)} \beta_{x,a} = 0$ (see Remark 9.1 and Lemma 9.1). In this case X can be represented as follows:

$$X = (X \setminus X_0) \cup X_0,$$

where

$$X_0 = \{x \in X | \sum_{a \in A(x)} \alpha_{x,a} = 0; \sum_{a \in A(x)} \beta_{x,a} = 0\}.$$

For $x \in X \setminus X_0$ the convexity of $\hat{S}_x$ can be proved in the same way as for the case $\theta_x > 0, \forall x \in X$. If $X_0 \neq \emptyset$ then for $x \in X_0$ we have $\hat{S}_x = S_x$ and the convexity of $\hat{S}_x$ is evident. So, the theorem holds.

□

9.5 The Main Results

In this section we prove the existence of Nash equilibria in mixed stationary strategies for an arbitrary average stochastic positional game. To prove this result we show that such a game in normal form can be formulated as a game with quasi-monotonic and graph-continuous payoffs of the players.

9.5.1 A Normal Form of an Average Stochastic Positional Game in Mixed Stationary Strategies

Based on the results from Sect. 9.4 we can now formulate the average stochastic positional game in the terms of mixed stationary strategies as follows.

Let $\mathbf{S}^i$, $i \in \{1, 2, \dots m\}$ be the set of solutions of the system

$$\begin{cases} \sum\limits_{a \in A(x)} s^i_{x,a} = 1, \quad \forall x \in X_i; \\ \qquad s^i_{x,a} \geq 0, \quad \forall x \in X_i, \ a \in A(x) \end{cases} \tag{9.43}$$

that determines the set of stationary strategies of player i. Each $\mathbf{S}^i$ is a convex compact set and an arbitrary extreme point corresponds to a basic solution s^i of system (9.43), where $s^i_{x,a} \in \{0, 1\}$, $\forall x \in X_i$, $a \in A(x)$, i.e. each basic solution of this system corresponds to a pure stationary strategy of player i.

On the set $\mathbf{S} = \mathbf{S}^1 \times \mathbf{S}^2 \times \cdots \times \mathbf{S}^m$ we define m payoff functions

$$\psi^i_\theta(s^1, s^2, \dots, s^m) = \sum_{k=1}^{m} \sum_{x \in X_k} \sum_{a \in A(x)} s^k_{x,a} f^i(x,a) q_x, \qquad i = 1, 2, \dots, m, \tag{9.44}$$

where q_x for $x \in X$ are determined uniquely from the following system of linear equations

$$\begin{cases} q_y - \sum\limits_{k=1}^{m} \sum\limits_{x \in X_k} \sum\limits_{a \in A(x)} s^k_{x,a}\, p^a_{x,y}\, q_x = 0, & \forall y \in X; \\ q_y + w_y - \sum\limits_{k=1}^{m} \sum\limits_{x \in X_k} \sum\limits_{a \in A(x)} s^k_{x,a}\, p^a_{x,y}\, w_x = \theta_y, & \forall y \in X \end{cases} \tag{9.45}$$

for an arbitrary fixed profile $\mathbf{s} = (s^1, s^2, \dots, s^m) \in \mathbf{S}$. The functions $\psi^i_\theta(s^1, s^2, \dots, s^m)$, $i = 1, 2, \dots, m$, represent the payoff functions for the average stochastic game in normal form $\langle \{\mathbf{S}^i\}_{i=\overline{1,m}}, \{\psi^i_\theta(\mathbf{s})\}_{i=\overline{1,m}} \rangle$. This game is determined by the tuple $(\{X_i\}_{i=\overline{1,m}}, \{A(x)\}_{x \in X}, \{f^i(x,a)\}_{i=\overline{1,m}}, p, \{\theta_y\}_{y \in X})$ where θ_y for $y \in X$ are given nonnegative values such that $\sum_{y \in X} \theta_y = 1$.

If $\theta_y = 0,\ \forall y \in X \setminus \{x_0\}$ and $\theta_{x_0} = 1$, then we obtain an average stochastic game in normal form $\langle \{\mathbf{S}^i\}_{i=\overline{1,m}},\ \{\omega^i_{x_0}(\mathbf{s})\}_{i=\overline{1,m}} \rangle$ when the starting state x_0 is fixed, i.e. $\psi^i_\theta(s^1, s^2, \ldots, s^m) = \omega^i_{x_0}(s^1, s^2, \ldots, s^m)$, $i = 1, 2, \ldots, m$. So, in this case the game is determined by $(\{X_i\}_{i=\overline{1,m}},\ \{A(x)\}_{x\in X},\ \{f^i(x,a)\}_{i=\overline{1,m}},\ p,\ x_0)$.

If $\theta_y > 0,\ \forall y \in X$ and $\sum_{y\in X}\theta_y = 1$, then we obtain an average stochastic game when the play starts in the states $y \in X$ with probabilities θ_y. In this case for the payoffs of the players in the game in normal form we have

$$\psi^i_\theta(s^1, s^2, \ldots, s^m) = \sum_{y\in X} \theta_y \omega^i_y(s^1, s^2, \ldots, s^m), \quad i = 1, 2, \ldots, m.$$

9.5.2 Existence of Mixed Stationary Nash Equilibria in Average Stochastic Positional Games

Let $\langle \{\mathbf{S}^i\}_{i=\overline{1,m}},\ \{\psi^i_\theta(\mathbf{s})\}_{i=\overline{1,m}} \rangle$ be the non-cooperative game in normal form that corresponds to the average stochastic positional game in stationary strategies determined by $(\{X_i\}_{i=\overline{1,m}},\ \{A(x)\}_{x\in X},\ \{f^i(x,a)\}_{i=\overline{1,m}},\ p,\ \{\theta_y\}_{y\in X})$. Hence, $\mathbf{S}^i$, $i = 1, 2, \ldots, m$, and $\psi^i_\theta(\mathbf{s})$, $i = 1, 2, \ldots, m$, are defined according to (9.43)–(9.45).

Theorem 9.6 *The game $\langle \{\mathbf{S}^i\}_{i=\overline{1,m}},\ \{\psi^i_\theta(s)\}_{i=\overline{1,m}} \rangle$ possesses a Nash equilibrium $\mathbf{s}^* = (s^{1*}, s^{2*}, \ldots, s^{m*}) \in \mathbf{S}$ which is a Nash equilibrium in mixed stationary strategies for the average stochastic positional game determined by $(\{X_i\}_{i=\overline{1,m}},\ \{A(x)\}_{x\in X},$ $\{f^i(x,a)\}_{i=\overline{1,m}},\ p,\ \{\theta_y\}_{y\in X})$. If $\theta_y > 0,\ \forall y \in X$, then $\mathbf{s}^* = (s^{1*}, s^{2*}, \ldots, s^{m*})$ is a Nash equilibrium in mixed stationary strategies for the average stochastic positional game $\langle \{\mathbf{S}^i\}_{i=\overline{1,m}}, \{\omega^i_y(\mathbf{s})\}_{i=\overline{1,m}} \rangle$ with an arbitrary starting state $y \in X$.*

Proof To prove the theorem we need to verify that $\langle \{\mathbf{S}^i\}_{i=\overline{1,m}},\ \{\psi^i_\theta(\mathbf{s})\}_{i=\overline{1,m}} \rangle$ satisfies the conditions of Theorem 9.3. So, we have to show that each payoff $\psi^i_\theta(s^i, \mathbf{s}^{-i})$ is quasi-monotonic with respect to s^i on convex and compact set $\mathbf{S}^i$, and each payoff function $\psi^i_\theta(s^i, \mathbf{s}^{-i})$ is graph-continuous.

Indeed, if players $1, 2, \ldots, i-1, i+1, \ldots, m$, fix their strategies $\hat{s}^k \in \mathbf{S}^k$, $k \neq i$ then we obtain an average Markov decision problem with respect to $s^i \in \mathbf{S}^i$ in which it is necessary to maximize the average reward function $\varphi^i(s^i) = \psi^i_\theta(s^i, \hat{\mathbf{s}}^{-i})$. According to Theorem 9.5 the function $\varphi^i(s^i) = \psi^i_\theta(s^i, \hat{\mathbf{s}}^{-i})$ possesses the property that it is quasi-monotonic with respect to s^i on $\mathbf{S}^i$. Additionally we can observe that if for the payoff $\psi^i(s^i, \mathbf{s}^{-i})$ we consider the function $F^i : \mathbf{S}^{-i} \to \mathbf{S}^i$ such that

$$F^i(\mathbf{s}^{-i}) = \hat{s}^i \in \phi^i(\mathbf{s}^{-i}) \text{ for } \mathbf{s}^{-i} \in \mathbf{S}^{-i},\ i \in \{1, 2, \ldots, m\}$$

where

$$\phi^i(\mathbf{s}^{-i}) = \{\, \hat{s}^i \in \mathbf{S}^i \mid \psi_\theta^i(\hat{s}^i, \mathbf{s}^{-i})) = \max_{s^i \in \mathbf{S}^i} \psi_\theta^i(s^i, \mathbf{s}^{-i})\},$$

then the function $\psi_\theta^i(F^i(\mathbf{s}^{-i}), \mathbf{s}^{-i})$ is continuous at $\mathbf{s}^{-i} = \overline{\mathbf{s}}^{-i}$ for an arbitrary $(\overline{s}^i, \overline{\mathbf{s}}^{-i}) \in \mathbf{S}$. So, $\psi_\theta^i(\mathbf{s})$ is graph-continuous and according to Theorem 9.3 the game $\langle\{\mathbf{S}^i\}_{i=\overline{1,m}}, \{\psi_\theta^i(\mathbf{s})\}_{i=\overline{1,m}}\rangle$ possesses a Nash equilibrium $\mathbf{s}^* \in \overline{\mathbf{S}}$. This Nash equilibrium is a Nash equilibrium in mixed stationary strategies for the average stochastic positional game determined by $(\{X_i\}_{i=\overline{1,m}}, \{A(x)\}_{x\in X}, \{f^i(x,a\}_{i=\overline{1,m}}, p, \{\theta_y\}_{y\in X})$. □

Thus, for an arbitrary average stochastic positional game a Nash equilibrium in mixed stationary strategies exists and the optimal stationary strategies of the players can be found using the game $\langle\{\mathbf{S}^i\}_{i=\overline{1,m}}, \{\psi_\theta^i(\mathbf{s})\}_{i=\overline{1,m}}\rangle$, where $\mathbf{S}^i$ and $\psi_\theta^i(\mathbf{s})$, $i = 1, 2, \ldots, m$, are defined according to (9.43)–(9.45).

9.6 Conclusion

Average stochastic positional games represent an important class of average stochastic games with finite state and action spaces that generalizes deterministic positional games with mean payoffs from [3, 6, 7]. For this class of games Nash equilibria exist in the set of mixed stationary strategies and the optimal mixed stationary strategies of the players can be found using the game model in normal form from Sect. 9.5.

Acknowledgements The author is grateful to the referee for interesting suggestions and remarks contributing to improve the presentation of the paper.

References

1. Boyd, S., Vandenberghe, L.: Convex Optimization. Cambridge University Press, Cambridge (2004)
2. Dasgupta, P., Maskin, E.: The existence of equilibrium in discontinuous economic games. Rev. Econ. Stud. **53**, 1–26 (1986)
3. Ehrenfeucht, A., Mycielski, J.: Positional strategies for mean payoff games. Int. J. Game Theory **8**, 109–113 (1979)
4. Fan, K.: Application of a theorem concerned sets with convex sections. Math. Ann. **1963**, 189–203 (1966)
5. Flesch, J., Thuijsman, F., Vrieze, K.: Cyclic Markov equilibria in stochastic games. Int. J. Game Theory **26**, 303–314 (1997)
6. Gurvich, V., Karzaniv, A., Khachyan, L.: Cyclic games and an algorithm to find minimax mean cycles in directed graphs. USSR Comput. Math. Math. Phys. **28**, 85–91 (1988)
7. Lozovanu, D., Pickl, S.: Nash equilibria conditions for cyclic games with p players. Electron. Notes Discrete Math. **25**, 117–124 (2006)

8. Lozovanu, D., Pickl, S.: Nash equilibria conditions for stochastic positional games. Contribution Game Theory Manag. **8**, 201–213 (2014)
9. Lozovanu, D., Pickl, S.: Determining the optimal strategies for zero-sum average stochastic positional games. Electron. Notes Discrete Math. **55**, 155–159 (2016)
10. Nash, J.: Non-cooperative games. Ann. Math. **54**, 286–293 (1953)
11. Puterman, M.: Markov Decision Processes: Discrete Dynamic Programming. Wiley, Hoboken (2005)
12. Rogers, P.: Nonzero-sum stochastic games. PhD thesis, University of California, Berkeley, Report ORC 69-8 (1966)
13. Shapley, L.: Stochastic games. Proc. Natl. Acad. Sci. USA **39**, 1095–1100 (1953)
14. Vieille, N.: Equilibrium in 2-person stochastic games I, II. Isr. J. Math. **119**(1), 55–126 (2009)

Chapter 10
Game Equilibria and Transition Dynamics in Networks with Heterogeneous Agents

Vladimir Matveenko, Maria Garmash, and Alexei Korolev

Abstract We study game equilibria in a model of production and externalities in network with two types of agents who possess different productivities. Each agent may invest a part of her endowment (for instance, time or money) at the first stage; consumption at the second period depends on her own investment and productivity as well as on the investments of her neighbors in the network. Three ways of agent's behavior are possible: passive (no investment), active (a part of endowment is invested) and hyperactive (the whole endowment is invested). We introduce adjustment dynamics and study consequences of junction of two regular networks with different productivities of agents. In particular, we study how the behavior of nonadopters (passive agents) changes when they connect to adopters (active or hyperactive) agents.

10.1 Introduction

Models of network economics and network games take into account structure of socio-economic systems, social settings, interactions of agents (see e.g. [1–4, 7–11, 13]). Such models assume that agents in a network act as rational decision makers, and the profile of actions of all agents in the network is a game equilibrium. Decision of each agent is influenced by behavior (or by knowledge) of her neighbors in the network. Usually, in such models the agents are assumed to be homogeneous except differences in their positions in the network; a specific problem is to study how the network structure relates to behavior of agents in the game equilibrium. However, since diversity and heterogeneity become an important aspect of contemporary social and economic life (international working teams is a typical example), an important task is to account for heterogeneity of agents as a factor

V. Matveenko (✉) · M. Garmash · A. Korolev
National Research University Higher School of Economics, St. Petersburg, Russia
e-mail: vmatveenko@hse.ru

L. A. Petrosyan et al. (eds.), *Frontiers of Dynamic Games*,
Static & Dynamic Game Theory: Foundations & Applications,
https://doi.org/10.1007/978-3-319-92988-0_10

defining differences in their behavior and well-being. This direction of research is only forming in the literature.

We add agents' heterogeneity into the two-period consumption-investment game model (see [16] for a special case of complete network and [14] for a general network case). In the model, in the first stage each agent in network, at the expense of diminishing current consumption, makes investment of some resource (such as money or time). Consumption in the second stage depends not only on her own investment and productivity but also on her 'environment'—the sum of investments of both herself and her neighbors. Total utility of each agent depends on her consumption in both stages. Such situations are typical for families, communities, international organizations, innovative industries etc.

The fact that other players influence the payoff only through the environment makes applicable the concept of 'Nash equilibrium with externalities', similar to the one introduced by Romer [16] and Lucas [12]. Under this concept, the agent is attached, in some degree, to the equilibrium of the game. Namely, it is assumed that the agent in the moment of decision-making considers the environment as exogenously given.

For this model Matveenko et al. [15] assume that there are two types of agents characterized by different productivities. It is shown that, in dependence on the type of agent and externality which she receives, three ways of agent's behavior are possible: passive (no investment), active (a part of endowment is invested) and hyperactive (the whole endowment is invested). In other terms, the equilibrium behavior of agents is defined by their generalized α-centralities which take into account their productivities. Matveenko et al. [15] confine themselves by studying the case of complete networks.

In the present paper we consider a more general case of regular (equidegree) networks with two types of agents. A central question studied in the paper is consequences of junction of two networks with different productivities of agents. We introduce a continuous adjustment dynamics and study the process of transition to a new equilibrium. The dynamics pattern and the nature of the resulting equilibrium depend on the parameters characterizing the heterogeneous agents. We find conditions under which the initial equilibrium holds after unification, as well as conditions under which the equilibrium changes. In particular, we study how the behavior of nonadopters (passive agents) changes when they connect to adopters (active or hyperactive) agents. Thus, our paper contributes a new approach to the literature on diffusion in networks (typical examples are technology adoption and development of provision of local public goods—see e.g. [5, 6, 17]).

The paper is organized in the following way. The game model is formulated in Sect. 10.2. Agent's behavior in equilibrium is characterized in Sect. 10.3. Section 10.4 studies equilibria with heterogeneous agents in regular network of a special class. Section 10.5 introduces and studies the adjustment dynamics which may start after a small disturbance of initial inner equilibrium or after a junction of networks. Section 10.6 studies consequences of junction of two regular networks with different types of agents. Section 10.7 concludes.

10.2 The Model

There is a network (undirected graph) with n nodes $i = 1, 2, ..., n$; each node represents an agent. At the first stage each agent i possesses initial endowment of good, e (it may be, for instance, time or money) and uses it partially for consumption at the first stage, c_1^i, and partially for investment into knowledge, k_i:

$$c_1^i + k_i = e, i = 1, 2, ..., n.$$

Investment immediately transforms one-to-one into knowledge which is used in production of good for consumption at the second stage, c_2^i.

The consumption of the second period is equal to production: $c_2^i = F(k_i, K_i)$. Production in node i is described by production function:

$$F(k_i, K_i) = g_i k_i K_i, \ g_i > 0,$$

which depends on the state of knowledge in ith node, k_i, and on *environment*, K_i. The environment is the sum of investments by the agent himself and her neighbors:

$$K_i = k_i + \tilde{K}_i, \tilde{K}_i = \sum_{j \in N(i)} k_j$$

where $N(i)$—is the set of incident nodes of node i. The sum of investments of neighbors, $\tilde{K}_i$, will be referred as *pure externality*.

Preferences of agent i are described by the quadratic utility function:

$$U_i(c_1^i, c_2^i) = c_1^i(e - ac_1^i) + d_i c_2^i,$$

where d_i is the value of consumption at the second stage, $d_i > 0$; a is a satiation coefficient. It is assumed that $c_1^i \in [0, e]$, the utility increases in c_1^i and is concave (the marginal utility decreases) with respect to c_1^i. A sufficient condition leading to such shape of the utility is $0 < a < 1/2$; we assume that this inequality is satisfied.

We will denote the product $d_i g_i$ by b_i and assume that $a < b_i$. Since increase of any of parameters d_i, g_i promotes increase of the second stage consumption, we will call b_i *productivity*. We will assume that $b_i \neq 2a, \ i = 1, 2, ..., n$. If $b_i > 2a$, we will say that ith agent is *productive*, and if $b_i < 2a$—that the agent is *unproductive*.

Three ways of behavior are possible: agent i is called *passive* if she makes zero investment, $k_i = 0$ (i.e. consumes the whole endowment at the first stage); *active* if $0 < k_i < e$; *hyperactive* if she makes maximally possible investment e (i.e. consumes nothing on the first stage).

We consider a game in which possible actions (strategies) of player i are the values of investment $k_i \in [0, e]$. *The Nash equilibrium with externalities* (for shortness, *the equilibrium*) is a profile of actions $k_1^*, k_2^*, ..., k_n^*$, such that each k_i^* is a solution of the following problem of maximization of ith player's utility given environment K_i:

$$U_i(c_1^i, c_2^i) \underset{c_1^i, c_2^i}{\longrightarrow} \max$$

$$\begin{cases} c_1^i = e - k_i, \\ c_2^i = F(k_i, K_i), \\ c_1^i \geq 0, c_2^i \geq 0, k_i \geq 0, \end{cases}$$

where the environment K_i is defined by the profile $k_1^*, k_2^*, ..., k_n^*$:

$$K_i = k_i^* + \sum_{j \in N(i)} k_j^*.$$

Substituting the constraints-equalities into the objective function, we obtain a new function (*payoff function*):

$$\begin{aligned} V_i(k_i, K_i) = U_i(e - k_i, F_i(k_i, K_i)) &= (e - k_i)(e - a(e - k_i)) + b_i k_i K_i \\ &= e^2(1 - a) - k_i e(1 - 2a) - a k_i^2 + b_i k_i K_i. \end{aligned} \tag{10.1}$$

If all players' solutions are internal ($0 < k_i^* < e, i = 1, 2, ..., n$), i.e. all players are active, the equilibrium will be referred to as *inner* equilibrium. It is clear that, the inner equilibrium (if it exists for given values of parameters) is defined by the system

$$D_1 V_i(k_i, K_i) = 0, \ i = 1, 2, ..., n, \tag{10.2}$$

where D_1 is the differentiation operator for the first argument of function. Here

$$D_1 V_i(k_i, K_i) = e(2a - 1) - 2a k_i + b_i K_i. \tag{10.3}$$

The following theorem will serve as a tool for comparison of utilities.

Theorem 10.1 (Theorem 1.2 in [15]) *Let W^* and W^{**} be networks with the same endowment e; i, j be, correspondingly, two their nodes; b_i, b_j be productivities of the agents at these nodes; k_i^*, K_i^*, U_i^* and k_j^{**}, K_j^{**}, U_j^{**} be equilibrium values of knowledge, environment and utilities in these two nodes; $k_i^* \in (0, e]$, $k_j^{**} \in (0, e]$. In such case*

1. *if $b_i K_i^* < b_j K_j^{**}$, then $U_i^* < U_j^{**}$;*
2. *if $b_i K_i^* \leq b_j K_j^{**}$, then $U_i^* \leq U_j^{**}$;*
3. *if $b_i K_i^* = b_j K_j^{**}$, then $U_i^* = U_j^{**}$.*

If $k_i^* = 0$, $K_j^{**} > 0$, then $U_i^* = U(e, 0) < U_j^{**}$.

10.3 Indication of Agent's Ways of Behavior

Definition 10.1 We will denote by $\tilde{k_i^S}$ the root of the equation

$$D_1 V_i(k_i, K_i) = (b_i - 2a)k_i + b_i \tilde{K_i} - e(1 - 2a) = 0.$$

Thus,

$$\tilde{k_i^S} = \frac{e(2a-1) + b_i \tilde{K_i}}{2a - b_i},$$

where $\tilde{K_i}$ is the pure externality received by the agent.

Remark 10.1 Evidently, if in equilibrium the agent is active, her investment is equal to $\tilde{k_i^S}$. In other cases this value has only "informative" role.

Remark 10.2 Lemma 2.1 and Corollary 2.1 in [15] imply that a profile of actions is an equilibrium only if for every agent i, $i = 1, 2, \ldots, n$:

1) If $k_i = 0$, then $\tilde{K_i} \le \frac{e(1-2a)}{b_i}$;
2) If $0 < k_i < e$, then $k_i = \tilde{k_i^S}$;
3) If $k_i = e$, then $\tilde{K_i} \ge \frac{e(1-b_i)}{b_i}$.

Definition 10.2 A network in which each node has the same degree (number of neighbors) is referred as *regular*. A regular network is denoted (n, m), where n is the number of nodes, and m is degree.

Corollary 10.1 *In regular network (n, m) in equilibrium, in which all the agents make the same investments*

1) if $b_i < \frac{1}{m+1}$, $i = 1, 2, ..., n$, then agents are passive;
2) if $b_i = \frac{1}{m+1}$, $i = 1, 2, ..., n$, then agents are passive or hyperactive;
3) if $b_i > \frac{1}{m+1}$, $i = 1, 2, ..., n$, then agents are passive, active or hyperactive.

Proof It follows immediately from Remark 10.2. In fact, the condition of passivity

$$0 \le \frac{e(1-2a)}{b_i}$$

is already valid. The condition of activity

$$0 < \tilde{k_i^S} = \frac{e(1-2a)}{(m+1)b_i} < e$$

is equivalent to

$$b_i > \frac{1}{m+1}.$$

The condition of hyperactivity

$$\tilde{K}_i = me \geq \frac{e(1-b_i)}{b_i}$$

is equivalent to

$$b_i \geq \frac{1}{m+1}.$$

□

Remark 10.3 Evidently, $\tilde{k_i^S}$ can be presented as

$$\tilde{k_i^S} = \frac{b_i K_i - e(1-2a)}{2a}. \tag{10.4}$$

Lemma 10.1 (Lemma 2.2 in [15]) *In equilibrium ith agent is passive iff*

$$K_i \leq \frac{e(1-2a)}{b_i}; \tag{10.5}$$

ith agent is active iff

$$\frac{e(1-2a)}{b_i} < K_i < \frac{e}{b_i}; \tag{10.6}$$

ith agent is hyperactive iff

$$K_i \geq \frac{e}{b_i}. \tag{10.7}$$

Remark 10.4 In any network, in which all agents have the same environment, there cannot be equilibrium in which an agent with a higher productivity is active while an agent with a lower productivity is hyperactive, or when an agent with a higher productivity is passive while an agent with a lower productivity is active or hyperactive.

10.4 Equilibria in Regular Network with Heterogeneous Types of Agents

Definition 10.3 Let us consider a regular network consisting of n_1 agents with productivity b_1 (agents of type 1) and n_2 agents with productivity b_2 (agents of type 2); $b_1 > b_2$. Let each agent of type 1 have m_1 neighbors of her own type and $m_2 + 1$ neighbors of type 2; and let each agent of type 2 have m_2 neighbors of her own type and m_1+1 neighbors of type 1. Such network will be referred as *biregular*.

Remark 10.5 A special case of biregular network is a complete network with n_1+n_2 agents, which is received in result of unification of two complete networks with n_1 and n_2 agents. In such network $m_i = n_i - 1, i = 1, 2$.

Definition 10.4 Equilibrium (or any other situation) is called *symmetric*, if all players of the same type choose the same action (make the same investment).

Let a biregular network be in a symmetric equilibrium, in which each 1st type agent makes investment k_1, and each 2nd type agent makes investment k_2. Then, for each agent environment is equal to $K = k_1(m_1 + 1) + k_2(m_2 + 1)$. According to Remark 10.4, only six symmetric equilibria are possible. The following proposition lists these possible symmetric equilibria and provides conditions of their existence.

Proposition 10.1 *In biregular network the following symmetric equilibria are possible.*

1) Equilibrium with all hyperactive agents is possible iff

$$b_1 > b_2 \geq \frac{1}{m_1 + m_2 + 2}. \tag{10.8}$$

2) Equilibrium in which 1st type agents are hyperactive and 2nd type agents are active is possible if

$$0 < \frac{1 - 2a - (m_1 + 1)b_2}{(m_2 + 1)b_2 - 2a} < 1. \tag{10.9}$$

$$m_1 + 1 + (m_2 + 1)\frac{1 - 2a - (m_1 + 1)b_2}{(m_2 + 1)b_2 - 2a} \geq \frac{1}{b_1}. \tag{10.10}$$

3) Equilibrium in which 1st type agents are hyperactive and 2nd type agents are passive is possible iff

$$b_1 \geq \frac{1}{m_1 + 1}, b_2 \leq \frac{1 - 2a}{m_1 + 1}. \tag{10.11}$$

4) *Equilibrium in which 1st type agents are active and 2nd type agents are passive is possible if*

$$b_1 > \frac{1}{m_1+1}, b_2 \le \frac{(m_1+1)b_1 - 2a}{m_1+1}. \tag{10.12}$$

5) *Equilibrium with all passive agents is always possible.*
6) *Equilibrium in which agents of both types are active is possible if*

$$(m_1+1)(b_1-b_2) < 2a, 2ab_1(m_1+m_2+2) > 2a + (m_2+1)(b_1-b_2).$$

Proof

1) Follows from Lemma 10.1.
2) This equilibrium is possible if inequality (10.9) is checked. According to (10.7), the equilibrium is possible under (10.10).
3) Since in this case the environment is $K = (m_1+1)$, according to (10.5) and (10.7), the equilibrium is possible iff (10.11) is checked.
4) According to (10.5) and (10.6), the equilibrium is possible if

$$b_1 > \frac{1}{m_1+1}, \ (m_1+1)k_1 \le \frac{e(1-2a)}{b_2},$$

where

$$k_i = \frac{e(1-2a)}{(m_1+1)b_1 - 2a}.$$

5) Follows from Remark 10.2.
6) The system of Eq. (10.2) turns into

$$\begin{cases} ((m_1+1)b_1 - 2a)k_1 + (m_2+1)b_1k_2 = e(1-2a), \\ (m_1+1)b_2k_1 + ((m_2+1)b_2 - 2a)k_2 = e(1-2a). \end{cases}$$

We solve this system by Kramer method and obtain

$$k_1^S = \frac{e(1-2a)\big((m_2+1)(b_2-b_1) - 2a\big)}{2a\big(2a - (m_1+1)b_1 - (m_2+1)b_2\big)},$$

$$k_2^S = \frac{e(1-2a)\big((m_1+1)(b_1-b_2) - 2a\big)}{2a\big(2a - (m_1+1)b_1 - (m_2+1)b_2\big)}.$$

It is clear that $k_1^S > k_2^S$; hence, the necessary and sufficient conditions of existence of the inner equilibrium are

$$k_2^S > 0, k_1^S < e,$$

i.e.

$$(m_1 + 1)(b_1 - b_2) < 2a, \tag{10.13}$$

$$2ab_1(m_1 + m_2 + 2) > 2a + (m_2 + 1)(b_1 - b_2). \tag{10.14}$$

Under inequalities (10.13), (10.14), the inner equilibrium is

$$k_1 = k_1^S, k_2 = k_2^S$$

□

Remark 10.6 The signs of the following derivatives show how a change in the types' productivities b_1, b_2 influences volumes of investments k_1, k_2:

$$(k_1)'_{b_1} = C_1((m_2 + 1)b_2 - 2a) \ (where\ C_1 > 0)$$

$$(k_1)'_{b_2} = C_2(-(m_1 + 1)(m_2 + 1)b_1 - (m_2 + 1)^2 b_1) < 0 \ (where\ C_2 > 0)$$

$$(k_2)'_{b_1} = C_3(-(m_1 + 1)(m_2 + 1)b_2 - (m_1 + 1)^2 b_2) < 0 \ (where\ C_3 > 0)$$

$$(k_2)'_{b_2} = C_4((m_1 + 1)b_1 - 2a) > 0 \ (where\ C_4 > 0)$$

Thus, with an increase in productivity of ith type agents, their equilibrium investments increase, while the equilibrium investments of jth type agents decrease ($j \neq i, i = 1, 2$).

10.5 Adjustment Dynamics and Dynamic Stability of Equilibria

Now we introduce adjustment dynamics which may start after a small deviation from equilibrium or after junction of networks each of which was initially in equilibrium. We model the adjustment dynamics in the following way.

Definition 10.5 In the adjustment process, each agent maximizes her utility by choosing a level of her investment; at the moment of decision-making she considers her environment as exogenously given. Correspondingly, in continuous time if

$k_i(t_0) = 0$, where t_0 is an arbitrary moment of time, and $D_1 V_i(k_i, K_i)|_{k_i=0} \le 0$, then $k_i(t) = 0$ for any $t > t_0$, and if $k_i(t_0) = e$ and $D_1 V_i(k_i, K_i)|_{k_i=e} \ge 0$, then $k_i(t) = e$ for any $t > t_0$; in all other cases, $k_i(t)$ satisfies the difference equation:

$$\dot{k_i} = \frac{b_i}{2a}\tilde{K_i} + \frac{b_i - 2a}{2a}k_i - \frac{e(1-2a)}{2a}.$$

Definition 10.6 The equilibrium is called *dynamically stable* if, after a small deviation of one of the agents from the equilibrium, dynamics starts which returns the equilibrium back to the initial state. In the opposite case, the equilibrium is called *dynamically unstable*.

In biregular network, let in initial time moment each 1st type agent invest k_{01} and each 2nd type agent invest k_{02}. Correspondingly, the environment (common for all agents) in the initial moment is $K = k_{01}(m_1 + 1) + k_{02}(m_2 + 1)$.

Assume that either $k_{01} = 0$ and $D_1 V_1(k_1, K)|_{k_1=0} > 0$, or $k_{01} = e$ and $D_1 V_1(k_1, K)|_{k_1=e} < 0$, or $k_{01} \in (0, e)$, and ether $k_{02} = 0$ and $D_1 V_2(k_2, K)|_{k_2=0} > 0$, or $k_{02} = e$ and $D_1 V_2(k_2, K)|_{k_2=e} < 0$, or $k_{02} \in (0, e)$. Then Definition 10.5 implies that the dynamics is described by the system of differential equations.

$$\begin{cases} \dot{k_1} = \frac{(m_1+1)b_1-2a}{2a}k_1 + \frac{(m_2+1)b_1}{2a}k_2 + \frac{e(2a-1)}{2a}, \\ \dot{k_2} = \frac{(m_1+1)b_2}{2a}k_1 + \frac{(m_2+1)b_2-2a}{2a}k_2 + \frac{e(2a-1)}{2a} \end{cases} \tag{10.15}$$

with initial conditions

$$\begin{cases} k_1^0 = k_{01}, \\ k_2^0 = k_02. \end{cases} \tag{10.16}$$

Proposition 10.2 *The general solution of the system of differential equations* (10.15) *has the form*

$$k(t) = C_1 \cdot exp\{-t\}\begin{pmatrix} -(m_2+1) \\ m_1+1 \end{pmatrix}$$

$$+ C_2 \cdot exp\Big\{\Big(\frac{(m_1+1)b_1 + (m_2+1)b_2}{2a} - 1\Big)t\Big\}\begin{pmatrix} b_1 \\ b_2 \end{pmatrix} + \begin{pmatrix} D_1 \\ D_2 \end{pmatrix}, \tag{10.17}$$

where $(D_1, D_2)^T$ *is the steady state of* (10.15),

$$D_1 = \frac{e(1-2a)\big((m_2+1)(b_2-b_1) - 2a\big)}{2a\big(2a - (m_1+1)b_1 - (m_2+1)b_2\big)}, \tag{10.18}$$

$$D_2 = \frac{e(1-2a)\big((m_1+1)(b_1-b_2)-2a\big)}{2a\big(2a-(m_1+1)b_1-(m_2+1)b_2\big)}. \tag{10.19}$$

The solution of the Cauchy differential problem (10.15)–(10.16) *has the form*

$$k(t) = \frac{b_1k_2^0 - b_2k_1^0 + b_2D_1 - b_1D_2}{(m_1+1)b_1+(m_2+1)b_2} \cdot exp\{-t\} \begin{pmatrix} -(m_2+1) \\ m_1+1 \end{pmatrix}$$

$$+ \frac{(m_1+1)k_1^0 + (m_2+1)k_2^0 - \tilde{D}}{(m_1+1)b_1+(m_2+1)b_2} \cdot exp\left\{\left(\frac{(m_1+1)b_1+(m_2+1)b_2}{2a} - 1\right)t\right\} \begin{pmatrix} b_1 \\ b_2 \end{pmatrix}$$

$$+ \begin{pmatrix} D_1 \\ D_2 \end{pmatrix}, \tag{10.20}$$

where

$$\tilde{D} = \frac{e(1-2a)(m_1+m_2+2)}{(m_1+1)b_1+(m_2+1)b_2-2a}. \tag{10.21}$$

Proof The characteristic equation of system (10.15) is

$$\begin{vmatrix} \frac{(m_1+1)b_1}{2a} - (\lambda+1) & \frac{(m_2+1)b_1}{2a} \\ \frac{(m_1+1)b_2}{2a} & \frac{(m_2+1)b_2}{2a} - (\lambda+1) \end{vmatrix}$$

$$= -(\lambda+1)\left(\frac{(m_1+1)b_1}{2a} + \frac{(m_2+1)b_2}{2a}\right) + (\lambda+1)^2 = 0.$$

Thus, the eigenvalues are

$$\lambda_1 = -1, \quad \lambda_2 = \frac{(m_1+1)b_1+(m_2+1)b_2}{2a} - 1.$$

An eigenvector corresponding λ_1 is

$$e_1 = \begin{pmatrix} -(m_2+1) \\ m_1+1 \end{pmatrix},$$

while an eigenvector corresponding λ_2 can be found as a solution of the system of equations

$$\begin{cases} \frac{-(m_2+1)b_2}{2a}x_1 + \frac{(m_2+1)b_1}{2a}x_2 = 0, \\ \frac{(m_1+1)b_2}{2a}x_1 - \frac{(m_1+1)b_1}{2a}x_2 = 0. \end{cases}$$

We find

$$e_2 = \begin{pmatrix} b_1 \\ b_2 \end{pmatrix}.$$

The general solution of the homogeneous system of differential equations corresponding (10.15) has the form

$$(k(t))_g = C_1 \cdot exp\{-t\} \begin{pmatrix} -(m_2+1) \\ m_1+1 \end{pmatrix}$$

$$+C_2 \cdot exp\left\{\left(\frac{(m_1+1)b_1 + (m_2+1)b_2}{2a} - 1\right)t\right\} \begin{pmatrix} b_1 \\ b_2 \end{pmatrix}.$$

As a partial solution of the system (10.15) we take its steady state, i.e. the solution of the linear system

$$\begin{cases} 0 = \frac{(m_1+1)b_1}{2a}(D_1 - 1) + \frac{(m_2+1)b_1}{2a} D_2 + \frac{e(2a-1)}{2a}, \\ 0 = \frac{(m_1+1)b_2}{2a} D_1 + \frac{(m_2+1)b_2}{2a}(D_2 - 1) + \frac{e(2a-1)}{2a}. \end{cases}$$

The solution is (10.18)–(10.19); hence, the general solution of the system (10.15) has the form (10.17). In solution of the Cauchy problem (10.15)–(10.16), constants of integration are defined from the initial conditions:

$$\begin{pmatrix} k_1^0 \\ k_2^0 \end{pmatrix} = C_1 \begin{pmatrix} -(m_2+1) \\ m_1+1 \end{pmatrix} + C\frac{1}{b_1+b_2} \begin{pmatrix} b_1 \\ b_2 \end{pmatrix} + \begin{pmatrix} D_1 \\ D_2 \end{pmatrix}. \tag{10.22}$$

Multiplying by $(-b_2, b_1)$ we obtain

$$C_1(b_2(m_2+1) + b_1(m_1+1)) + b_1 D_2 - b_2 D_1 = b_1 k_2^0 - b_2 k_1^0.$$

Thus,

$$C_1 = \frac{b_1 k_2^0 - b_2 k_1^0 + b_2 D_1 - b_1 D_2}{(m_1+1)b_1 + (m_2+1)b_2}.$$

However, since one of the eigenvalues is zero, we need only constant C to write the solution. Multiplying by (m_1+1, m_2+1) we obtain

$$(m_1+1)k_1^0 + (m_2+1)k_2^0 = C_2((m_1+1)b_1 + (m_2+1)b_2) + (m_1+1)D_1 + (m_2+1)D_2.$$

We denote $\tilde{D} = (m_1+1)D_1 + (m_2+1)D_2$ and derive expression (10.21). Thus,

$$C_2 = \frac{(m_1+1)k_1^0 + (m_2+1)k_2^0 - \tilde{D}}{(m_1+1)b_1 + (m_2+1)b_2}.$$

Substituting for C_1 and C_2 into (10.17) we obtain (10.20). □

Let us find conditions of dynamic stability/instability for the equilibria in biregular network, which are listed in Proposition 10.1.

Proposition 10.3

1. *The equilibrium with all hyperactive agents is stable iff*

$$b_1 > \frac{1}{m_1+m_2+2}, b_2 > \frac{1}{m_1+m_2+2}. \tag{10.23}$$

2. *The equilibrium, in which agents of 1st type are hyperactive and agents of 2nd type are active, is stable iff*

$$m_1 + 1 + \frac{(m_2+1)\big(1-2a-(m_1+1)b_2\big)}{(m_2+1)b_2 - 2a} > \frac{1}{b_1}, \tag{10.24}$$

$$\frac{(m_2+1)b_2}{2a} < 1.$$

3. *The equilibrium, in which agents of 1st type are hyperactive and agents of 2nd type are passive, is stable iff*

$$b_1 > \frac{1}{m_1+1}, b_2 < \frac{1-2a}{m_1+1}.$$

4. *The equilibrium, in which agents of 1st type are active and agents of 2nd type are passive, is always unstable.*
5. *The equilibrium with all passive agents is always stable.*
6. *The equilibrium with all active agents is always unstable.*

Proof

1. According to Definition 10.5 and Eq. (10.3),

$$D_1V_1(k_1, K)|_{k_1=e} = b_1(m_1+m_2+2)e - e, D_1V_2(k_2, K)|_{k_2=e}$$
$$= b_2(m_1+m_2+2)e - e.$$

Both derivatives are positive iff (10.23) is checked.

2. According to Definition 10.5 and Eqs. (10.3), (10.10),

$$D_1V_1(k_1, K)|_{k_1=e} = b_1\left((m_1+1)e + \frac{(m_2+1)e(1-2a-(m_1+1)b_2)}{(m_2+1)b_2-2a}\right)e - e$$
$$\geq 0.$$

However, for dynamic stability, the strict inequality is needed. Let (10.24) be checked and $k_1 = e$. The differential equation describing dynamics of each of the 2nd group agents is

$$\dot{k_2} = \frac{(m_2+1)b_2-2a}{2a}k_2 + \frac{(m_1+1)eb_2}{2a} + \frac{e(2a-1)}{2a}. \tag{10.25}$$

For stability it is necessary and sufficient that $\frac{(m_2+1)b_2}{2a} < 1$.

3. According to Definition 10.5 and Eqs. (10.3), (10.11),

$$D_1V_1(k_1, K)|_{k_1=e} = b_1e - e \geq 0,$$

$$D_1V_2(k_2, K)|_{k_2=e} = e(2a-1) + b_2(m_1+1)e \leq 0.$$

For stability the strict inequalities are needed.

4. According to Definition 10.5 and Eqs. (10.3), (10.12),

$$D_1V_1(k_2, K)|_{k_2=0} = e(2a-1) + b_2\frac{(m_1+1)e(1-2a)}{(m_1+1)b_1-2a}$$
$$= \frac{e(1-2a)\big((m_1+1)(b_2-b_1)+2a\big)}{(m_1+1)b_1-2a} \leq 0$$

For stability the strict inequalities are needed. Let the second inequality in (10.12) be satisfied strictly. The differential equation for any of the 1st group agents is

$$\dot{k_1} = \frac{(m_1+1)b_1-2a}{2a}k_1 + \frac{e(2a-1)}{2a}.$$

According to the first inequality in (10.12),

$$(m_1+1)b_1 > 1 > 2a;$$

hence, the equilibrium is unstable.

5. According to Definition 10.5 and Eq. (10.3),

$$D_1 V_1(k_1, K)|_{k_1=0} = e(2a - 1) < 0,$$

$$D_1 V_2(k_2, K)|_{k_2=0} = e(2a - 1) < 0.$$

6. One of the eigenvalues of the system (10.15) is

$$\lambda_2 = \frac{(m_1 + 1)b_1 + (m_2 + 1)b_2}{2a} - 1 > 0;$$

hence, the equilibrium is unstable. □

10.6 Biregular Junction of Two Similar Networks

Definition 10.7 Two regular networks (n_1, m_1) and (n_2, m_2) will be called *similar* with *similarity coefficient* q, if

$$\frac{n_1}{m_1 + 1} = \frac{n_2}{m_2 + 1} = q.$$

Remark 10.7 Since for complete network $n = m + 1$, any two complete networks are similar with similarity coefficient $q = 1$.

Definition 10.8 Let there be two regular networks (n_1, m_1) and (n_2, m_2) with same similarity coefficient q, and let them unify, creating a biregular network. Each agent of the first network (1st type agent in the unified network) establishes $m_2 + 1$ links with agents of the second network (2nd type agents in the unified network), while each agent of the second network establishes $m_1 + 1$ links with agents of the first network. (In special case of complete networks this means that each agent of the first network establishes links with each agent of the second network). Such junction of similar networks will be called *biregular*.

Thus, after a biregular junction of similar networks, each of $n_1 + n_2$ agents of the unified network has degree $m_1 + m_2 + 1$, i.e. regular network $(n_1 + n_2, m_1 + m_2 + 1)$ is formed. This regular network has the same similarity coefficient q as each of the initial unified networks: it is easy to check that

$$\frac{n_1 + n_2}{m_1 + m_2 + 2} = q.$$

Let all agents of the first of the networks have productivity b_1, and all agents of the second network—productivity b_2, and let the networks initially be in a symmetric equilibrium, in which agents in the networks make investments k_{01} and k_{02}, correspondingly. Then, evidently, after junction the symmetry still holds: all agents of the same type behave similarly. It follows from the fact that at time moment t environment of each agent is equal to

$$k_1(t)(m_1+1)+k_2(t)(m_2+1),$$

i.e. environment is the same for all agents.

Will the agents hold their initial behavior? Or, will a transition dynamics start after the junction?

Proposition 10.4 *After biregular junction of similar networks, all agents of the unified network hold their initial behavior (make the same investments as before the junction) in the following four cases:*

1) *if initially agents in both networks are hyperactive;*
2) *if*

$$b_2 \leq \frac{1-2a}{m_1+1},$$

and initially agents in the 1st network are hyperactive, and agents in the 2nd network are passive;

3) *if*

$$b_1 > \frac{1}{m_1+1}, \quad b_2 \leq b_1 - \frac{2a}{m_1+1},$$

and initially agents in the 1st network are active, and agents in the 2nd network are passive;

4) *if initially agents in both networks are passive.*

In all other cases the equilibrium changes.

Proof

1) According to Corollary 10.1,

$$b_1 \geq \frac{1}{m_1+1}, \quad b_2 \geq \frac{1}{m_2+1},$$

Substituting $k_1 = e$ and $k_2 = e$ into (10.3) we obtain, correspondingly,

$$D_1V_1(k_1, K)|_{k_1=e} = b_1(m_1+m_2+2)e - e \geq 0,$$

$$D_2V_2(k_2, K)|_{k_2=e} = b_2(m_1+m_2+2)e - e \geq 0.$$

2) According to Corollary 10.1,

$$b_1 \geq \frac{1}{m_1+1}.$$

Substituting $k_2 = 0$ and $k_1 = e$ into (10.3) we obtain

$$D_1V_1(k_1, K)|_{k_1=e} = b_1(m_1+1)e - e \geq 0,$$
$$D_1V_2(k_2, K)|_{k_2=0} = b_2(m_1+1)e - e(1-2a) \leq 0.$$

3) Substituting $k_1 = \frac{e(1-2a)}{(m_1+1)b_1-2a}$ and $k_2 = 0$ into (10.3) we obtain

$$D_1V_1(k_1, K) = \frac{e(1-2a)\big(2a-(m_1+1)b_1-2a+(m_1+1)b_1\big)}{(m_1+1)b_1-2a} = 0,$$
$$D_1V_2(k_2, K) = \frac{e(2a-1)\big((m_1+1)(b_1-b_2)-2a\big)}{(m_1+1)b_1-2a} \leq 0.$$

4) Substituting $k_1 = 0, k_2 = 0$ into (10.3) we obtain

$$D_1V_1(k_1, K)|_{k_1=0} = D_1V_2(k_2, K)|_{k_2=0} = e(2a-1) \leq 0.$$

In all other cases the initial values of investments of agents will not be equilibrium in the unified network, and there will be a transition dynamics. □

Proposition 10.4 shows, in particular, that passive agents (nonadopters), when connected with adopters, can remain nonadopters only if their productivity, b_2, is relatively low.

A pattern of transition process after the junction depends on initial conditions and parameters values. If adjustment dynamics of the unified regular network starts, it is described by the system of difference equations (10.15) with initial conditions (10.16).

Proposition 10.5 *We consider a biregular junction of two similar regular networks. Let the agents in the 1st network before junction be hyperactive (hence, $b_1 \geq \frac{1}{m_1+1}$ by Corollary 10.1) and agents in the 2nd network be passive. Then the following cases are possible.*

1. *If $b_2 \leq \frac{1-2a}{m_1+1}$, then after junction all agents hold their initial behavior, and there is no transition process in the unified network. The unified network is in equilibrium with $\{k_1 = e, \quad k_2 = 0\}$.*

2. *If* $b_2 > \frac{1-2a}{m_1+1}$ *and* $b_2 \geq \frac{2a}{m_2+1}$, *then the 1st group agents stay hyperactive; investments of the 2nd group agents increase, until they also become hyperactive. The unified network comes to equilibrium* $\{k_1 = e, \quad k_2 = e\}$.[1]
3. *If* $\frac{2a}{m_2+1} > b_2 > \frac{1-2a}{m_1+1}$, *then the 1st group agents stay hyperactive; investments of the 2nd group agents increase. The unified network comes to equilibrium* $\left\{k_1 = e, \quad k_2 = \frac{e\left((m_1+1)b_2+2a-1\right)}{2a-(m_2+1)b_2}\right\}$ *if* $b_2 < \frac{1}{m_1+m_2+2}$, *and to equilibrium* $\{k_1 = e, \quad k_2 = e\}$ *if* $b_2 \geq \frac{1}{m_1+m_2+2}$.

In cases 2 and 3, utilities of all agents in the unified network increase. In case 1, the utilities do not change.

Proof

1. Follows from Proposition 10.4, point 2.

2–3. If for agents of the 2nd group

$$D_1 V_2(k_2, K)|_{k_2=0} = b_2(m_1 + 1)e - e(1 - 2a) > 0,$$

they change their investments according to the differential equation (10.25). The general solution of Eq. (10.25) is:

$$k_2(t) = C \exp\left(\frac{(m_2 + 1)b_2 - 2a}{2a}\right) + D, \tag{10.26}$$

where

$$D = \frac{e\left((m_1 + 1)b_2 + 2a - 1\right)}{2a - (m_2 + 1)b_2}, \tag{10.27}$$

and $(m_1 + 1)b_2 > 1 - 2a$. The initial conditions imply

$$C = k_2^0 - D = -D.$$

The partial solution satisfying initial conditions is

$$k_2(t) = D\left(1 - \exp\left(\frac{(m_2 + 1)b_2 - 2a}{2a}\right)\right).$$

[1] Notice that conditions $b_2 > \frac{1-2a}{m_1+1}$ and $b_2 \geq \frac{2a}{m_2+1}$ imply $b_2 > \frac{1}{m_1+m_2}$, i.e. condition of existence of equilibrium $\{k_1 = e, \quad k_2 = e\}$.

If $b_2 > \frac{2a}{m_2+1}$, then $\frac{(m_2+1)b_2}{2a}$, $D < 0$, and $k_2(t)$ converges to e. After the value e is achieved, $k_2(t) = e$, since

$$D_1 V_2(k_2, K)|_{k_2=e} = b_2(m_1 + m_2 + 2)e - e \geq (1 - 2a + 2a)e - e = 0.$$

If $b_2 < \frac{2a}{m_2+1}$, then $\frac{(m_2+1)b_2}{2a} < 1$, $D > 0$, and $k_2(t)$ converges to D if $D < e$, i.e. if $b_2 < \frac{1}{m_1+m_2+2}$. In the opposite case, if $b_2 \geq \frac{1}{m_1+m_2+2}$, $k_2(t)$ converges to e.

It is clear that both the equilibria, $\{k_1 = e, \quad k_2 = e\}$ and $\left\{k_1 = e, k_2 = \frac{e\left((m_1+1)b_2+2a-1\right)}{2a-(m_2+1)b_2}\right\}$, possible in result of junction, are stable.

In the "resonance" case,

$$b_2 = \frac{2a}{m_2 + 1},$$

we are looking for the solution of differential equation (10.25) in form tD, where

$$D = \frac{e\left((m_1 + 1)b_2 + 2a - 1\right)}{2a}.$$

Since $(m_1 + 1)b_2 > 1 - 2a$, the value of investment $k_2(t)$ converges to e and, since this value is achieved, stays equal to e, because

$$D_1 V_2(k_2, K)|_{k_2=e} = b_2(m_1 + +m_2 + 2)e - e \geq (1 - 2a + 2a)e - e = 0.$$

The last statement (concerning utilities) follows directly from Theorem 10.1. □

Proposition 10.6 *We consider a biregular junction of two similar regular networks. Let agents of the 1st network before junction be hyperactive (which implies $b_1 \geq \frac{1}{m_1+1}$ by Corollary 10.1), and agents of the 2nd network be active (which implies $b_2 > \frac{1}{m_2+1}$). The unified network moves to the equilibrium with all hyperactive agents. The utilities of all agents increase.*

Proof The 1st group agents stay hyperactive, because, by (10.3),

$$D_1 V(k_1, b_1, K)|_{k_1=e} = e(2a - 1) - 2ae + b_1(m_1 + 1)e + b_2(m_2 + 1)k_2 \geq 0.$$

For the 2nd group agents we have Eq. (10.25). Its general solution is (10.26), where D is defined by (10.26). From the initial conditions we find

$$C = k_2^0 - D = \frac{e(1 - 2a)}{(m_2 + 1)b_2 - 2a} - \frac{e(1 - 2a - (m_1 + 1)b_2)}{(m_2 + 1)b_2 - 2a} \frac{(m_1 + 1)b_2}{(m_2 + 1)b_2 - 2a} > 0.$$

Hence, $k_2(t)$ achieves the value e.

The statement concerning utilities follows directly from Theorem 10.1. □

Proposition 10.7 *We consider a biregular junction of two similar regular networks. If before junction agents of both networks are hyperactive (this implies* $b_1 \geq \frac{1}{m_1+1}$*,* $b_2 \geq \frac{1}{m_2+1}$ *by Corollary 10.1), they stay hyperactive after junction: there is no transition dynamics, and utilities of all agents do increase.*

Proof It follows from Proposition 10.4, item 1. The increase of utilities follows from Theorem 10.1. □

Proposition 10.8 *We consider a biregular junction of two similar regular networks. If before junction agents of both networks are passive, they stay passive after junction: there is no transition dynamics, and agents' utilities do not change.*

Proof It follows from Proposition 10.4, item 4. Utilities do not change according to Theorem 10.1. □

The following two propositions show how, depending on the relation between the heterogeneous productivities, passive agents (nonadopters) may change their behavior (become adopters).

Proposition 10.9 *We consider a biregular junction of two similar regular networks. Let agents of 1st network before junction be active (which implies* $b_1 > \frac{1}{m_1+1}$ *by Corollary 10.1),* $k_1^0 = \frac{e(1-2a)}{(m_1+1)b_1-2a}$*, and agents of the 2nd network be passive. Then the following cases are possible.*

1. *Under* $(m_1+1)b_1 \geq (m_1+1)b_2 + 2a$*, all agents hold their initial behavior, and there is no transition process.*
2. *Let* $(m_1+1)b_1 < (m_1+1)b_2 + 2a$*. If* $b_2 \geq \frac{2a}{m_2+1}$ *and* $\frac{e-D_1-k_1^0}{b_1} < \frac{e-D_2}{b_2}$*, then the network moves to the equilibrium with all hyperactive agents. If* $b_2 < \frac{2a}{m_2+1}$ *and* $\frac{e-D_1-k_1^0}{b_1} < \frac{e-D_2}{b_2}$*, then the network moves to the equilibrium, in which the 1st group agents are hyperactive and the 2nd group agents are active,*

$$k_2 \frac{e(1-2a-(m_1+1)b_2)}{(m_2+1)b_2-2a}.$$

3. *If* $\frac{e-D_1-k_1^0}{b_1} \geq \frac{e-D_2}{b_2}$*, then the network moves to the equilibrium with all hyperactive agents.*

In case 1, utilities of all agents do not change; in case 2, utilities of all agents increase.

Proof For the 2nd group agents, initially

$$D_1V(k_2, b_2, K) = e(2a-1) + b_2\frac{(m_1+1)e(1-2a)}{(m_1+1)b_1-2a}$$

$$= \frac{e(1-2a)\big((m_1+1)(b_2-b_1)+2a\big)}{(m_1+1)b_1-2a}.$$

Thus, $D_1V(k_2, b_2, K) \le 0$ if $(m_1+1)b_1 \ge (m_1+1)b_2+2a$. In this case the 2nd group agents stay passive. The 1st group agents also hold their behavior unchanged, because their environment does not change.

Now, let $(m_1+1)b_1 < (m_1+1)b_2+2a$. The 2nd group agents increase their investments, and so do agents of the 1st group, because their environment increases. Conditions $(m_1+1)b_1 < (m_1+1)b_2+2a$ and $b_1 > \frac{1}{m_1+1}$ imply $b_2 > \frac{1-2a}{m_1+1}$. Hence, by Lemma 10.1, the equilibrium with hyperactive agents of one of the groups and active agents of another group is always possible, as well as the equilibrium with all hyperactive agents.

Agents of one of the groups may achieve the investment level e earlier than the agents of another group. Let it be the 1st group, i.e. $\frac{e-D_1-k_1^0}{b_1} < \frac{e-D_2}{b_2}$. The investment level of the 2nd group agents in this moment is some $\tilde{k}_2^0$. After that investments of the 2nd group agents follow Eq. (10.25). The general solution of (10.25) is (10.26), where D has the form (10.27). From the initial conditions we have $C = \tilde{k}_2^0 - D$. Thus, if $b_2 > \frac{2a}{m_2+1}$, then $D < 0$, which implies $C > 0$; hence, investments of the 2nd group agents will achieve level e. If $b_2 < \frac{2a}{m_2+1}$, then investments of agents of the 2nd group will become equal to $D > 0$.

In the "resonance" case, $b_2 = \frac{2a}{m_2+1}$, as previously, the general solution of Eq. (10.25) has the form

$$k_2(t) = C + t\frac{\big((m_1+1)b_2+2a-1\big)}{2a}.$$

It follows from initial conditions that $C = \tilde{k}_2^0$, so the partial solution of (10.25) satisfying the initial conditions is

$$k_2(t) = \tilde{k}_2^0 + t\frac{e\big((m_1+1)b_2+2a-1\big)}{2a}.$$

Since $(m_1+1)b_2 > 1-2a$, the value of investments of the 2nd group agents in this case also achieves e.

Suppose now, that the 2nd group agents have received the investment level e first, i.e. $\frac{e-D_1-k_1^0}{b_1} > \frac{e-D_2}{b_2}$, while the investment level of the 1st group agents was equal to some $\tilde{k}_1^0$. It is possible only if $b_2 > b_1$. From that moment the investments of the

1st group agents follow equation

$$\dot{k_1} = \frac{(m_1+1)b_1 - 2a}{2a}k_1 + \frac{(m_2+1)eb_1}{2a} + \frac{e(2a-1)}{2a}, \tag{10.28}$$

whose general solution is

$$k_1(t) = C\exp\left(\frac{(m_1+1)b_1-2a}{2a}\right)^n + D,$$

where

$$D = \frac{e(1-2a-(m_2+1)b_1)}{(m_1+1)b_1-2a}.$$

From the initial condition we have $C = \tilde{k_1^0} - D$. Moreover, $\tilde{k_1^0} > k_1^0 = \frac{e(1-2a)}{(m_1+1)b_1-2a}$, which implies

$$\begin{aligned} C = \tilde{k_1^0} - D &> \frac{e(1-2a)}{(m_1+1)b_1-2a} - \frac{e(1-2a-(m_2+1)b_1)}{(m_1+1)b_1-2a} \\ &= \frac{(m_2+1)b_1}{(m_1+1)b_1-2a} > 0. \end{aligned}$$

Hence, investments of the 1st group agents achieve e.

In case when agents of both groups achieve investment level e simultaneously, i.e. $\frac{e-D_1-k_1^0}{b_1} = \frac{e-D_2}{b_2}$, the network, evidently, turns to the equilibrium with all hyperactive agents. □

Proposition 10.10 *We consider a biregular junction of two similar regular networks. If before junction agents of both networks are active (this implies* $b_1 > \frac{1}{m_1+1}$, $b_2 > \frac{1}{m_2+1}$ *by Corollary 10.1), then after junction all agents become hyperactive; their utilities increase.*

Proof The initial conditions are $k_1^0 = \frac{e(1-2a)}{(m_1+1)b_1-2a}$, $k_2^0 = \frac{e(1-2a)}{(m_2+1)b_2-2a}$. According to (10.3),

$$\begin{aligned} &D_1V(k_1, b_1, K) \\ &= e(2a-1) + \frac{\big(b_1(m_1+1)-2a\big)e(1-2a)}{(m_1+1)b_1-2a} + \frac{b_1(m_2+1)e(1-2a)}{(m_2+1)b_2-2a} > 0, \end{aligned}$$

$$D_1 V(k_2, b_2, K)$$
$$= e(2a-1) + \frac{\big(b_2(m_2+1) - 2a\big)e(1-2a)}{(m_2+1)b_2 - 2a} + \frac{b_2(m_1+1)e(1-2a)}{(m_1+1)b_1 - 2a} > 0.$$

Thus, agents of both groups will increase their investments following Eq. (10.26). Agents of one of the group will achieve investment level e first. Let it be the 1st group, and let investments of the 2nd group agents in this moment be $\tilde{k_2^0}$. Then investments of the 2nd group agents follow differential equation (10.25), whose general solution is (10.26), where D has the form (10.27). The initial conditions imply $C = \tilde{k_2^0} - D$, but

$$\tilde{k_2^0} > k_2^0 = \frac{e(1-2a)}{(m_2+1)b_2 - 2a} > D,$$

Hence, $C > 0$. Thus, investments of the 2nd group agents will also achieve level e. Absolutely similar argument is for the case when the 2nd group achieves the investment level e first. □

Remark 10.8 In all cases considered in Propositions 10.4–10.10, agents' utilities in result of junction do increase or, at least, do not decrease. Thus, all the agents have an incentive to unify, or, at least, have no incentive not to unify.

10.7 Conclusion

Research on the role of heterogeneity of agents in social and economic networks is rather new in the literature. In our model we assume presence of two types of agents possessing different productivities. At the first stage each agent in network may invest some resource (such as money or time) to increase her gain at the second stage. The gain depends on her own investment and productivity, as well as on investments of her neighbors in the network. Such situations are typical for various social, economic, political and organizational systems. In framework of the model, we consider relations between network structure, incentives, and agents' behavior in the game equilibrium state in terms of welfare (utility) of the agents.

We touch some questions of network formation and identify agents potentially interested in particular ways of enlarging the network. We introduce continuous adjustment dynamics which may start after a deviation from equilibrium or after a junction of networks initially being in equilibrium.

Earlier, a special case of complete networks was considered in [15]. Here we introduce a more general case of regular networks. In particular, we study behavior of agents with different productivities in two biregular networks after junction. Specifically, we show that if a network consisting of non-adopters (passive agents) does unify with a network consisting of adopters (active or hyperactive agents), and the non-adopters possess a low productivity, then there is no transition process, and the non-adopters stay passive. Under somewhat higher productivity, the non-adopters become adopters (come to active state), and under even higher productivity they become hyperactive.

Agents, who are initially active in a symmetric equilibrium in regular network (which implies that their productivities are sufficiently high), also may increase their level of investment in result of unification with another regular network with hyperactive or active agents. The unified network comes into equilibrium in which all agents are hyperactive.

A natural task for future research is to expand the results to broader classes of networks.

Acknowledgements The research is supported by the Russian Foundation for Basic Research (project 17-06-00618). The authors are grateful to two anonymous referees for valuable comments.

References

1. Acemoglu, D., Robinson, J.A.: Why Nations Fall: The Origins of Power, Prosperity, and Poverty. Crown Publishers, New York (2012)
2. Ballester, C., Calvo-Armengol, A., Zenou, Y.: Who's who in networks. Wanted: the key player. Econometrica **74**(5), 1403–1417 (2006)
3. Bramoullé, Y., Kranton, R.: Public goods in networks. J. Econ. Theory **135**, 478–494 (2007)
4. Bramoullé, Y., Kranton, R., D'Amours, M.: Strategic interaction and networks. Am. Econ. Rev. **104**(3), 898–930 (2014)
5. Eliott, M., Golub, B.: A network approach to public goods. Cambridge Working Papers in Economics 1813. Faculty of Economics, University of Cambridge (2018)
6. Estrada, E.: The Structure of Complex Networks. Theory and Applications. Oxford University Press, Oxford (2011)
7. Galeotti, A., Goyal, S., Jackson, M.O., Vega-Redondo, F., Yariv, L.: Network games. Rev. Econ. Stud. **77**, 218–244 (2010)
8. Goyal, S.: Connections: An Introductions to the Economics of Networks. Princeton University Press, Princeton (2010)
9. Jackson, M.O.: Social and Economic Networks. Princeton University Press, Princeton (2008)
10. Jackson, M.O., Zenou, Y.: Games on networks. In: Young, P., Zamir, S. (eds.) Handbook of Game Theory, vol. 4, pp. 95–163. Elsevier, Amsterdam (2014)
11. Jackson, M.O., Rogers, B.W., Zenou, Y.: The economic consequences of social network structure. J. Econ. Lit. **55**(1), 49–95 (2017)
12. Lucas, R.: On the mechanics of economic development. J. Monet. Econ. **2**(1), 3–42 (1988)
13. Martemyanov, Y.P., Matveenko, V.D.: On the dependence of the growth rate on the elasticity of substitution in a network. Int. J. Process Manag. Benchmarking **4**(4), 475–492 (2014)
14. Matveenko, V.D., Korolev, A.V.: Knowledge externalities and production in network: game equilibria, types of nodes, network formation. Int. J. Comput. Econ. Econ. **7**(4), 323–358 (2017)
15. Matveenko, V., Korolev, A., Zhdanova, M.: Game equilibria and unification dynamics in networks with heterogeneous agents. Int. J. Eng. Bus. Manag. **9**, 1–17 (2017)
16. Romer, P.M.: Increasing returns and long-run growth. J. Polit. Econ. **94**, 1002–1037 (1986)
17. Scharf, K.: Private provision of public goods and information diffusion in social groups. Int. Econ. Rev. **55**(4), 1019–1042 (2014)

Chapter 11
Non-cooperative Differential Game Model of Oil Market with Looking Forward Approach

Ovanes Petrosian, Maria Nastych, and Dmitrii Volf

Abstract The paper applies Looking Forward Approach to analyze the world oil market with the framework of a differential game model of quantity competition oligopoly. Namely Looking Forward Approach is used to take into account dynamically updating information. Under the information we understand the forecast of the oil demand dynamics. We focus on the period from December 2015 to November 2016 and suppose that during this time interval countries did not cooperate officially on the amounts of oil to be produced. Therefore, their behavior can be modeled using the non-cooperative game model. As a solution concept for this conflict-controlled process we use feedback Nash equilibrium. In order to define the parameters of model open source data is used, results of numerical simulations and comparison with the historical data are presented.

11.1 Introduction

The paper is devoted to constructing a game theoretical model for the world oil market using the Looking Forward Approach. Game models with Looking Forward Approach allow taking into account the variability of market demand, the adaptation of participants actions to a changing environment and the actual planning horizons for demand. That is the intuition to apply the approach to the oil market which has highly volatile prices.

The object of this paper is to simulate the oil market dynamic during the period from December 2015 to November 2016. The largest oil exporters reached an agreement about the reduction of oil production to raise the prices in the aftermath of November 30, 2016 and the game started to be cooperative. We suppose that

O. Petrosian (✉) · D. Volf
Saint Petersburg State University, Saint Petersburg, Russia

M. Nastych
National Research University Higher School of Economics, Saint Petersburg, Russia
e-mail: manastych@hse.ru

L. A. Petrosyan et al. (eds.), *Frontiers of Dynamic Games*,
Static & Dynamic Game Theory: Foundations & Applications,
https://doi.org/10.1007/978-3-319-92988-0_11

countries did not cooperate officially on the amounts of oil to be produced before this date. Therefore their behavior can be simulated using the non-cooperative game model. As an optimality principal the feedback Nash equilibrium is used. OPEC countries, which we call player one, and eleven non-OPEC countries, which we call player two, produce more than 60% of oil in the world together. The model includes US shale and non-shale oil producing companies as players three and four respectively as main market rivals. All the other oil exporting countries are united under player five. To obtain data on the world oil market we used open sources. Namely, we used International Energy Agency for monthly data on crude oil supply from January 2015 till November 2016, Finam agency for monthly data on brent and light oil prices from January 2015 till November 2016, Rystad Energy Ucube and oil market news for the cost of producing a barrel of oil in 2016.

Following well-established tradition, we use oligopoly quantity setting to model the oil market. Likewise, Moran in [12] and Krasner in [10] analyze main features of the oil oligopoly. The authors in [11] give a selective survey of oligopoly models for energy production. In the paper [20] the author examines cartel formation in the world oil market under Cournot setting. The authors in [3] also use the quantity competitive environment to model collisions and proportionate adjustment of production levels.

Existing differential games often rely on the assumption of time-invariant game structures for the derivation of equilibrium solutions. However, many events in the considerably far future are intrinsically unknown. In this paper, information about the players' future payoffs will be revealed as the game proceeds. Making use of the newly obtained information, the players revise their strategies accordingly, and the process will continue indefinitely. Looking Forward Approach for differential games provides a more realistic and practical alternative to the study of classical differential games. Looking Forward Approach enables us to construct game theoretical models in which the game structure can change or update over time (time-dependent formulation) and players do not have full information about the change of the game structure, but they have full information about the game structure on the truncated time interval. We understand the information about the motion equation and the payoff functions as information about the game structure. The duration of the period of this information is known in advance. At defined instants information about the game structure is being updated. Looking Forward Approach was mainly developed for cooperative games with transferable utility [7, 15–17], but there are also papers on non-cooperative differential games [22], dynamic games [23] and games with non-transferable utility [18].

The concept of the Looking Forward Approach is new in game theory, especially in differential games, and it gives the foundation for further study of differential games with dynamic updating. At the moment there were no known attempts of constructing approaches for modeling conflict-controlled processes where information about the process updates in time dynamically. The first time the Looking Forward Approach was presented in the paper [15], it was applied to the cooperative differential game with finite-horizon. The paper [17] on the subject is focused on studying Looking Forward Approach with stochastic forecast and dynamic adaptation in the case when information about the conflicting process can change during the game. In the paper [7] the Looking Forward Approach was applied to a cooperative differential game of pollution control. The aim of the paper was to study dependency of the resulting solution upon the value of the informational horizon; the corresponding optimization problem was formulated and solved. In the paper [16] the Looking Forward Approach was applied to the cooperative differential game with infinite-horizon. Papers [22] and [23] are devoted to study of cooperative differential games and non-cooperative dynamic games with infinite horizon where information about the process updates dynamically. The focus of these papers is a profound formulation of Hamilton–Jacobi–Bellman equations for a different types of forecasts and information structures. Paper [18] is devoted to studying the Looking Forward Approach for cooperative differential games with non-transferable utility, and current paper is the continuation of that paper. Here we try to construct non-cooperative game model and apply it to the real historical data. Future steps are to construct game model describing the six-month long agreement which was signed by the OPEC countries and eleven non-OPEC countries about the reduction of oil production at the summit in Vienna on November 30, 2016.

Looking Forward Approach has the common ground as the Model Predictive Control theory worked out within the framework of numerical optimal control. We analyze [5, 8, 19, 21] to get recent results in this area. Model predictive control is a method of control when the current control action is achieved by solving, at each sampling instant, a finite horizon open-loop optimal control problem using the current state of an object as the initial state. This type of control is able to cope with hard limitations on controls and states, which is definitely its strong point over the rest of methods. It has got, therefore, a wide application in petro-chemical and related industries where key operating points are located close to the set of admissible states and controls. Mathematically, the main problem that the Model Predictive Control solves is the provision of movement along the target trajectory under the conditions of random perturbations and unknown dynamical system. At each time step the optimal control problem is solved for defining controls which will lead the system to the target trajectory. Looking Forward Approach, on the other hand, solves the problem of modeling behavior of players when information about the process updates dynamically. It means that Looking Forward Approach does not use a target trajectory, but answers the question of composing trajectory which will be used by the players of the process and allocating a cooperative payoff along the trajectory.

In the current paper the Looking Forward Approach is applied to the non-cooperative oligopoly differential model [4] of the oil market with the largest oil exporters and other oil producing countries. Considered game model is defined by the set of linear differential equations and the quadratic utility functions, which means that it is a linear quadratic differential game. These type of game models are very popular in the literature and a recent exposition of this theory can be found in [2, 6]. The popularity of these games is caused on the one hand by practical considerations. To some extent these kinds of differential games are analytically and numerically solvable. Important references that contributed to this research during the last decennium are Basar and Bernhard [2], Basar and Olsder [1]. Noncooperative game theory deals with strategic interactions among multiple decision makers with the objective functions depending on choices of all the players and suggests solution concepts for a case when players do not cooperate or make any arrangements about players actions. Players cannot simply optimize their own objective function independently of the choices of the other players. In 1950 and 1951 in [13, 14] by John Nash such a solution concept was introduced and called Nash equilibrium. It has a property that if all players but one stay put, then the player who has the option of moving away from the solution point should not have any incentive to do so because she cannot improve her payoff. It suggests that none of the players can improve their payoff by a unilateral move. The Nash equilibrium solution concept provides a reasonable noncooperative equilibrium solution for nonzero-sum games when the roles of the players are symmetric, that is to say when no single player dominates the decision process.

The paper is structured as follows: In Sect. 11.2 we describe the initial game model. In Sect. 11.3 we define the notion of truncated subgame. In Sect. 11.4 we define the feedback Nash equilibrium for each truncated subgame. In Sect. 11.5 we describe the process of information updating, define a notion of a resulting trajectory and payoffs. In Sect. 11.6 we present the results of numerical simulation of oil price on the market on the time interval from December 2015 to November 2016.

11.2 Initial Game Model

Consider the differential game model of Cournot oligopoly [4] on the oil market. An oligopolistic market of n asymmetrical counties (players) belonging to the set $N = \{1, \dots, n\}$, producing oil, and competing for the quantity produced q_i under price stickiness is given by the differential game $\Gamma(p_0, T - t_0)$ with prescribed duration $T - t_0$ and initial state $p(t_0) = p_0 \in P \subset R$.

According to the model market price p_i evolves according to the differential equation:

$$\dot{p}(t) = s(\hat{p}(t) - p(t)), \;\; p(t_0) = p_0, \tag{11.1}$$

where $\hat{p}(t) \in P \subset R$ is the notional level of price at time t, $p(t)$ is its current level, and the parameter $s : 0 < s < 1$, is the speed of adjustment. Thus, prices adjust to the deference between its notional level and its current level.

Further, we assume that the notional prices at any time t are defined by linear inverse demand function

$$\hat{p}(t) = a - d \sum_{i \in N} q_i(t). \tag{11.2}$$

Players $i \in N$ choose quantity $q_i(t) \in U_i \subset R$ produced in order to maximize their profits:

$$K_i(p_0, T - t_0; q_1, \dots, q_n) = \int_{t_0}^{T} e^{-\rho(t-t_0)} \left[q_i(t)(p(t) - c_i - g_i q_i(t))\right] dt, \tag{11.3}$$

here, $0 \leq \rho \leq 1$ represents the positive discount rate, which is the same for all the periods and all the players to simplify the model and to equalize players as symmetrical participants in the global capital market. $C_i(t) = c_i q_i(t) + g_i q_i^2(t)$ is the total cost function for each player i.

11.3 Truncated Subgame

Suppose that the information for players is updated at fixed time instants $t = t_0 + j\Delta t$, $j = 0, \dots, l$, where $l = \frac{T}{\Delta t} - 1$. During the time interval $[t_0 + j\Delta t, t_0 + (j+1)\Delta t]$, players have certain information about the dynamics of the game (11.1) and payoff function (11.5) on the time interval $[t_0 + j\Delta t, t_0 + j\Delta t + \overline{T}]$, where $\Delta t \leq \overline{T} \leq T$. At the instant $t = t_0 + (j+1)\Delta t$ information about the game is being updated and the same procedure repeats for time interval with number $j + 1$.

To model this kind of behavior we introduce the following definition (Fig. 11.1). Denote vectors $p_{j,0} = p(t_0 + j\Delta t)$, $p_{j,1} = p(t_0 + (j+1)\Delta t)$.

Definition 11.1 Let $j = 0, \dots, l$. A truncated subgame $\bar{\Gamma}_j(p_{j,0}, t_0 + j\Delta t, t_0 + j\Delta t + \overline{T})$ is defined on the time interval $[t_0 + j\Delta t, t_0 + j\Delta t + \overline{T}]$. The motion equation

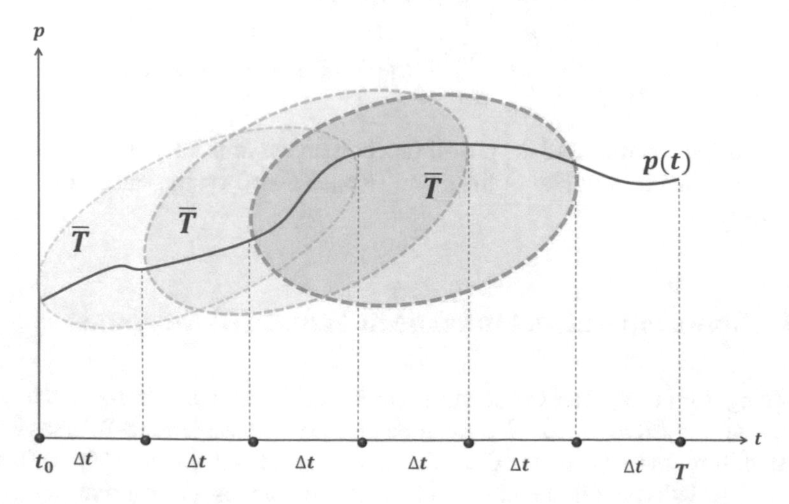

Fig. 11.1 Each oval represents random truncated information, which is known to players during the time interval $[t_0 + j\Delta t, t_0 + (j+1)\Delta t]$, $j = 0, \dots, l$

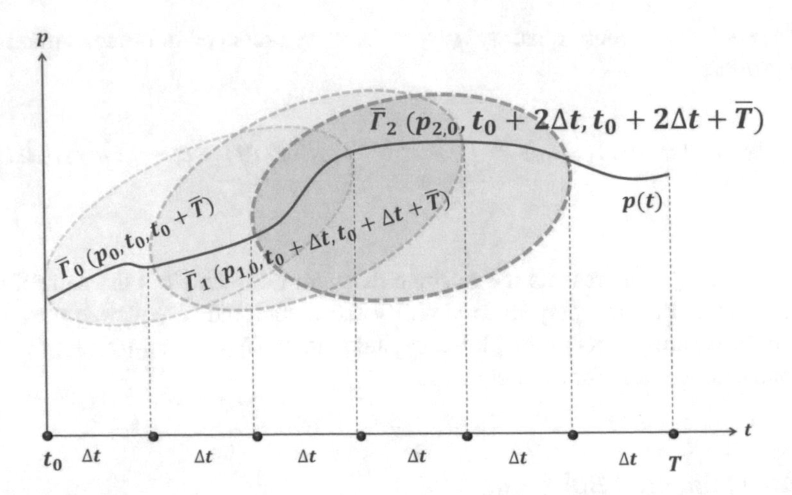

Fig. 11.2 Behaviour of players in the game with truncated information can be modeled using the truncated subgames $\bar{\Gamma}_j(p_{j,0}, t_0 + j\Delta t)$, $j = 0, \ldots, l$

and the initial condition of the truncated subgame $\bar{\Gamma}_j(p_{j,0}, t_0 + j\Delta t, t_0 + j\Delta t + \overline{T})$ have the following form:

$$\dot{p}(t) = s\Big(a_j - d_j \sum_{i \in N} q_i^j(t) - p(t)\Big), \;\; p(t_0 + j\Delta t) = p_{j,0}. \tag{11.4}$$

The payoff function of the player i in truncated subgame j is equal to

$$K_i^j(p_{j,0}, t_0 + j\Delta t, t_0 + j\Delta t + \overline{T}; q_1^j, \ldots, q_n^j) =$$
$$= \int_{t_0 + j\Delta t}^{t_0 + j\Delta t + \overline{T}} e^{-\rho(t - t_0)} \Big[q_i^j(t)(p(t) - c_i - g_i q_i^j(t))\Big] dt. \tag{11.5}$$

The motion equation and the payoff function on the time interval $[t_0 + j\Delta t, t_0 + j\Delta t + \overline{T}]$ coincide with that of the game $\Gamma(p_0, T - t_0)$ on the same time interval (Fig. 11.2).

11.4 Non-cooperative Outcome in Truncated Subgame

According to [1, 9] non-cooperative Nash equilibrium solution of the game $\bar{\Gamma}_j(p_{j,0}, t_0 + j\Delta t, t_0 + j\Delta t + \overline{T})$ can be defined by the Fleming-Bellman-Isaacs partial differential equations. Consider a family of subgames $\bar{\Gamma}_j(p(t), t, t_0 + j\Delta t + \overline{T})$ with payoff structure (11.3) and dynamics (11.4), starting at the time $t \in [t_0 + j\Delta t, t_0 + j\Delta t + \overline{T}]$ with initial state $p(t)$. Let $q_j^{NE}(t, p) =$

$(q_1^{jNE}(t,p),\ldots,q_n^{jNE}(t,p))$ for $t \in [t_0 + j\Delta t, t_0 + j\Delta t + \overline{T}]$ denote a set of feedback strategies that constitutes a Nash equilibrium solution to the game $\bar{\Gamma}_j(p(t), t, t_0 + j\Delta t + \overline{T})$ and $V_i^j(\tau, p) : [t, T] \times R^n \to R$ denote the value function of player $i \in N$ that satisfy the corresponding Bellman-Isaacs-Fleming equations [1, 9].

Theorem 11.1 *Assume there exists a continuously differential function $V_i^j(t,p)$: $[t_0 + j\Delta t, t_0 + j\Delta t + \overline{T}] \times R \to R$ satisfying the partial differential equation*

$$V_t^{j,i}(t,p) = \max_{q_i^j}\left\{e^{-\rho(t-t_0)}\left[q_i^j(p - c_i - g_i q_i^j)\right] + \right.$$

$$\left. + V_p^{j,i}(t,p)s\left(a - d\left[q_i^j(t) + \sum_{k\neq i} q_k^{jNE}(t)\right] - p(t)\right)\right\},\quad i = 1,\ldots,n. \qquad (11.6)$$

where $V_i^j(t_0 + j\Delta t + \overline{T}, p) = 0$. Denote by $q_j^{NE}(t)(t,p)$ controls which maximize right hand side of (11.6). Then $q_j^{NE}(t)(t,p)$ provides a feedback Nash equilibrium in the truncated subgame $\bar{\Gamma}_j(p_{j,0}, t_0 + j\Delta t, t_0 + j\Delta t + \overline{T})$.

Since the considered differential game is a LQ differential game, then the feedback Nash equilibrium is unique (see [1]).

In this game model Bellman function $V_i^j(t,p)$ can be obtained in the form:

$$V_i^j(t,p) = e^{-\rho(t-t_0)}\left[A_i^j(t)p^2 + B_i^j(t)p + C_i^j(t)\right],\quad i = \overline{1,n}. \qquad (11.7)$$

Substituting (11.7) in (11.6) we can determine Nash equilibrium strategies in the following:

$$q_i^{jNE}(t,p) = \frac{(c_i - p) + ds\left[B_i^j(t) + 2A_i^j(t)p\right]}{2g_i},\quad i = \overline{1,n}, \qquad (11.8)$$

where functions $A_i^j(t)$, $B_i^j(t)$, $C_i^j(t)$, $t \in [t_0 + j\Delta t, t_0 + j\Delta t + \overline{T}]$ are defined by the system of differential equations:

$$\dot{A}_i^j(t) = A_i^j(t)\left[\rho + 2s\right] + \frac{(2A_i^j(t)d_j s - 1)^2}{4g_i} - \sum_{k\neq i}\frac{A_i^j(t)d_j s - 2A_i^j(t)A_k^j(t)d^2s^2}{g_k}$$

$$\dot{B}_i^j(t) = B_i^j(t)\left[\rho + s\right] - \frac{c_i}{2g_i} - 2A_i^j(t)a_j s - \sum_{k\neq i}\frac{A_i^j(t)B_k^j(t)d_j^2s^2}{g_k} -$$

$$-\sum_{k\in N}\frac{B_i^j(t)d_j s - A_i^j(t)c_k d_j s - A_k^j(t)B_i^j(t)d_j^2s^2}{g_k}$$

$$\dot{C}_i^j(t) = C_i^j(t)\rho - B_i^j(t)a_j s + \frac{c_i^2 + (B_i^j(t)d_j s)^2}{4g_i} + \sum_{k \neq i} \frac{B_i^j(t)B_k^j(t)d_j^2 s^2}{2g_k} +$$

$$+ \sum_{k \in N} \frac{B_i^j(t)c_k d_j s}{2g_k}$$

with the boundary conditions $A_i^j(t_0 + j\Delta t + \overline{T}) = 0$, $B_i^j(t_0 + j\Delta t + \overline{T}) = 0$ and $C_i^j(t_0 + j\Delta t + \overline{T}) = 0$.

Substituting $q_j^{NE}(t, p)$ (11.8) into (11.4) yields the dynamics of Nash equilibrium trajectory:

$$\dot{p}(t) = s\Big(a - d\sum_{i \in N} q_i^{jNE}(t, p) - p(t)\Big), \; p(t_0 + j\Delta t) = p_{j,0}. \tag{11.9}$$

Denote by $p_j^{NE}(t)$ the solution of system (11.9).

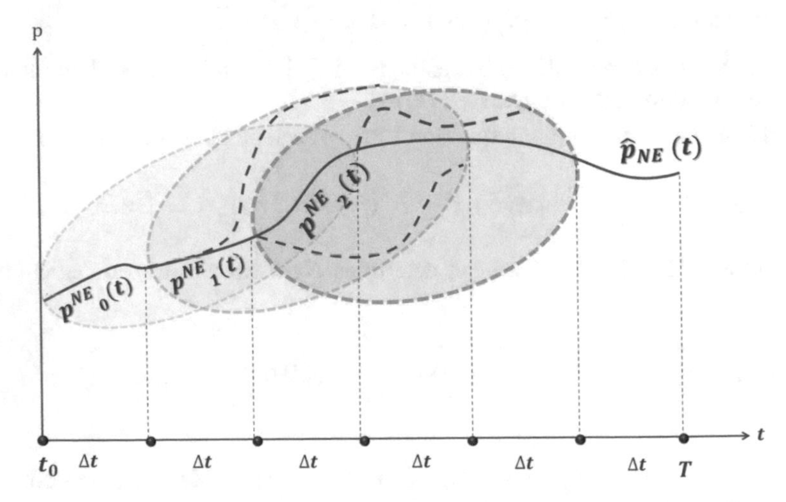

Fig. 11.3 Solid line represents the conditionally non-cooperative trajectory $\{\hat{p}_{NE}(t)\}_{t=t_0}^T$. Dashed lines represent parts of non-cooperative trajectories that are not used in the composition, i.e., each dashed trajectory is no longer optimal in the current truncated subgame

11.5 Information Updating

Suppose that each truncate subgame $\bar{\Gamma}_j(p_{j,0}, t_0+j\Delta t, t_0+j\Delta t+\overline{T})$ develops along $p_j^{NE}(t)$ then the whole non-cooperative game with Looking Forward Approach develops along:

Definition 11.2 Conditionally non-cooperative trajectory $\{\hat{p}_{NE}(t)\}_{t=t_0}^{T}$ is a combination of $p_j^{NE}(t)$ for each truncated subgame $\bar{\Gamma}_j(p_{j,0}^{NE}, t_0 + j\Delta t, t_0 + j\Delta t + \overline{T})$ (Fig. 11.3):

$$\{\hat{p}_{NE}(t)\}_{t=t_0}^{T} = \begin{cases} p_0^{NE}(t),\ t \in [t_0, t_0 + \Delta t), \\ \dots, \\ p_j^{NE}(t),\ t \in [t_0 + j\Delta t, t_0 + (j+1)\Delta t), \\ \dots, \\ p_l^{NE}(t),\ t \in [t_0 + l\Delta t, t_0 + (l+1)\Delta t]. \end{cases} \quad (11.10)$$

Along the conditionally non-cooperative trajectory players receive payoff according to the following formula:

Definition 11.3 Resulting non-cooperative outcome for player $i = 1, \dots, n$ in the game $\Gamma(p_0, T - t_0)$ with Looking Forward Approach has the following form:

$$\hat{V}_i(t, \hat{p}_{NE}(t)) = \sum_{m=j+1}^{l} \left[V_i^m(t_0 + m\Delta t, p_{m,0}^{NE}) - V_i^m(t_0 + (m+1)\Delta t, p_{m,1}^{NE})\right] + \quad (11.11)$$
$$+\left[V_i^j(t, p_j^{NE}(t)) - V_i^j(t_0 + (j+1)\Delta t, p_{j,1}^{NE})\right],\ i \in N.$$

11.6 Numerical Simulation

Game starts from December 2015 and lasts till the summit in Vienna at November 2016. We consider the oil as homogeneous product, we appraise the demand function with parameters of average world oil price and total world oil supply. We calculate the average oil prices for each period based just on the two major trading classifications of brent crude and light crude which are accessible on Finam agency data source. As an initial price we take the average price in December 2015 which is equal to $p_0 = 34.51$.

Date	Total world supply, million barrels per day	Brent, $ for barrel	Light, $ for barrel	Average price, $ for barrel
12.2015	96.411	35.910	33.110	34.510
01.2016	95.875	36.640	33.990	35.315
02.2016	95.420	40.140	37.820	38.980
03.2016	95.294	47.320	45.990	46.655
04.2016	95.400	49.520	48.750	49.135
05.2016	95.187	49.740	48.640	49.190
06.2016	95.954	43.270	41.760	42.515
07.2016	96.891	46.970	45.000	45.985
08.2016	95.894	49.990	48.050	49.020
09.2016	96.001	48.510	46.970	47.740
10.2016	97.362	44.520	43.120	43.820

To estimate a demand function one needs to mine several observations for the same period. Unfortunately, available petroleum statistics cannot provide data with such frequency. As a result, we fixed the parameter of a choke price of the demand function to ensure its linear form and to simplify our work. Being a minimum price with a zero demand by the definition of a choke price, this quantity must be higher than the historical maximum of the oil price. The oil price reached its maximum of $122 for the barrel in May of 2008. It is equal to $282 at the end of October 2016 with adjust for inflation of 10$ as we assumed above. We suppose that fixation the parameter a_j on the level of 300 gives us the approximate choke price. Parameter d_j can be obtained as

$$d_j = (a_j - \hat{p}(t-1)) \sum_{i \in N} q_i(t-1). \tag{11.12}$$

The length of each Δt-time interval is 1 month. Players use appraised demand with parameters a_j and d_j as the forecast for the next $\overline{T} = 3$ periods. We set a value upon the parameters of cost function by using total cost of producing a barrel and average volumes of oil production for our players in 2016 and by fixing the parameter g_i on the level 0.7 for each player and each period. Both the parameters of c_i and g_i remains unchanged during the game. We assume that the speed of adjustment $s = 0.2$ and discount factor $r = 10\%$.

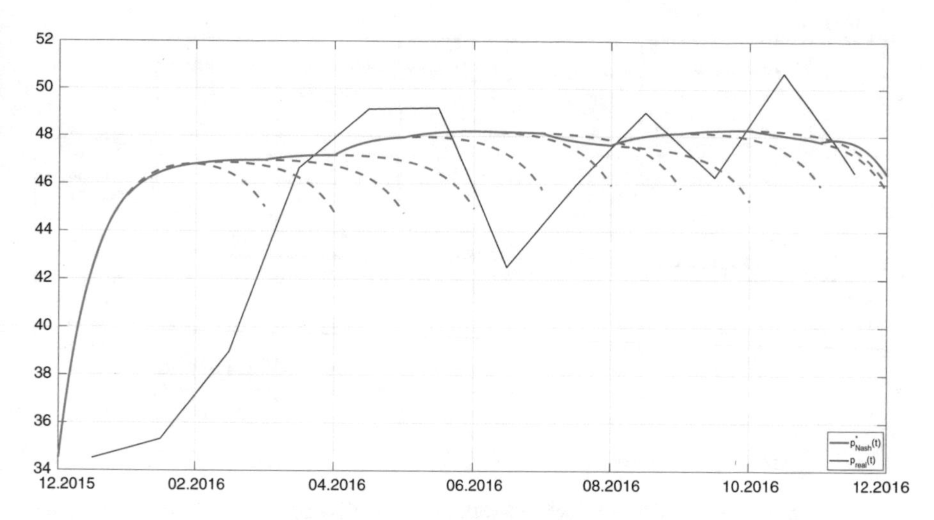

Fig. 11.4 Conditionally non-cooperative trajectory of the oil price $\hat{p}_{NE}(t)$ (thick solid line) with Looking Forward Approach, historical oil price trajectory (thick dotted line)

Date	a	d
12.2015	300	2.717
01.2016	300	2.753
02.2016	300	2.760
03.2016	300	2.735
04.2016	300	2.658
05.2016	300	2.629
06.2016	300	2.634
07.2016	300	2.683
08.2016	300	2.621
09.2016	300	2.617
10.2016	300	2.627

i	Producer	c	g
1	OPEC	3.169	0.7
2	Non-OPEC	17.333	0.7
3	US shale	20.238	0.7
4	US non-shale	18.182	0.7
5	Others	20.867	0.7

On the Fig. 11.4 comparison of conditionally non-cooperative trajectory $\{\hat{p}_{NE}(t)\}_{t=t_0}^{T}$ and historical average oil price dynamics is presented.

On the Fig. 11.5 Nash feedback strategies (11.8) corresponding to the $\{\hat{p}_{NE}(t)\}_{t=t_0}^{T}$ and historical quantities of production of oil are presented for each group of countries.

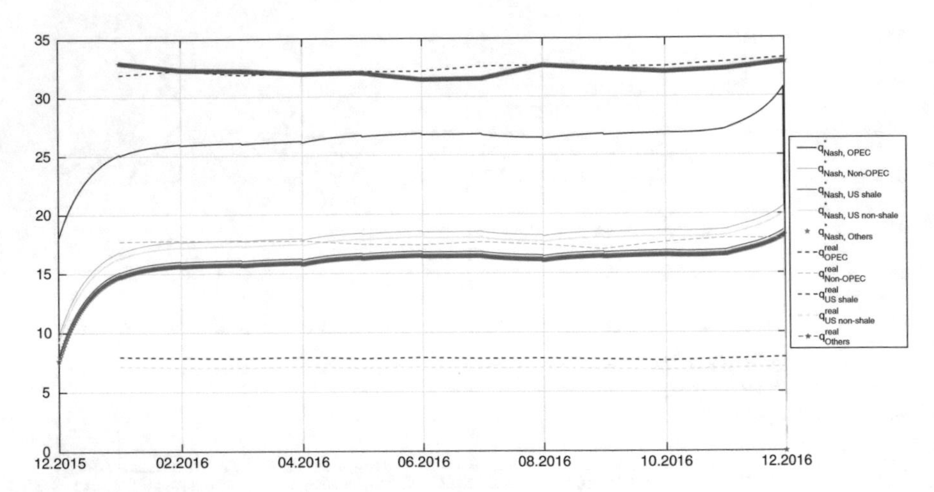

Fig. 11.5 Nash feedback strategies defined with Looking Forward Approach (solid lines), and corresponding historical quantities of production of oil (dashed lines)

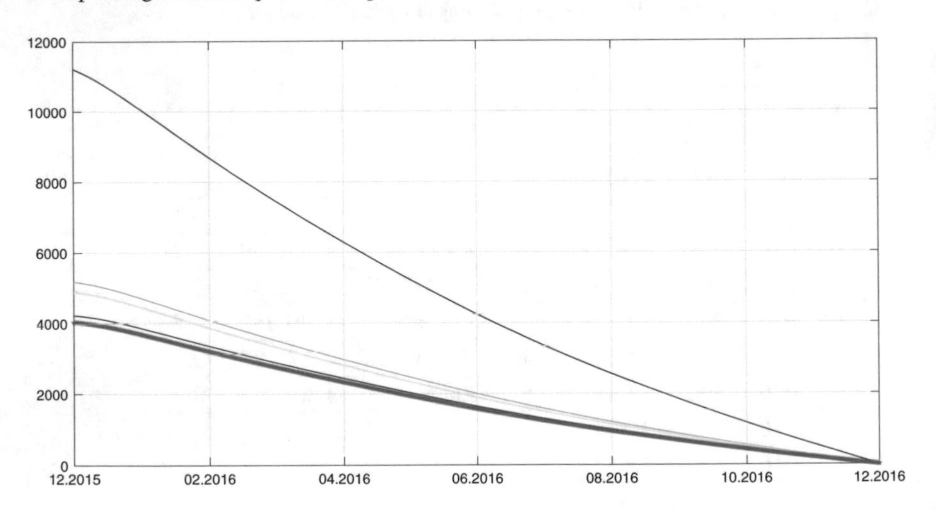

Fig. 11.6 Payoffs of players corresponding to the Nash equilibrium $\hat{V}_i(t, \hat{p}_{NE}(t))$

According to the Figs. 11.4 and 11.5 we can suggest that parameters from the table below c_i and g_i can be used for constructing other models based on this approach.

On the Fig. 11.6 we can see the payoff of players $\hat{V}_i(t, \hat{p}_{NE}(t))$, $i \in N$ corresponding to the Nash equilibrium strategies along the trajectory $\hat{p}_{NE}(t)$.

11.7 Conclusion

Differential game model of oil market is considered. Looking Forward Approach is used for constructing model where information about the process updates dynamically. An attempt has been made for constructing oil market model for time period from December 2015 to November 2016. Namely, we assume that from December 2015 to November 2016 a non-cooperative oligopoly structure was on the oil market. Therefore, we construct non-cooperative game model and adapt it to the real oil price data. Thereafter, at the end of November 2016, the largest oil exporters signed an agreement for reduction of oil extraction to rise the prices. For this case we will continue our research in future papers. Numerical results shows high applicability of Looking Forward Approach for modelling and simulating real life conflict-controlled processes. But there are still a lot of open questions such as defining appropriate value for information horizon, defining form of payoff functions and type of forecast for parameters which is used by the players. Future steps are to construct game model describing the six-month long agreement which was signed by the OPEC countries and eleven non-OPEC countries about the reduction of oil production at the summit in Vienna on November 30, 2016.

References

1. Basar, T., Olsder, G.J.: Dynamic Noncooperative Game Theory. Society for Industrial and Applied Mathematics, Philadelphia (1999)
2. Basar, T., Pierre, B.: H∞-optimal Control and Related Minimax Design Problems. Birkhauser, Basel (1995)
3. Berger, K., Hoel, M., Holden, S., Olsen, O.: The oil market as an oligopoly. Discussion paper. Centr. Bureau Stat. **32**, 1–19 (1988)
4. Cellini, R., Lambertini, L.: A differential oligopoly game with differentiated goods and sticky prices. Eur. J. Oper. Res. **176**, 1131–1144 (2007)
5. Dona, J., Goodwin, G., Seron, M.: Constrained Control and Estimation: An Optimisation Approach. Springer, New York (2005)
6. Engwerda, J.: LQ Dynamic Optimization and Differential Games. Chichester, Wiley (2005)
7. Gromova, E., Petrosian, O.: Control of informational horizon for cooperative differential game of pollution control. In: 2016 International Conference Stability and Oscillations of Nonlinear Control Systems (Pyatnitskiy's Conference), pp. 1–4 (2016)
8. Han, S., Kwon, W.: Receding Horizon Control: Model Predictive Control for State Models. Springer, New York (2005)
9. Isaacs, R.: Differential Games. Wiley, New York (1965)
10. Krasner, S.D.: The great oil sheikdown. Foreign Policy **13**, 123–138 (1973)
11. Ludkovski, M., Sircar, R.: Game theoretic models for energy production. commodities, energy and environmental finance. Fields Inst. Commun. **74**, 317–333 (2015)
12. Moran, T.H.: Managing an oligopoly of would-be sovereigns: the dynamics of joint control and self-control in the international oil industry past, present, and future. Int. Organ. **41**(4), 575–607 (1987)
13. Nash, J.F. Jr.: Equilibrium points in n-person games. Proc. Natl. Acad. Sci. **36**(1), 48–49 (1950)
14. Nash, J.F. Jr.: Non-cooperative games. Ann. Math. **54**(2), 286–295 (1951)

15. Petrosian, O.: Looking forward approach in cooperative differential games. Int. Game Theory Rev. **18**(2), 1–14 (2016)
16. Petrosian, O.: Looking forward approach in cooperative differential games with infinite-horizon. Vestnik Leningrad State Univ. **4**, 18–30 (2016)
17. Petrosian, O., Barabanov, A.: Looking forward approach in cooperative differential games with uncertain-stochastic dynamics. J. Optim. Theory Appl. **172**(1), 328–347 (2017)
18. Petrosian, O., Nastych, M., Volf, D.: Differential game of oil market with moving informational horizon and nontransferable utility. In: Constructive Nonsmooth Analysis and Related Topics (dedicated to the memory of V.F. Demyanov) (CNSA), 2017, pp. 1–4. Springer, New York (2017)
19. Rawlings, J., Mayne, D.: Model Predictive Control: Theory and Design. Nob Hill Publishing, Madison (2009)
20. Salant, S.: Exhaustible resources and industrial structure: a Nash-Cournot approach to the world oil market. J. Polit. Econ. **84**, 1079–1094 (1976)
21. Wang, L.: Model Predictive Control System Design and Implementation Using MATLAB. Springer, New York (2005)
22. Yeung, D., Petrosian, O.: Cooperative stochastic differential games with information adaptation. In: International Conference on Communication and Electronic Information Engineering (CEIE 2016) (2017)
23. Yeung, D., Petrosian, O.: Infinite horizon dynamic games: a new approach via information updating. Int. Game Theory Rev. **20**(1), 1–23 (2017)

Chapter 12
S-strongly Time-Consistency in Differential Games

Leon A. Petrosyan and Ekaterina V. Gromova

Abstract In the paper the definition of S-strongly time-consistency in differential games is introduced. The approach of the construction of S-strong time-consistent subcore of the classical core on the base of characteristic function obtained by normalization of classical characteristic function is formulated. Its relation to another characteristic function obtained by an integral extension of the original characteristic function is studied.

12.1 Introduction

Dynamic games theory has many applications in different areas (see [1, 2, 11, 13]). Particularly important are cooperative differential games that are widely used for modeling the situations of joint decision taking by many agents. When considering such problems, the realizability of cooperative solution in time turns out to be one of the central issues.

As it was mentioned earlier, [9, 13], an attempt to transfer the optimality principles (cooperative solution) from the static cooperative game theory to n-persons differential games leads to dynamically unstable (time inconsistent) optimality principles that renders meaningless their use in differential games. Hence, the notion of time consistent cooperative solution and an approach to determining such cooperative solution was proposed in [9].

A strong time-consistent optimality principle has even more attractive property. Namely, strong time consistency of the core considered as a cooperative solution implies that a single deviation from the chosen imputation taken from the core in favor of another imputation from the core does not lead to non-realizability of the cooperative agreement (the core) defined for the whole duration of the game, [7]. This implies that the overall payment for players will also be contained in the core.

L. A. Petrosyan · E. V. Gromova (✉)
Saint Petersburg State University, Saint Petersburg, Russia
e-mail: l.petrosyan@spbu.ru; e.v.gromova@spbu.ru

L. A. Petrosyan et al. (eds.), *Frontiers of Dynamic Games*,
Static & Dynamic Game Theory: Foundations & Applications,
https://doi.org/10.1007/978-3-319-92988-0_12

In this paper, a cooperative differential game with the set of players N is studied in general setting on the finite time horizon. The work is of fundamental character, but may potentially have a big practical impact because it proposes a constructive approach to the definition of a new cooperative solution which satisfies the condition of strong time-consistency.

In the paper, we study different approaches to constructing a strongly time-consistent cooperative solution, which are based on the use of additional procedures for the imputation distribution on the time interval $[t_0, T]$ (IDP) for classical cooperative solution, i.e., the core, and on the transformations of the classical characteristic function $V(S, \cdot)$, $S \subseteq N$. Furthermore, we present results illustrating the relationship between the introduced concepts.

In [7, 8], it was shown that it is possible to define a new type of characteristic function $\bar{V}(S, \cdot)$ on the base of integral transformation of the classical characteristic function $V(S, \cdot)$ such that the resulting optimality principles are strongly time-consistent.

In [10], another approach to the construction of the characteristic function $\hat{V}(S, \cdot)$ on the base of normalizing transformation of $V(S, \cdot)$ had been suggested and it was shown that the core constructed on the base of the new $\hat{V}(S, \cdot)$ belongs to the classical core.

In this contribution we track the connection between the optimality principles constructed on the basis of classical characteristic function and the constructions resulted from the new types of characteristic function. We study the property of strong-time consistency for all constructed optimality principles and suggest a modification of the notion of strong time-consistency as described below.

The notion of S-strong time-consistency can be considered as a weakening of the strong time-consistency and means the following: after a single deviation from the chosen imputation from the optimality principle $\hat{M}(x_0, t_0)$ in favor of another imputation from the same optimality principle $\hat{M}(x^*(t), t)$ the resulting imputation will belong to a larger set $M(x_0, t_0) \supset \hat{M}(x_0, t_0)$ even if the resulting solution does not belong to the initial set $\hat{M}(x_0, t_0)$. Note that S-strong time-consistency of the cooperative solution is considered with respect to another (bigger) set, hence the prefix S-.

The construction of a S-strongly dynamically stable subcore on the base of all described approaches is presented.

12.2 List of Key Notations

x	trajectory of the system
u	control vector $u = \{u_1, \ldots, u_n\}$
$K_i(x, t, u)$	payoff of the player i in a subgame starting at t from x
N	set of players (the grand-coalition)
S	subset of players (a coalition), $S \subseteq N$
$V(S, x, t)$	basic characteristic function (c.f.)

$\bar{V}(S, x, t)$	an integral extension of the c.f. V
$\hat{V}(S, x, t)$	a normalized c.f. V
$L(x, t)$	set of imputations associated with V
$\bar{L}(x, t)$	set of imputations associated with $\bar{V}$
$C(x, t)$	core associated with V
$\bar{C}(x, t)$	core associated with $\bar{V}$
$\hat{C}(x, t)$	core associated with $\hat{V}$

12.3 Basic Game

Consider the differential game $\Gamma(x_0, t_0)$ starting from the initial position x_0 and evolving on time interval $[t_0, T]$. The equations of the system's dynamics have the form

$$\dot{x} = f(x, u_1, \ldots, u_n), \; x(t_0) = x_0,$$
$$u_i \in U_i \subset \text{Comp}R^m, \; x \in R^l, \; i = 1, \ldots, n. \tag{12.1}$$

The players' payoffs are

$$K_i(x, t_0; u_1, \ldots, u_n) = \int_{t_0}^{T} h_i(x(t))dt, \; i = 1, \ldots, n, \; h_i(\cdot) \geq 0,$$

where $x(t)$ is the solution of system (12.1) with controls $u_1, \ldots, u_n$. The non-negativeness of the utility function $h_i(\cdot)$ is an important assumption of the model.

It is furthermore assumed that the system (12.1) satisfies all the conditions guaranteeing the existence and uniqueness of solution $x(t)$ on the time interval $[t_0, T]$ for all admissible measurable open loop controls $u_1(t), \ldots, u_n(t), \; t \in [t_0, T]$. Let there exist a set of controls

$$u^*(t) = \{u_1^*(t), \ldots, u_n^*(t)\}, \; t \in [t_0, T]$$

such that

$$\max_{u_1,\ldots,u_n} \sum_{i=1}^{n} K_i(x_0, t_0; u_1(t), \ldots, u_n(t)) = \sum_{i=1}^{n} \int_{t_0}^{T} h_i(x^*(t))dt = V(N; x_0, t_0). \tag{12.2}$$

The solution $x^*(t)$ of the system (12.1) corresponding to $u^*(t)$, is called the *cooperative trajectory*.

In cooperative game theory, [6], it is assumed that the players initially agree upon the use of the controls $u^*(t) = \{u_1^*(t), \ldots, u_n^*(t)\}$ and hence, in the cooperative formulation the differential game $\Gamma(x_0, t_0)$ always develops along the cooperative trajectory $x^*(t)$.

Let $N = \{1, \dots, i, \dots, n\}$ be the set of all players. Let $S \subseteq N$ and denote by $V(S; x_0, t_0)$ the characteristic function of the game $\Gamma(x_0, t_0)$, [6]. Note that $V(N; x_0, t_0)$ is calculated by the formula (12.2). Let $V(S; x^*(t), t)$, $S \subseteq N$, $t \in [t_0, T]$ be a (superadditive) characteristic function of the subgame $\Gamma(x_0, t_0)$ constructed by any relevant method [5].

So, we state the following properties for characteristic function:

$$V(\emptyset; x_0, t_0) = 0;$$

$$V(N; x_0, t_0) = \sum_{i=1}^{n} \int_{t_0}^{T} h_i(x^*(\tau))d\tau;$$

$$V(S_1 \cup S_2; x_0, t_0) \geq V(S_1; x_0, t_0) + V(S_2; x_0, t_0). \tag{12.3}$$

For the sake of definiteness we can assume that the characteristic function $V(S; x_0, t_0)$ is constructed as the value of a zero-sum differential game based on the game $\Gamma(x_0, t_0)$ and played between the coalition S (the first maximizing player) and the coalition $N \setminus S$ (the second minimizing player), and in each situation the payoff of coalition S is assumed to be the sum of players' payoffs from this coalition.

Consider the family of subgames $\Gamma(x^*(t), t)$ of game $\Gamma(x_0, t_0)$ along the cooperative trajectory $x^*(t)$, i.e. a family of cooperative differential games from the initial state $x^*(t)$ defined on the interval $[t, T]$, $t \in [t_0, T]$ and the payoff functions

$$K_i(x^*(t), t; u_1, \dots, u_n) = \int_{t}^{T} h_i(x(\tau))d\tau, \ i = 1, \dots, n,$$

where $x(\tau)$ is a solution of (12.1) from initial position $x^*(t)$ with controls $u_1, \dots, u_n$.

Let $V(S; x^*(t), t)$, $S \subseteq N$, $t \in [t_0, T]$ be the (superadditive) characteristic function of subgame $\Gamma(x^*(t), t)$, s.t. the properties (12.3) hold. For $V(N; x^*(t), t)$, the Bellman optimality condition along $x^*(t)$ holds, i.e.

$$V(N; x_0, t_0) = \int_{t_0}^{t} \sum_{i=1}^{n} h_i(x^*(\tau))d\tau + V(N; x^*(t), t).$$

12.4 Construction of a Core with a New Characteristic Function

Define the new characteristic function $\bar{V}(S; x_0, t_0)$, $S \subseteq N$, similar to [7, 8], by the formula

$$\bar{V}(S; x_0, t_0) = \int_{t_0}^{T} V(S; x^*(\tau), \tau) \frac{\sum_{i=1}^{n} h_i(x^*(\tau))}{V(N; x^*(\tau), \tau)} d\tau. \tag{12.4}$$

Similarly, we define for $t \in [t_0, T]$

$$\bar{V}(S; x^*(t), t) = \int_t^T V(S; x^*(\tau), \tau) \frac{\sum_{i=1}^n h_i(x^*(\tau))}{V(N; x^*(\tau), \tau)} d\tau. \tag{12.5}$$

One can readily see that the function $\bar{V}(S; x_0, t_0)$ has the all properties (12.3) of the characteristic function of the game $\Gamma(x_0, t_0)$. Indeed,

$$\bar{V}(\emptyset; x_0, t_0) = 0,$$

$$\bar{V}(N; x_0, t_0) = V(N; x_0, t_0) = \sum_{i=1}^n \int_{t_0}^T h_i(x^*(\tau)) d\tau,$$

$$\bar{V}(S_1 \cup S_2; x_0, t_0) \geq \bar{V}(S_1; x_0, t_0) + \bar{V}(S_2; x_0, t_0).$$

for $S_1,\ S_2 \subset N,\ S_1 \cap S_2 = \emptyset$ (here we use the superadditivity of function $V(S; x_0, t_0)$). The similar statement is true also for function $\bar{V}(S; x^*(t), t)$ which is defined as the characteristic function of $\Gamma(x^*(t), t)$.

Let $L(x_0, t_0)$ be the set of imputations in $\Gamma(x_0, t_0)$ determined by characteristic function of $V(S; x_0, t_0)$, $S \subseteq N$, i.e.

$$L(x_0, t_0) = \left\{ \xi = \{\xi_i\} : \sum_{i=1}^n \xi_i = V(N; x_0, t_0),\ \xi_i \geq V(\{i\}; x_0, t_0) \right\}. \tag{12.6}$$

Similarly, we define the set of imputations $L(x^*(t), t)$, $t \in [t_0, T]$ in the subgame $\Gamma(x^*(t), t)$:

$$\begin{aligned} L(x^*(t), t) = \big\{ \xi^t = \{\xi_i^t\} : \textstyle\sum_{i=1}^n \xi_i^t = V(N; x^*(t), t), \\ \xi_i^t \geq V(\{i\}; x^*(t), t), i \in N \big\}. \end{aligned} \tag{12.7}$$

We denote the set of imputations defined by characteristic functions $\bar{V}(S; x_0, t_0)$ and $\bar{V}(S; x^*(t), t)$ by $\bar{L}(x_0, t_0)$ and $\bar{L}(x^*(t), t)$, respectively. These imputations are defined in the same way as (12.6), (12.7).

Let $\xi(t) = \{\xi_i(t)\} \in L(x^*(t), t)$ be the integrable selector [9], $t \in [t_0, T]$, define

$$\bar{\xi}_i = \int_{t_0}^T \xi_i(\tau) \frac{\sum_{i=1}^n h_i(x^*(\tau))}{V(N; x^*(\tau), \tau)} d\tau, \tag{12.8}$$

$$\bar{\xi}_i^t = \int_t^T \xi_i(\tau) \frac{\sum_{i=1}^n h_i(x^*(\tau))}{V(N; x^*(\tau), \tau)} d\tau, \tag{12.9}$$

where $t \in [t, T]$ and $i = 1, \ldots, n$.

One can see that

$$\sum_{i=1}^{n} \bar{\xi}_i = V(N; x_0, t_0),$$

$$\sum_{i=1}^{n} \bar{\xi}_i^t = V(N; x^*(t), t).$$

Moreover, we have

$$\bar{\xi}_i \geq \int_{t_0}^{T} V(\{i\}; x^*(\tau), \tau) \frac{\sum_{i=1}^{n} h_i(x^*(\tau))}{V(N; x^*(\tau), \tau)} d\tau = \bar{V}(\{i\}; x_0, t_0)$$

and similarly

$$\bar{\xi}_i^t \geq \bar{V}(\{i\}; x^*(t), t), \ i = 1, \ldots, n, \ t \in [t_0, T],$$

i.e. the vectors $\bar{\xi} = \{\bar{\xi}_i\}$ and $\bar{\xi}^t = \{\bar{\xi}_i^t\}$ are imputations in the games $\Gamma(x_0, t_0)$ and $\Gamma(x^*(t), t), t \in [t_0, T]$, respectively, if the functions $\bar{V}(S; x_0, t_0)$ and $\bar{V}(S; x^*(t), t)$ are used as characteristic functions.

We have that $\bar{\xi} \in \bar{L}(x_0, t_0)$ and $\bar{\xi}^t \in \bar{L}(x^*(t), t)$.

Denote by $C(x_0, t_0) \subset L(x_0, t_0)$, $C(x^*(t), t) \subset L(x^*(t), t)$, $t \in [t_0, T]$, the core of the game $\Gamma(x_0, t_0)$ and of the subgame $\Gamma(x^*(t), t)$, respectively (it is assumed that the sets $C(x^*(t), t)$, $t \in [t_0, T]$, are not empty along the cooperative trajectory $x^*(t)$). For an application of the core in differential games see also [3].

So, we have

$$C(x_0, t_0) = \{\xi = \{\xi_i\}, s.t. \sum_{i \in S} \xi_i \geq V(S; x_0, t_0), \ \sum_{i \in N} \xi_i = V(N; x_0, t_0), \ \forall S \subset N\}.$$

Let further $\tilde{C}(x_0, t_0)$ and $\tilde{C}(x^*(t), t)$, $t \in [t_0, T]$ be the core of the game $\Gamma(x_0, t_0)$ and of $\Gamma(x^*(t), t)$, constructed using the characteristic function $\bar{V}(S; x, t_0)$, defined by the formulas (12.4) and (12.5). Thus, $\tilde{C}(x_0, t_0)$ is the set of imputations $\{\tilde{\xi}_i\}$ such that

$$\sum_{i \in S} \tilde{\xi}_i \geq \bar{V}(S; x_0, t_0), \ \ \forall S \subset N; \qquad \sum_{i \in N} \tilde{\xi}_i = \bar{V}(N; x_0, t_0) = V(N; x_0, t_0) \tag{12.10}$$

and $\tilde{C}(x^*(t), t)$ is the set of imputations $\{\tilde{\xi}_i^t\}$, s.t.

$$\sum_{i \in S} \tilde{\xi}_i^t \geq \bar{V}(S; x^*(t), t), \ \ \forall S \subset N; \qquad \sum_{i \in N} \tilde{\xi}_i^t = \bar{V}(N; x^*(t), t) = V(N; x^*(t), t).$$

Let in the formulas (12.8) and (12.9) $\xi(t)$ be an integrable selector, $\xi(t) \in C(x^*(t), t_0)$, $t \in [t_0, T]$. Define the set

$$\bar{C}(x_0, t_0) = \left\{\bar{\xi} : \bar{\xi} = \int_{t_0}^{T} \xi(\tau) \frac{\sum_{i=1}^{n} h_i(x^*(\tau))}{V(N; x^*(\tau), \tau)} d\tau\}, \ \forall \xi(\tau) \in C(x^*(\tau), \tau)\right\}.$$

Similarly, we define

$$\bar{C}(x^*(t), t) = \left\{\bar{\xi}^t : \bar{\xi}^t = \int_{t}^{T} \xi(\tau) \frac{\sum_{i=1}^{n} h_i(x^*(\tau))}{V(N; x^*(\tau), \tau)} d\tau\}, \ \forall \xi(\tau) \in C(x^*(\tau), \tau)\right\}.$$

We have the following lemma.

Lemma 12.1

$$\bar{C}(x_0, t_0) \subseteq \tilde{C}(x_0, t_0), \qquad \bar{C}(x^*(t), t) \subseteq \tilde{C}(x^*(t), t), \qquad \forall t \in [t_0, T].$$

Proof To prove this lemma, we use the necessary and sufficient conditions for imputations from the core (12.10).

We have $\forall \bar{\xi} \in \bar{C}(x_0)$:

$$\sum_{i \in S} \bar{\xi}_i = \sum_{i \in S} \int_{t_0}^{T} \xi_i(\tau) \frac{\sum_{i=1}^{n} h_i(x^*(\tau))}{V(x^*(\tau), \tau, N)} d\tau.$$

For imputations from the (basic) core $C(x^*(t), t)$ we have

$$\sum_{i \in S} \xi_i(t) \geq V(S, x^*(t), t), \quad \forall S \subset N.$$

Hence,

$$\sum_{i \in S} \bar{\xi}_i \geq \bar{V}(S, x_0, t_0), \quad \forall S \subset N,$$

and $\bar{C}(x_0) \subseteq \tilde{C}(x_0)$.

The inclusion $\bar{C}(x^*(t), t) \subseteq \tilde{C}(x^*(t), t)$, $\forall t \in [t_0, T]$ can be proved in a similar way. □

Moreover, we also have the converse result.

Lemma 12.2

$$\tilde{C}(x_0, t_0) \subseteq \bar{C}(x_0, t_0); \qquad \tilde{C}(x^*(t), t) \subseteq \bar{C}(x^*(t), t), \qquad \forall t \in [t_0, T].$$

Proof We show that for each imputation $\tilde{\xi}_i \in \tilde{C}(x_0, t_0)$, $\tilde{\xi}_i^t \in \tilde{C}(x^*(t), t)$ there exists an integrable selector $\xi(t) \in C(x^*(t), t)$, $t \in [t_0, T]$ such that

$$\tilde{\xi}_i = \int_{t_0}^{T} \frac{\xi_i(\tau) \sum_{i \in N} h_i(x^*(\tau))}{V(N; x^*(\tau), \tau)} d\tau,$$

$$\tilde{\xi}_i^t = \int_{t}^{T} \frac{\xi_i(\tau) \sum_{i \in N} h_i(x^*(\tau))}{V(N; x^*(\tau), \tau)} d\tau,$$

$$i = 1, \ldots, n.$$

Since $\tilde{\xi}^t$ is an imputation, we have

$$\tilde{\xi}_i^t \geq \bar{V}(\{i\}, x^*(t), t) = \int_t^T \frac{V(\{i\}; x^*(\tau), \tau)}{V(N; x^*(\tau), \tau)} \sum_{i \in N} h_i(x^*(\tau)) d\tau.$$

Moreover, by summing up we get

$$\bar{V}(N; x^*(t), t) = \sum_{i=1}^{n} \tilde{\xi}_i^t.$$

The non-negativeness of the utility functions $h_i(\cdot)$ implies that there exist $\alpha_i \geq 0$, $i = 1, \ldots, n$ such that

$$\tilde{\xi}_i^t = \int_t^T \frac{\alpha_i(\tau) + V(\{i\}; x^*(\tau), \tau))}{V(N; x^*(\tau), \tau)} \sum_{i \in N} h_i(x^*(\tau)) d\tau,$$

and

$$\frac{\sum_{i=1}^{n} (\alpha_i(\tau) + V(\{i\}; x^*(\tau), \tau))}{V(N; x^*(\tau), \tau)} = 1.$$

Obviously, that $\xi(\tau) = \{\xi_i(\tau) = \alpha_i(\tau) + V(\{i\}; x^*(\tau), \tau))\}$ is an imputation in the game with the characteristic function $V(S; x^*(\tau), \tau))$. But we can also prove that $\xi(\tau) = \{\xi_i(\tau) = \alpha_i(\tau) + V(\{i\}; x^*(\tau), \tau))\}$ belongs to the core $C(x^*(\tau), \tau)$. For $\tilde{\xi}^t \in \tilde{C}(x^*(t), t)$ we have

$$\sum_{i \in S} \tilde{\xi}_i^t = \int_t^T \frac{\sum_{i \in S} (\alpha_i(\tau) + V(\{i\}; x^*(\tau), \tau)))}{V(N; x^*(\tau), \tau)} \sum_{i \in N} h_i(x^*(\tau)) d\tau$$

$$\geq \bar{V}(S, x^*(t), t) = \int_t^T \frac{V(S; x^*(\tau), \tau)}{V(N; x^*(\tau), \tau)} \sum_{i \in N} h_i(x^*(\tau)) d\tau,$$

and hence we get

$$\sum_{i \in S}(\alpha_i(\tau) + V(\{i\}; x^*(\tau), \tau))) \geq V(S; x^*(\tau), \tau).$$

The lemma is proved. □

The preceding results imply that

$$\tilde{C}(x^*(t), t) \equiv \bar{C}(x^*(t), t), \ \forall t \in [t_0, T].$$

It means, that the core $\tilde{C}(x_0, t_0)$ constructed by using characteristic function $\bar{V}$ coincides with the set of imputations $\bar{C}(x_0, t_0)$ constructed by formula (12.8) for any imputation $\xi(t)$ from the initial core $C(x^*(t), t)$. Later on we will use the unified notation $\bar{C}(x_0, t_0)$ for both sets.

12.5 Strong Time-Consistency

The property of strong dynamic stability (strong time consistency) coincides with the property of dynamic stability (time consistency) for scalar-valued principles of optimality such as the Shapley value [8] or the "proportional solution". However, for set-valued principles of optimality it has significant and non-trivial sense, which is that any optimal behavior in the subgame with the initial conditions along the cooperative trajectory computed at some intermediate time $t \in [t_0, T]$, together with optimal behavior on the time interval $[t, T]$ is optimal in the problem with the initial condition t_0. This property is almost never fulfilled for such set-valued principles of optimality as the core or the NM-solution.

Let us formulate the definition of strong time-consistency for an arbitrary optimality principle $M(x_0, t_0)$ based on previous results, [9]. A slightly different definition was given in [4].

Introduce the subset $M(x_0, t_0)$ of the imputation set $L(x_0, t_0)$ as the optimality principle in the cooperative game $\Gamma(x_0, t_0)$. $M(x_0, t_0)$ can be a core, a NM-solution, a Shapley value or another one. Similarly, we define this set for all subgames $\Gamma(x^*(t), t)$ along the cooperative trajectory $x^*(t)$.

Definition 12.1 The solution (optimality principle) $M(x_0, t_0)$ is said to be strongly time-consistent in the game $\Gamma(x_0, t_0)$ if

1. $M(x^*(t), t) \neq \emptyset, t \in [t_0, T]$.
2. for any $\xi \in M(x_0, t_0)$ there exists a vector-function $\beta(\tau) \geq 0$ such that

$$M(x_0, t_0) \supset \int_{t_0}^{t} \beta(\tau) d\tau \oplus M(x^*(t), t),$$

$\forall t \in [t_0, T]$, $\int_{t_0}^{T} \beta(t) dt = \xi \in M(x_0, t_0)$.

Here symbol $\oplus$ is defined as follows. Let $a \in R^n$, $B \subset R^n$, then

$$a \oplus B = \{a + b : b \in B\}.$$

Let us consider the core $\bar{C}(x_0, t_0)$ as the set $M(x_0, t_0)$. Thus we have the following lemma.

Lemma 12.3 *$\bar{C}(x_0, t_0)$ is a strongly time-consistent optimality principle.*

Proof From the definition of the set $\bar{C}(x_0, t_0)$ we have that any imputation $\bar{\xi} \in \bar{C}(x_0, t_0)$ has the form (12.8). Then for any $\bar{\xi} \in \bar{C}(x_0, t_0)$ there exists

$$\bar{\beta}_i(t) = \xi_i(t) \frac{\sum_{i \in N} h_i(x^*(t)}{V(N; x^*(t), t)} \geq 0, \ i = 1, \ldots, , t \in [t_0, T]$$

such that $\bar{\xi} = \int_{t_0}^T \bar{\beta}(t)dt \in \bar{C}(x_0, t_0)$.

Let us take another imputation $\hat{\xi}^t$ from the core $\bar{C}(x^*(t), t)$. Then according to the definition of the set $\bar{C}(x^*(t), t)$ we have that there exists a selector $\hat{\xi}(t)$ from the initial basic core $C(x^*(t), t)$, i.e. $\hat{\xi}(t) \in C(x^*(t), t)$ such that

$$\hat{\beta}_i(t) = \hat{\xi}_i(t) \frac{\sum_{i \in N} h_i(x^*(t)}{V(N; x^*(t), T - t)} \geq 0, \ i = 1, \ldots, N, \ t \in [t_0, T],$$

such that $\hat{\xi}^t = \int_t^T \hat{\beta}(t)dt \in \bar{C}(x^*(t), t)$.

Let us consider the vector-function

$$\check{\xi}(\tau) = \begin{cases} \xi(\tau) & \tau \in [t_0, t], \\ \hat{\xi}(\tau), & \tau \in (t, T], \end{cases} \tag{12.11}$$

It is obvious that $\check{\xi}(\tau) \in C(x^*(\tau), \tau)$, $\forall \tau \in [t_0, T]$. Then we have a new vector

$$\check{\xi} = \int_{t_0}^t \bar{\beta}(\tau)d\tau + \hat{\xi}^t = \int_{t_0}^T \check{\xi}(\tau) \frac{\sum_{i \in N} h_i(x^*(\tau)}{V(N; x^*(\tau), \tau)} d\tau,$$

where $\check{\xi}(\tau) \in C(x^*(\tau), \tau)$, $\forall \tau \in [t_0, T]$.

From the definition of the set $\bar{C}(x_0, t_0)$ we have that new vector $\check{\xi} \in \bar{C}(x_0, t_0)$. The vector $\hat{\xi}^t$ had been taken from the core $\bar{C}(x^*(t), t)$ arbitrarily.

So, we have shown that

$$\bar{C}(x_0, t_0) \supset \int_{t_0}^T \xi(t) \frac{\sum_{i \in N} h_i(x^*(\tau))}{V(N; x^*(\tau), \tau)} d\tau \oplus \bar{C}(x^*(t), t),$$

$t \in [t_0, T]$.

The lemma is proved. □

The value

$$\xi_i(t)\frac{\sum_{i\in N} h_i(x^*(t)}{V(N; x^*(t), t)} \geq 0$$

is interpreted as the rate at which the ith player's component of the imputation, i.e., $\bar{\xi}_i$, is distributed over the time interval $[t_0, T]$.

12.6 S-strongly Time-Consistency

As above we consider the subset $M(x_0, t_0)$ of the imputation set $L(x_0, t_0)$ as the optimality principle in the cooperative game $\Gamma(x_0, t_0)$ which can be a core, a NM-solution, a Shapley value or another one. Similarly, we define this set for all subgames $\Gamma(x^*(t), t)$ along the cooperative trajectory $x^*(t)$.

Suppose we have two different optimality principles (cooperative solutions) $M(x_0, t_0)$ and $\hat{M}(x_0, t_0)$ such that

$$\hat{M}(x_0, t_0) \subseteq M(x_0, t_0),$$

$$\hat{M}(x^*(t), t) \subseteq M(x^*(t), t),$$

$\forall t \in [t_0, T]$. Again, we assume that these sets are non-empty during the whole game.

Definition 12.2 The cooperative solution $\hat{M}(x_0, t_0)$ is S-strongly time-consistent (dynamically stable) with respect to the set $M(x_0, t_0)$ if for any imputation $\xi \in \hat{M}(x_0, t_0)$ there exists $\beta(\tau) \geq 0$ such that

$$M(x_0, t_0) \supset \int_{t_0}^{t} \beta(\tau)d\tau \oplus \hat{M}(x^*(t), t),$$

$\forall t \in [t_0, T]$, $\int_{t_0}^{T} \beta(t)dt = \xi \in \hat{M}(x_0, t_0)$.

Here we introduce the definition of strong time-consistency of the optimality principle with respect to another (bigger) set, hence the prefix S-.

This definition means the following: even if the resulting solution will not belong to the initial set $\hat{M}(x_0, t_0)$ it will stay within the set $M(x_0, t_0)$ which includes $\hat{M}(x_0, t_0)$.

From Definition 12.2 we have the following proposition.

Lemma 12.4 *Let the optimality principle $M(x_0, t_0)$ such that $M(x^*(t), t) \neq \emptyset$, $\forall t \in [t_0, T]$ be strongly time-consistent. Then any subset $\hat{M}(x_0, t_0)$, $\hat{M}(x_0, t_0) \subseteq M(x_0, t_0)$ such that $\hat{M}(x^*(t), t) \neq \emptyset$, $\hat{M}(x^*(t), t) \subseteq M(x^*(t), t)$, $\forall t \in [t_0, T]$, is S-strongly time-consistent with respect to $M(x_0, t_0)$.*

12.7 The Construction of a *S*-strongly Dynamically Stable Subcore

In the following we identify a subset $\hat{C}(x_0, t_0)$ of the imputations in the set $\bar{C}(x_0, t_0)$, which would belong to the core $C(x_0, t_0)$, defined on the basis of the classical characteristic function $V(S; x_0, t_0)$.

Consider the value

$$\max_{t \leq \tau \leq T} \frac{V(S; x^*(\tau), \tau)}{V(N; x^*(\tau), \tau)} = \lambda(S, t_0), \tag{12.12}$$

then the following inequality holds

$$\bar{V}(S; x_0, t_0) \leq \lambda(S, t_0) \int_{t_0}^{T} \sum_{i \in N} h_i(x^*(\tau)) d\tau = \lambda(S, t_0) V(N; x_0, t_0). \tag{12.13}$$

We introduce a new characteristic function

$$\hat{V}(S; x_0, t_0) = \lambda(S, t_0) V(N; x_0, t_0). \tag{12.14}$$

Similarly, for $t \in [t_0, T]$ define the respective characteristic function $\hat{V}(S; x^*(t), t)$ as

$$\hat{V}(S; x^*(t), t) = \lambda(S, t) V(N; x^*(t), t), \tag{12.15}$$

where

$$\lambda(S, t) = \max_{t \leq \tau \leq T} \frac{V(S; x^*(\tau), \tau)}{V(N; x^*(\tau), \tau)}. \tag{12.16}$$

From (12.12), (12.13), (12.15) and (12.16) we get

$$\hat{V}(S; x_0, t_0) \geq \bar{V}(S; x_0, t_0),$$

$$\hat{V}(S; x^*(t), t) \geq \bar{V}(S; x^*(t), t).$$

Notice that

$$\bar{V}(N; x_0, t_0) = \hat{V}(N; x_0, t_0),$$

$$\bar{V}(N; x^*(t), t) = \hat{V}(N; x^*(t), t).$$

In addition, for all $S_1, S_2,\ S_1 \subset S_2$

$$\hat{V}(S_1; x^*(t), t) \leq \hat{V}(S_2; x^*(t), t),\ t \in [t_0, T].$$

Unfortunately, the property of superadditivity for the function $\hat{V}(S; x^*(t), t), t \in [t_0, T]$ does not hold in general. One can write

$$\hat{V}(S; x^*(t), t) = \lambda(S, t)V(N; x^*(t), t) =$$
$$= \max_{t \le \tau \le T} \frac{V(S; x^*(\tau), \tau)}{V(N; x^*(\tau), \tau)} V(N; x^*(t), t) \ge$$
$$\ge V(N; x^*(t), t) \frac{V(S; x^*(t), t)}{V(N; x^*(t), t)} \ge V(S; x^*(t), t), \quad S \subset N. \tag{12.17}$$

The preceding inequality leads to the following lemma.

Lemma 12.5 *The following inequality holds true:*

$$V(S; x^*(t), t) \le \hat{V}(S; x^*(t), t), \quad \forall t \in [t_0, T].$$

Denote by $\hat{C}(x_0, t_0)$ the set of imputations $\xi = (\xi_1, \dots \xi_n)$ such that

$$\sum_{i \in S} \xi_i \ge \hat{V}(S; x_0, t_0), \quad \forall S \subset N,$$
$$\sum_{i \in N} \xi_i = \hat{V}(N; x_0, t_0). \tag{12.18}$$

Assume that the set $\hat{C}(x^*(t), t)$ is not empty when $t \in [t_0, T]$. It is easy to see that it is analogous to the core $C(x_0, t_0)$, if the function $\hat{V}(S; x^*(t), t)$ is chosen as the characteristic function.

Thereby we have the statement.

Theorem 12.1 ([10]) *The following inclusion takes place:*

$$\hat{C}(x^*(\tau), \tau) \subset C(x^*(\tau), \tau) \cap \bar{C}(x^*(\tau), \tau), \quad \forall \tau \in [t_0, T]. \tag{12.19}$$

We can also formulate the following Theorem (see Fig. 12.1 for an illustration).

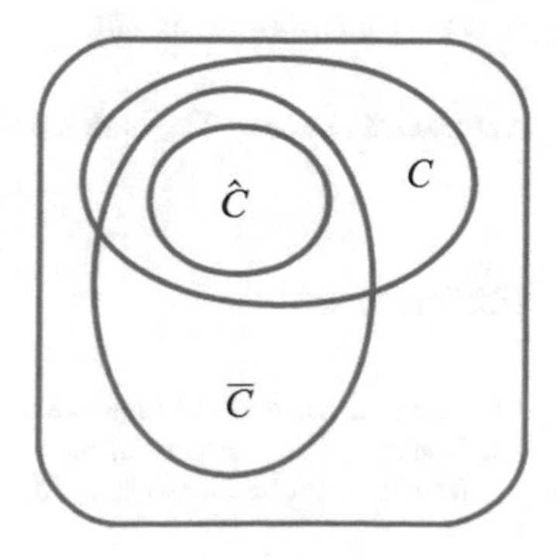

Fig. 12.1 The basic core and the associated cores

Theorem 12.2 *The subcore* $\hat{C}(x_0, t_0) \subset C(x_0, t_0)$ *is S-strongly time-consistent with respect to the set* $\bar{C}(x_0, t_0)$.

Proof From Theorem 12.1 we have that $\hat{C}(x_0, t_0) \subset C(x_0, t_0) \cap \bar{C}(x_0, t_0)$, and hence $\hat{C}(x_0, t_0) \subset \bar{C}(x_0, t_0)$. Lemma 12.3 implies that $\bar{C}(x_0, t_0)$ is strong-time consistent optimality principle.

Finally, the requested result follows from Lemma 12.4. □

The preceding theorem shows that using the new characteristic function (12.14) we constructed a subset of the classical core $C(x_0, t_0)$ (subcore) in the game $\Gamma(x_0, t_0)$ which is S-time-consistent with respect to $\bar{C}(x_0, t_0)$.

This gives an interesting practical interpretation of the subcore $\hat{C}(x_0, t_0)$. Selecting the imputation ξ from the subcore as a solution, we guarantee that if the players—when evolving along the cooperative trajectory in subgames—change their mind by switching to another imputation within the current subcore $\hat{C}(x^*(\tau), \tau)$, the resulting imputation will not leave the set $\bar{C}(x_0, t_0)$ which is also a core in $\Gamma(x_0, t_0)$, but with the characteristic function of the form $\bar{V}(S, \cdot)$ (12.3) obtained by an integral transformation of classical characteristic function $V(S, x^*(\tau), \tau)$ in the games $\Gamma(x^*(\tau), \tau)$.

From Theorem 12.1 it follows that the imputations of type $\hat{C}(x^*(t), t)$ belong to the classical core of the game $\Gamma(x^*(t), t)$ for all $t \in [t_0, T]$. In this sense, Theorem 12.1 establishes a new principle of optimality (cooperative solution).

12.8 Conclusion

In the paper we introduced the definition of S-strong time-consistency in differential games. The approach to the construction of an S-strong time-consistent subcore of the classical core is based on the use of normalized initial characteristic function. We also considered its relation to another characteristic function obtained by an integral extension of the original characteristic function.

We shown that the computed subset of the classical core can be considered as a new optimality principle (cooperative solution) in differential games.

In the future we plan to study the relationship of proposed approach with another constructive approach [12] which allows to identify another subset of the core which is strongly time-consistent.

Acknowledgements The work has been supported by the grant RSF 17-11-01079.

References

1. Basar, T., Olsder, G.J.: Dynamic Noncooperative Game Theory. Academic, London (1982)
2. Dockner, E.J., Jorgensen, S., Long, N.V., Sorger, G.: Differential Games in Economics and Management Science. Cambridge University Press, Cambridge (2000)

3. Gromova, E.: The Shapley value as a sustainable cooperative solution in differential games of three players. In: Recent Advances in Game Theory and Applications. Springer, Berlin (2016)
4. Gromova, E.V., Petrosyan, L.A.: Strongly time-consistent cooperative solution for a differential game of pollution control. UBS **55**, 140–159 (2015)
5. Gromova, E.V., Petrosyan, L.A.: On an approach to constructing a characteristic function in cooperative differential games. Autom. Remote Control **78**(1680), (2017). https://doi.org/10.1134/S0005117917090120
6. Neumann, J., Morgenstern, O.: Theory of Games and Economic Behavior. Princeton University Press, Princeton (1947)
7. Petrosyan, L.A.: Strongly dynamically stable differential optimality principles. Vestnik SPb. Univ. Vyp. 1: Mat. Mekh. Astronom. **4**, 40–46 (1993)
8. Petrosyan, L.A.: Characteristic function in cooperative differential games. Vestnik SPb. Univ. Vyp. 1: Mat. Mekh. Astronom. **1**, 48–52 (1995)
9. Petrosyan, L.A., Danilov, N.N.: Stability of the solutions in nonantagonistic differential games with transferable payoffs. Vestnik Leningrad. Univ. Vyp. 1: Mat. Mekh. Astronom. **1**, 52–59 (1979)
10. Petrosyan, L.A., Pankratova, Y.B.: Construction of strongly time-consistent subcores in differential games with prescribed duration. Trudy Inst. Mat. i Mekh. UrO RAN **23**(1), 219–227 (2017)
11. Petrosyan, L.A., Zenkevich, N.A.: Game Theory. World Scientific, Singapore (2016)
12. Petrosyan, O.L., Gromova, E.V., Pogozhev, S.V.: Strong time-consistent subset of core in cooperative differential games with finite time horizon. Mat. Teor. Igr Pril. **8**(4), 79–106 (2016)
13. Yeung, D.W.K., Petrosyan, L.A.: Subgame Consistent Economic Optimization, p. 395. Birkhauser, New York (2012)

Chapter 13
Characteristic Functions in a Linear Oligopoly TU Game

Artem Sedakov

Abstract We consider a linear oligopoly TU game without transferable technologies in which the characteristic function is determined from different perspectives. In so-called γ-, δ-, and ζ-games, we study the properties of characteristic functions such as monotonicity, superadditivity, and supermodularity. We also show that these games have nonempty cores of a nested structure when the δ-characteristic function is supermodular.

13.1 Introduction

In the definition of a TU game, the characteristic function plays an important role as it measures the worth of any coalition of players, which, in turn, influences players' cooperative payoffs. When the game is initially formulated as a normal-form game, the characteristic function of the corresponding TU game has to be determined. The first study on this problem was done in [12] in which the concepts of so-called α- and β-characteristic functions were proposed. Later in [1], TU games based on these characteristic functions were called α- and β-games, respectively. When transiting from a normal-form game to the corresponding TU game, other studies devoted to the definition of the characteristic function include the concepts of γ-, δ-, and ζ-games proposed in [8, 9], and [5], respectively.[1] All these definitions of the corresponding TU games proceed from the assumption that any coalition of players maximizes the sum of the payoffs of its members.

---

[1] Characteristic functions considered in [9] and [8] were called later the γ- and δ-characteristic functions, in [2] and [10], respectively.

A. Sedakov (✉)
Saint Petersburg State University, Saint Petersburg, Russia
e-mail: a.sedakov@spbu.ru

L. A. Petrosyan et al. (eds.), *Frontiers of Dynamic Games*,
Static & Dynamic Game Theory: Foundations & Applications,
https://doi.org/10.1007/978-3-319-92988-0_13

In this paper, we study the properties of the aforementioned characteristic functions applicable to linear oligopoly games where a finite number of firms producing a homogeneous product compete in a market. The literature on this topic covers two means of determining the cost of a group of firms (coalition): games with transferable technologies (weak synergy) [6, 13, 14], and games without transferable technologies [3, 4, 7]. Here, we follow the second approach as it is consistent with [12] in determining the profit of a coalition.

For the class of oligopoly TU games under consideration, the properties of α- and β-games have already been studied in [3, 7]. We continue studying the properties of γ-, δ-, and ζ-games such as monotonicity, superadditivity, and convexity. The remainder of the paper has the following structure. In Sect. 13.2, we consider a basic linear oligopoly game for which both noncooperative and cooperative solutions are presented. Next, Sect. 13.3 provides closed-form expressions for α-, β-, γ-, δ-, and ζ-characteristic functions, while their properties are examined in Sect. 13.4. The existence of the cores of linear oligopoly TU games based on the aforementioned characteristic functions is discussed in Sect. 13.5. Section 13.6 concludes.

13.2 The Model

We consider a market consisting of firms–competitors producing a homogeneous product. Denote the set of the firms by $N = \{1, \dots, n\}$ with $n \geqslant 2$. Each firm decides on its output, i.e., the quantity it must produce, $q_i \in Q_i = [0, a]$ with $a > 0$, thus the output is the firm's strategy. The market price for the product is determined by the profile of quantities $q = (q_1, \dots, q_n)$ according to the inverse demand function $P(q) = \left(a - \sum_{i \in N} q_i\right)_+ = \max\{0, a - \sum_{i \in N} q_i\}$. Under the assumption of linearity of the cost function $C_i(q_i) = c_i q_i$ with $c_i < a$ for any firm $i \in N$, we obtain the following expression of firm i's profit: $\pi_i(q) = (P(q) - c_i)q_i$. Thus we have a noncooperative normal-form game $(N, \{Q_i\}_{i \in N}, \{\pi_i\}_{i \in N})$. We note that π_i is not concave on $\prod_{j \in N} Q_j$ for any $i \in N$.

For any subset $S \subseteq N$, let $I_S = \{i \in N : i = \arg\min_{j \in S} c_j\}$, a firm belonging to I_S be denoted by i_S, and $c_S = \sum_{j \in S} c_j$.

13.2.1 Nash Equilibrium

A *Nash equilibrium* in the game $(N, \{Q_i\}_{i \in N}, \{\pi_i\}_{i \in N})$ is the profile $q^* = (q_1^*, \dots, q_n^*)$ such that $\pi_i(q^*) \geqslant \pi_i(q_i, q_{-i}^*)$ for any $i \in N$ and $q_i \in [0, a]$, where q_{-i}^* denotes the profile of outputs of all firms except firm i in q^*. For practical reasons, we suppose that the price $P(q)$ is positive under the equilibrium. It is well-known that the Nash equilibrium profile q^* has the form:

$$q_i^* = \frac{a + c_N}{n + 1} - c_i, \quad i \in N. \tag{13.1}$$

From the expression of equilibrium outputs, it follows that $q_i^* + c_i = q_j^* + c_j$ and therefore $q_i^* - q_j^* = c_j - c_i$ for any two firms i and j. To meet a positive equilibrium profile q^*, we additionally require that

$$(n+1)c_i < a + c_N \quad \text{for all} \quad i \in N. \tag{13.2}$$

Under the Nash equilibrium profile q^* we notice the following: $\sum_{i \in N} q_i^* < a$, the profit of firm $i \in N$ is positive and it equals $\pi_i(q^*) = (q_i^*)^2$; the equilibrium price for the product becomes $P(q^*) = \frac{a+c_N}{n+1}$ what exceeds the unit cost of any firm owning to inequality (13.2).

13.2.2 Cooperative Agreement

Now we shall consider the case when firms aim at maximizing the sum of their profits without being restricted in forming one alliance. This means that one must consider the following optimization problem:

$$\max_q \sum_{i \in N} \pi_i(q) \quad \text{subject to } q_i \in [0, a], \quad i \in N. \tag{13.3}$$

For practical reasons, we isolate the case when the price $P(q)$ is positive under the solution. Otherwise, when this price equals zero, the sum to be maximized will be nonpositive. The optimal solution of problem (13.3) will be denoted by $\bar{q} = (\bar{q}_1, \dots, \bar{q}_n)$ and called the *cooperative agreement*. The optimization problem (13.3) may be written in an alternative form:

$$\max_{q_1,\dots,q_n} \sum_{i \in N} (a - c_i) q_i - \left(\sum_{i \in N} q_i \right)^2 \tag{13.4}$$

$$\text{subject to} \quad q_i \in [0, a], \quad i \in N.$$

To analyze both the solution and the value of problem (13.4), we will use the following statement.

Proposition 13.1 *Let $z, z_1, \dots, z_k$ be real numbers such that $z \geqslant z_1 \geqslant \dots \geqslant z_k > 0$. The value of the constrained optimization problem*

$$\max_{x_1,\dots,x_k} \sum_{i=1}^k z_i x_i - \left(\sum_{i=1}^k x_i \right)^2 \tag{13.5}$$

$$\textit{subject to} \quad x_i \in [0, z], \quad i = 1, \dots, k,$$

equals $z_1^2/4$. This value is attained at $x = (z_1/2, 0, \dots, 0)$.

Note that Proposition 13.1 does not list all optimal solutions. For example, when $z_1 = z_2$, the aforementioned optimization problem (13.5) admits any optimal solution $(\frac{wz_1}{2}, \frac{(1-w)z_1}{2}, 0, \ldots, 0)$ where $w \in [0, 1]$. When $z_1 = \ldots = z_{k_0}$ for some integer $k_0 \leqslant k$, the profile $x = (x_1, \ldots, x_k)$ where

$$x_i = \begin{cases} \dfrac{z_1}{2k_0}, & \text{if } i \leqslant k_0, \\ 0, & \text{if } k_0 < i \leqslant k, \end{cases}$$

also appears to be the optimal solution of the optimization problem (13.5) as it gives the same value of $z_1^2/4$. For the analysis that we carry out below, we are interested only in the value of the constrained optimization problem, therefore the optimal solution/solutions are not of much relevance.

The cooperative agreement $\bar{q} = (\bar{q}_1, \ldots, \bar{q}_n)$ can directly be found from Proposition 13.1:

$$\bar{q}_i = \begin{cases} \dfrac{a - c_{i_N}}{2|I_N|}, & \text{if } i \in I_N, \\ 0, & \text{otherwise.} \end{cases} \tag{13.6}$$

Under the cooperative agreement, only firms with the lowest unit cost produce positive output. Firm i's profit under this agreement equals

$$\pi_i(\bar{q}) = \begin{cases} \dfrac{(a - c_{i_N})^2}{4|I_N|}, & \text{if } i \in I_N, \\ 0, & \text{otherwise,} \end{cases} \tag{13.7}$$

and the sum of firms' profits is $\sum_{i \in N} \pi_i(\bar{q}) = (a - c_{i_N})^2/4$. The price on the product will be $P(\bar{q}) = (a + c_{i_N})/2$. Comparing the equilibrium and cooperative policies, we conclude that $P(\bar{q}) > P(q^*)$. Moreover since $\bar{q}$ maximizes the sum of firms' profits, it immediately follows that $\sum_{i \in N} \pi_i(\bar{q}) \geqslant \sum_{i \in N} \pi_i(q^*)$, yet there may exist a firm $j \in N$ that $\pi_j(\bar{q}) < \pi_j(q^*)$. At the same time, $\sum_{i \in N} \bar{q}_i < \sum_{i \in N} q_i^*$. Indeed,

$$\begin{aligned} \sum_{i \in N} q_i^* - \sum_{i \in N} \bar{q}_i &= \frac{na - c_N}{n+1} - \frac{a - c_{i_N}}{2} = \frac{(n-1)a - 2c_N + (n+1)c_{i_N}}{2(n+1)} \\ &> \frac{(n+1)c_{N \setminus i_N} - (n-1)c_N - 2c_N + (n+1)c_{i_N}}{2(n+1)} \\ &= \frac{(n+1)c_N - (n-1)c_N - 2c_N}{2(n+1)} = 0. \end{aligned}$$

The inequality is true because $(n-1)(a + c_N) > (n+1)c_{N \setminus i_N}$ owning to (13.2). Summarizing the above, under the cooperative agreement firms produce less product, its price is higher, and firms get more profit in total with respect to the Nash equilibrium agreement.

13.3 Characteristic Functions in a Linear Oligopoly TU Game

We have observed that under the cooperative agreement, a firm may receive zero profit, however its profit is always positive under the Nash equilibrium. To encourage firms to cooperate with each other, the joint profit of $\sum_{i\in N}\pi_i(\bar{q})$ should be allocated in an alternative way, differing from (13.7). For this reason, we first move from a noncooperative game $(N, \{Q_i\}_{i\in N}, \{\pi_i\}_{i\in N})$ to its cooperative version called a *cooperative game* or TU game, and then allocate that joint profit with the use of an appropriate cooperative solution. We denote the cooperative game by (N, v) where $v : 2^N \mapsto \mathbb{R}$ is the characteristic function assigning the worth $v(S)$ to any subset $S \subseteq N$ called coalition with $v(\varnothing) = 0$. In this section, we consider different approaches for determining the characteristic function v. To emphasize a particular approach, we will use a superscript for v, however for any of the approaches $v(N) = (a - c_{i_N})^2/4$ will denote the joint profit to be allocated.

13.3.1 α-Characteristic Function

The first measure determining the worth of any coalition $S \subset N$ and considered in the game-theoretic literature was the α-characteristic function v^α introduced in [12]. The value $v^\alpha(S)$ is interpreted as the maximum value that coalition S can get in the worst-case scenario, i.e., when the complement $N \setminus S$ acts against S:

$$v^\alpha(S) = \max_{q_i\in[0,a], i\in S}\ \min_{q_j\in[0,a], j\in N\setminus S} \sum_{i\in S}\pi_i(q). \tag{13.8}$$

From [3] it follows that $v^\alpha(S) = 0$ for any coalition $S \subset N$, and the profile of outputs that solves (13.8) is of the form:

$$q_i^{\alpha,S} = \begin{cases} 0, & \text{if } i \in S, \\ \dfrac{a}{|N \setminus S|}, & \text{if } i \in N \setminus S, \end{cases} \tag{13.9}$$

with $\sum_{i\in N} q_i^{\alpha,S} = a$.

13.3.2 β-Characteristic Function

Another measure determining the worth of coalition $S \subset N$ was also considered in [12]. The value $v^{\beta}(S)$ amounts to the smallest value that the complement $N \setminus S$ can force S to receive, without knowing its actions, and this value is defined as

$$v^{\beta}(S) = \min_{q_j \in [0,a], j \in N \setminus S} \max_{q_i \in [0,a], i \in S} \sum_{i \in S} \pi_i(q). \tag{13.10}$$

In [3] it was shown that $v^{\beta}(S) = 0$ for any coalition $S \subset N$ thus $v^{\alpha}(S) = v^{\beta}(S) = 0$, and the profile of outputs that solves (13.10) is the same: $q_i^{\beta,S} = q_i^{\alpha,S}$, $i \in N$ with $\sum_{i \in N} q_i^{\beta,S} = a$.

13.3.3 γ-Characteristic Function

Considered in [2, 9], the γ-characteristic function v^{γ} for any coalition $S \subset N$ assigns its equilibrium payoff in a noncooperative game played between S acting as one player and players from $N \setminus S$ acting as singletons. Hence we get the following result.

Proposition 13.2 *For any coalition $S \subset N$, it holds that*

$$v^{\gamma}(S) = \left(q_{i_S}^* + \frac{1}{n - s + 2} \sum_{j \in S \setminus i_S} q_j^* \right)^2. \tag{13.11}$$

Proof According to the definition of the γ-characteristic function, coalition $S \subset N$ aims at maximizing the profit $\sum_{i \in S} \pi_i(q)$ over $q_i \in [0, a]$ for all $i \in S$, whereas each firm $j \in N \setminus S$ seeks to maximize its own profit $\pi_j(q)$ over $q_j \in [0, a]$. Maximizing $\sum_{i \in S} \pi_i(q)$ with respect to the profile of quantities of firms from S, we get the reaction of S (by Proposition 13.1):

$$q_i^{\gamma,S} = \begin{cases} \dfrac{a - \sum\limits_{j \in N \setminus S} q_j - c_{i_S}}{2|I_S|}, & \text{if } i \in S \cap I_S, \\ 0, & \text{if } i \in S \setminus I_S. \end{cases} \tag{13.12}$$

At the same time for any $j \in N \setminus S$, maximizing $\pi_j(q)$ with respect to the q_j, the first-order conditions imply $q_j = a - \sum_{i \in S} q_i - \sum_{i \in N \setminus S} q_i - c_j$. Summing these equalities over all $j \in N \setminus S$ and substituting expression (13.12) into this sum, we obtain that

$$\sum_{i\in N\setminus S} q_i^{\gamma,S} = \frac{(n-s)(a+c_{i_S})-2c_{N\setminus S}}{n-s+2},$$

where $s=|S|$. Thus

$$v^{\gamma}(S)=\frac{1}{4}\left(a-\sum_{i\in N\setminus S} q_i^{\gamma}(S)-c_{i_S}\right)^2=\left(\frac{a-(n-s+1)c_{i_S}+c_{N\setminus S}}{n-s+2}\right)^2$$
$$=\left(q_{i_S}^*+\frac{1}{n-s+2}\sum_{j\in S\setminus i_S} q_j^*\right)^2.$$

The equilibrium profile of outputs which is used to find the value $v^{\gamma}(S)$ for $S\subset N$ is of the form:

$$q_i^{\gamma,S}=\begin{cases}\frac{1}{|I_S|}\left(q_{i_S}^*+\frac{1}{n-s+2}\sum_{j\in S\setminus i_S} q_j^*\right), & \text{if } i\in S\cap I_S,\\ 0, & \text{if } i\in S\setminus I_S,\\ q_i^*+\frac{1}{n-s+2}\sum_{j\in S\setminus i_S} q_j^*, & \text{if } i\in N\setminus S,\end{cases} \tag{13.13}$$

and $\sum_{i\in N} q_i^{\gamma,S}\leqslant\sum_{i\in N} q_i^*$. □

13.3.4 δ-Characteristic Function

Motivated by the computational complexity of α-, β-, and γ-characteristic functions known for that moment, Petrosjan and Zaccour [8] introduced the δ-characteristic function v^{δ} which for any coalition $S\subset N$ was determined as its best response against the Nash equilibrium output of singletons from $N\setminus S$, i.e.:

$$v^{\delta}(S)=\max_{q_i\in[0,a],i\in S}\sum_{i\in S}\pi_i(q_S,q_{N\setminus S}^*). \tag{13.14}$$

Here the equilibrium profile of outputs of firms from coalition $N\setminus S$ is given by (13.1).

Proposition 13.3 *For any coalition $S \subset N$, it holds that*

$$v^{\delta}(S) = \left(q_{i_S}^* + \frac{1}{2} \sum_{j \in S \setminus i_S} q_j^* \right)^2. \tag{13.15}$$

Proof By the definition of the δ-characteristic function, the expression of the Nash equilibrium output (13.1), and the result of Proposition 13.1, we obtain

$$\begin{aligned}
v^{\delta}(S) &= \max_{q_i \in [0,a], i \in S} \sum_{i \in S} \left(\left(a - \sum_{j \in S} q_j - \sum_{j \in N \setminus S} q_j^* \right)_+ - c_i \right) q_i \\
&= \max_{q_i \in [0,a], i \in S} \sum_{i \in S} \left(a - \sum_{j \in N \setminus S} q_j^* - c_i - \sum_{j \in S} q_j \right) q_i \\
&= \max_{q_i \in [0,a], i \in S} \sum_{i \in S} \left(a - \sum_{j \in N \setminus S} q_j^* - c_i \right) q_i - \left(\sum_{j \in S} q_j \right)^2 \\
&= \frac{1}{4} \left(a - \sum_{j \in N \setminus S} q_j^* - c_{i_S} \right)^2 \\
&= \frac{\left((s+1) q_{i_S}^* - c_S + s c_{i_S} \right)^2}{4} = \left(q_{i_S}^* + \frac{1}{2} \sum_{j \in S \setminus i_S} q_j^* \right)^2.
\end{aligned}$$

Here we assumed that the total output does not exceed a, otherwise the maximum in (13.14) would be negative. Note that one of the profiles of quantities that solves maximization problem (13.14) is of the form:

$$q_i^{\delta,S} = \begin{cases} \frac{1}{|I_S|} \left(q_{i_S}^* + \frac{1}{2} \sum_{j \in S \setminus i_S} q_j^* \right), & \text{if } i \in S \cap I_S, \\ 0, & \text{if } i \in S \setminus I_S, \\ q_i^*, & \text{if } i \in N \setminus S, \end{cases} \tag{13.16}$$

and $\sum_{i \in N} q_i^{\delta,S} \leqslant \sum_{i \in N} q_i^{\gamma,S}$, but $\sum_{i \in S} q_i^{\delta,S} > \sum_{i \in S} q_i^{\gamma,S}$. Hence the proposition is proved. □

We notice a relationship between the γ- and δ-characteristic functions. For any S, it holds that

$$v^{\gamma}(S) = \left(\frac{2\sqrt{v^{\delta}(S)} + (n-s)\sqrt{v^{\delta}(i_S)}}{n-s+2} \right)^2,$$

i.e., $v^{\gamma}(S)$ is the square of the weighted average of $\sqrt{v^{\delta}(S)}$ and equilibrium output $q^*_{i_S}$ of firm i_S having the smallest unit cost in coalition S.

13.3.5 ζ-Characteristic Function

An approach for determining the worth of a coalition by means of the so-called ζ-characteristic function v^{ζ} was presented in [5]. For coalition $S \subset N$, the value $v^{\zeta}(S)$ measures the worst profit that S can achieve following the cooperative agreement $\bar{q}$ given by (13.6). In other words, $v^{\zeta}(S)$ is the value of the minimization problem

$$v^{\zeta}(S) = \min_{q_j \in [0,a], j \in N \setminus S} \sum_{i \in S} \pi_i(\bar{q}_S, q_{N \setminus S}). \tag{13.17}$$

Proposition 13.4 *For any coalition $S \subset N$, it holds that*

$$v^{\zeta}(S) = -|S \cap I_N| c_{i_N} \bar{q}_{i_N}. \tag{13.18}$$

Proof By the definition of the ζ-characteristic function and the expression of the cooperative output (13.6), we get

$$\begin{aligned} v^{\zeta}(S) &= \min_{q_j \in [0,a], j \in N \setminus S} \sum_{i \in S} \left(\left(a - \sum_{j \in S} \bar{q}_j - \sum_{j \in N \setminus S} q_j \right)_+ - c_i \right) \bar{q}_i \\ &= \sum_{i \in S \cap I_N} \left(\left(- \sum_{j \in S} \bar{q}_j \right)_+ - c_i \right) \bar{q}_i = -|S \cap I_N| c_{i_N} \bar{q}_{i_N}. \end{aligned}$$

A profile of firms' outputs that solves minimization problem (13.17) is given by:

$$q_i^{\zeta,S} = \begin{cases} \bar{q}_i, & \text{if } i \in S, \\ \dfrac{a}{|N \setminus S|}, & \text{if } i \in N \setminus S, \end{cases} \tag{13.19}$$

and $\sum_{i \in N} q_i^{\zeta,S} > a$. Thus the statement of the proposition is proved. □

13.4 Properties of the Characteristic Functions

In this section, we study properties of the characteristic functions that have been introduced and the relationships between them. Characteristic function v is *monotonic* if $v(R) \leqslant v(S)$ for any coalitions $R \subset S$. Characteristic function v is *superadditive* if $v(S \cup R) \geqslant v(S) + v(R)$ for any disjoint coalitions $S, R \subseteq N$. Characteristic function v is *supermodular* if $v(S \cup R) + v(S \cap R) \geqslant v(S) + v(R)$ for any coalitions $S, R \subseteq N$. When v is supermodular, the game (N, v) is *convex*. The properties of v^{α}, v^{β} such as monotonicity, superadditivity, supermodularity were examined in [3, 7]. Some results for v^{γ} were presented in [6, 9], for example, the existence of the γ-core for an oligopoly game either with transferable technologies or without transferable technologies but with $n \leqslant 4$. In the present section we study the properties of v^{γ}, v^{δ}, and v^{ζ} for the linear oligopoly game without transferable technologies.

Proposition 13.5 *Characteristic functions v^{γ} and v^{δ} are monotonic whereas v^{ζ} is not.*

Proof Let $R \subset S \subseteq N$ with $|R| = r$ and $|S| = s$. Therefore, $c_{i_R} \geqslant c_{i_S}$ and $q^*_{i_R} \leqslant q^*_{i_S}$. First, prove the monotonicity of v^{δ}. Since v^{δ} is nonnegative, it suffices to show that $\sqrt{v^{\delta}}$ is monotonic. Indeed,

$$\sqrt{v^{\delta}(S)} - \sqrt{v^{\delta}(R)} = q^*_{i_S} - q^*_{i_R} + \frac{1}{2} \sum_{j \in (S \setminus i_S) \setminus (R \setminus i_R)} q^*_j > 0.$$

Second, prove the monotonicity of v^{γ}. Again, since v^{γ} is nonnegative, it suffices to show that $\sqrt{v^{\gamma}}$ is monotonic. We have:

$$\sqrt{v^{\gamma}(S)} - \sqrt{v^{\gamma}(R)} = q^*_{i_S} + \frac{1}{n-s+2} \sum_{j \in S \setminus i_S} q^*_j - q^*_{i_R} - \frac{1}{n-r+2} \sum_{j \in R \setminus i_R} q^*_j$$

$$\geqslant q^*_{i_S} - q^*_{i_R} + \left(\frac{1}{n-s+2} - \frac{1}{n-r+2} \right) \sum_{j \in S \setminus i_S} q^*_j > 0.$$

Finally, show that v^{ζ} is not monotonic. Indeed, for $R \subset S \subset N$, it holds that $v^{\zeta}(S) - v^{\zeta}(R) = (|R \cap I_N| - |S \cap I_N|) c_{i_N} \bar{q}_{i_N} \leqslant 0$, but $v^{\zeta}(N) - v^{\zeta}(S) = (a - c_{i_N})^2/4 + |S \cap I_N| c_{i_N} \bar{q}_{i_N} > 0$, and this completes the proof. □

Proposition 13.6 *Characteristic functions v^{δ} and v^{ζ} are superadditive whereas v^{γ} is superadditive only in case of duopoly; i.e., when $n = 2$.*

Proof Let $S, R \subseteq N$ be two disjoint coalitions with $|S| = s$ and $|R| = r$. Without loss of generality, we suppose that $c_{i_S} \leqslant c_{i_R}$, therefore $q^*_{i_S} \geqslant q^*_{i_R}$. First, prove the superadditivity of v^δ. From (13.15), it can be easily verified that

$$\sqrt{v^\delta(S\cup R)}=\sqrt{v^\delta(S)}+\sqrt{v^\delta(R)}-\frac{1}{2}\sqrt{v^\delta(i_R)}. \tag{13.20}$$

Using (13.20), we obtain:

$$\begin{aligned}
&v^\delta(S\cup R)-v^\delta(S)-v^\delta(R)\\
&=\frac{1}{4}v^\delta(i_R)+2\sqrt{v^\delta(S)v^\delta(R)}-\sqrt{v^\delta(i_R)}\left(\sqrt{v^\delta(S)}+\sqrt{v^\delta(R)}\right)\\
&=\frac{1}{4}v^\delta(i_R)+\sqrt{v^\delta(S)}\left(\sqrt{v^\delta(R)}-\sqrt{v^\delta(i_R)}\right)+\sqrt{v^\delta(R)}\left(\sqrt{v^\delta(S)}-\sqrt{v^\delta(i_R)}\right)\\
&\geqslant\frac{1}{4}v^\delta(i_R)+\sqrt{v^\delta(S)}\left(\sqrt{v^\delta(R)}-\sqrt{v^\delta(i_R)}\right)+\sqrt{v^\delta(R)}\left(\sqrt{v^\delta(S)}-\sqrt{v^\delta(i_S)}\right)>0.
\end{aligned}$$

Second, to prove the superadditivity of v^ξ, we note that for $S\cup R\subset N$,

$$\begin{aligned}
v^\xi(S\cup R)-v^\xi(S)-v^\xi(R)&=-c_{i_N}\bar{q}_{i_N}(|(S\cup R)\cap I_N|-|S\cap I_N|-|R\cap I_N|)\\
&=-c_{i_N}\bar{q}_{i_N}\left(|(S\cap I_N)\cup(R\cap I_N)|-|S\cap I_N|-|R\cap I_N|\right)\\
&=-c_{i_N}\bar{q}_{i_N}\left(|S\cap I_N|+|R\cap I_N|\right.\\
&\quad\left.-|(S\cap I_N)\cap(R\cap I_N)|-|S\cap I_N|-|R\cap I_N|\right)\\
&=c_{i_N}\bar{q}_{i_N}|(S\cap I_N)\cap(R\cap I_N)|=0,
\end{aligned}$$

because $(S\cap I_N)\cap(R\cap I_N)=\varnothing$ when S and R are disjoint coalitions. At the same time, when $S\cup R=N$, we have $v^\xi(S\cup R)-v^\xi(S)-v^\xi(R)=v^\xi(N)+c_{i_N}\bar{q}_{i_N}(|S\cap I_N|+|R\cap I_N|)>0$.

And finally, consider v^γ. The superadditivity of v^γ in case of duopoly is obvious. Let $S=i_S$ and $R=i_R$. Using (13.11), it follows that

$$\begin{aligned}
v^\gamma(i_S\cup i_R)-v^\gamma(i_S)-v^\gamma(i_R)&=\left(q^*_{i_S}+\frac{q^*_{i_R}}{n}\right)^2-\left(q^*_{i_S}\right)^2-\left(q^*_{i_R}\right)^2\\
&=q^*_{i_S}q^*_{i_R}\left(\frac{2}{n}-\frac{q^*_{i_R}}{q^*_{i_S}}\left(1-\frac{1}{n^2}\right)\right),
\end{aligned}$$

which becomes negative when $n>q^*_{i_S}/q^*_{i_R}+\sqrt{1+\left(q^*_{i_S}/q^*_{i_R}\right)^2}\geqslant 1+\sqrt{2}>2$. The statement is proved. □

Proposition 13.7 *Characteristic function v^ξ is supermodular; v^γ is supermodular only in case of duopoly, and v^δ is supermodular either when $n\leqslant 4$, or when firms are symmetrical.*

Proof We first prove the supermodularity of v^ζ. Using results of Proposition 13.6, we obtain: $v^\zeta(S \cup R) + v^\zeta(S \cap R) - v^\zeta(S) - v^\zeta(R) = c_{i_N}\bar{q}_{i_N}(|(S \cap I_N) \cap (R \cap I_N)| - |(S \cap R) \cap I_N|) = 0$ when $S \cup R \subset N$. If $S \cup R = N$, then $v^\zeta(S \cup R) + v^\zeta(S \cap R) - v^\zeta(S) - v^\zeta(R) = v^\zeta(N) + c_{i_N}\bar{q}_{i_N}(|S \cap I_N| + |R \cap I_N| - |(S \cap R) \cap I_N|) > 0$ because the expression in the brackets is nonnegative.

Second, supermodularity implies superadditivity, and if v^γ is not superadditive, it cannot also be supermodular. By Proposition 13.6, in case of duopoly, v^γ is superadditive and therefore supermodular.

Finally, consider v^δ. Let $S, R \subseteq N$ be two coalitions with $|S| = s$ and $|R| = r$. Without loss of generality, we suppose that $c_{i_S} \leqslant c_{i_R} \leqslant c_{i_{S\cap R}}$, therefore $q^*_{i_S} \geqslant q^*_{i_R} \geqslant q^*_{i_{S\cap R}}$. It can be verified that

$$\sqrt{v^\delta(S \cup R)} = \sqrt{v^\delta(S)} + \sqrt{v^\delta(R)} - \sqrt{v^\delta(S \cap R)} - \frac{1}{2}\sqrt{v^\delta(i_R)} + \frac{1}{2}\sqrt{v^\delta(i_{S\cap R})}, \tag{13.21}$$

which is an extension of (13.20) when coalitions S and R are not necessarily disjoint. Using (13.21) and recalling that $\sqrt{v^\delta(i_R)} = q^*_{i_R}$ and $\sqrt{v^\delta(i_{S\cap R})} = q^*_{i_{S\cap R}}$, we have:

$$\begin{aligned} v^\delta(S \cup R) + v^\delta(S \cap R) - v^\delta(S) - v^\delta(R) &= \frac{1}{4}\left(q^*_{i_R} - q^*_{i_{S\cap R}}\right)^2 \\ &+2\left(\sqrt{v^\delta(S)} - \sqrt{v^\delta(S \cap R)}\right)\left(\sqrt{v^\delta(R)} - \sqrt{v^\delta(S \cap R)}\right) \\ &- \left(q^*_{i_R} - q^*_{i_{S\cap R}}\right)\left(\sqrt{v^\delta(S)} + \sqrt{v^\delta(R)} - \sqrt{v^\delta(S \cap R)}\right). \end{aligned}$$

Due to the monotonicity of v^δ, the latter expression is positive when $q^*_{i_R} = q^*_{i_{S\cap R}}$, i.e., the supermodularity condition holds. This is also the case when firms are symmetrical, hence v^δ will be supermodular.

Now we show that v^δ is supermodular in a general case for $n \leqslant 4$. In case of duopoly supermodularity is obvious. Let $n = 3$, and without loss of generality, we suppose $c_1 \leqslant c_2 \leqslant c_3$, thus here there is only one case of our interest: $S = \{1, 3\}$, $R = \{2, 3\}$. The case when $S = \{1, 2\}$, $R = \{2, 3\}$ is not of much interest since $q^*_{i_R} = q^*_{i_{S\cap R}}$ and therefore $i_R = i_{S\cap R} = 2$. Similarly, when $S = \{1, 2\}$, $R = \{1, 3\}$, we have $q^*_{i_R} = q^*_{i_{S\cap R}}$ and therefore $i_R = i_{S\cap R} = 1$. Other cases lead either to the inequality for superadditivity or to the case when the supermodularity inequality becomes an equality. Consider the aforementioned case. Let $S = \{1, 3\}$, $R = \{2, 3\}$. We obtain: $v^\delta(S \cup R) + v^\delta(S \cap R) - v^\delta(S) - v^\delta(R) = \frac{1}{4}(-3(q^*_2)^2 + 3(q^*_3)^2 + 4q^*_1q^*_2 - 2q^*_2q^*_3) \geqslant \frac{1}{4}((q^*_2)^2 + 3(q^*_3)^2 - 2q^*_2q^*_3) = \frac{1}{4}((q^*_2 - q^*_3)^2 + 2(q^*_3)^2) > 0$.

Let now $n = 4$ and $c_1 \leqslant c_2 \leqslant c_3 \leqslant c_4$. There are ten cases when the inequality guaranteeing supermodularity should be verified (when coalitions S and R intersect, but $q^*_{i_R} \neq q^*_{i_{S\cap R}}$): $S = \{1, 3\}$, $R = \{2, 3\}$; $S = \{1, 4\}$, $R = \{2, 4\}$; $S = \{1, 4\}$, $R = \{3, 4\}$; $S = \{2, 4\}$, $R = \{3, 4\}$; $S = \{1, 3\}$, $R = \{2, 3, 4\}$; $S = \{1, 4\}$, $R = \{2, 3, 4\}$;

$S = \{1, 3, 4\}$, $R = \{2, 3\}$; $S = \{1, 3, 4\}$, $R = \{2, 4\}$; $S = \{1, 2, 4\}$, $R = \{3, 4\}$; and $S = \{1, 3, 4\}$, $R = \{2, 3, 4\}$. Prove for the case when $S = \{1, 3, 4\}$, $R = \{2, 3, 4\}$. We have: $v^\delta(S \cup R) + v^\delta(S \cap R) - v^\delta(S) - v^\delta(R) = \frac{1}{4}(-3(q_2^*)^2 + 3(q_3^*)^2 + 4q_1^*q_2^* - 2q_2^*q_3^* - 2q_2^*q_4^* + 2q_3^*q_4^*) \geqslant \frac{1}{4}((q_2^* - q_3^*)^2 + 2(q_3^*)^2 - 2q_4^*(q_2^* - q_3^*)) = \frac{1}{4}((q_2^* - q_3^* - q_4^*)^2 + 2(q_3^*)^2 - (q_4^*)^2) > 0$. All other cases can be examined in a similar way. Hence the proposition is now proved. □

Example 13.1 Consider an oligopoly with $N = \{1, 2, 3, 4, 5\}$ and the following values of parameters: $a = 10$, $c_1 = c_2 = 1$, $c_3 = c_4 = c_5 = 2$. From (13.1) we get: $q_1^* = q_2^* = 2$, $q_3^* = q_4^* = q_5^* = 1$. Let $S = \{1, 3, 4, 5\}$ and $R = \{2, 3, 4, 5\}$, therefore $i_S = 1$, $i_R = 2$, $i_{S \cup R} \in \{1, 2\}$, and $i_{S \cap R} \in \{3, 4, 5\}$. Using (13.15), we obtain $v^\delta(S) = v^\delta(R) = 12.25$, $v^\delta(S \cup R) = 20.25$, and $v^\delta(S \cap R) = 4$ which means that $v(S \cup R) + v(S \cap R) < v(S) + v(R)$ and v^δ is not supermodular.

Proposition 13.8 *For any coalition $S \subset N$, the condition $v^\zeta(S) \leqslant v^\alpha(S) = v^\beta(S) \leqslant v^\gamma(S) \leqslant v^\delta(S)$ is satisfied.*

Proof The fulfillment of two inequalities $v^\zeta(S) \leqslant v^\alpha(S)$ and $v^\beta(S) \leqslant v^\gamma(S)$ is obvious. Prove that $v^\gamma(S) \leqslant v^\delta(S)$. Since values $v^\gamma(S)$ and $v^\delta(S)$ are positive for all S, it suffices to show that $\sqrt{v^\gamma(S)} \leqslant \sqrt{v^\delta(S)}$. We have $\sqrt{v^\delta(S)} - \sqrt{v^\gamma(S)} = \frac{n-s}{2(n-s+2)} \sum_{j \in S \setminus i_S} q_j^*$ which is positive. The statement of the proposition is hence proved. □

13.5 Cooperative Solutions for a Linear Oligopoly TU Game

An imputation set of cooperative game (N, v) is the set $\mathscr{I}[v] = \{(\xi_1, \ldots, \xi_n) : \sum_{i \in N} \xi_i = v(N);\ \xi_i \geqslant v(\{i\})\}$. A cooperative solution is a rule that maps v into a subset of $\mathscr{I}[v]$. In particular, the core of the game (N, v) is defined as the set $\mathscr{C}[v] = \{(\xi_1, \ldots, \xi_n) \in \mathscr{I}[v] : \sum_{i \in S} \xi_i \geqslant v(S),\ \ S \subset N\}$. The Shapley value $\Phi[v] = (\Phi_1[v], \ldots, \Phi_n[v])$ is an imputation whose components are defined as $\Phi_i[v] = \sum_{S \subseteq N} \frac{(n-|S|)!(|S|-1)!}{n!}(v(S) - v(S \setminus \{i\}))$, $i \in N$. The core of the game (N, v^α) will be called the α-core and denoted by $\mathscr{C}[v^\alpha]$. Similarly, we determine β-, γ-, δ-, and ζ-cores and denote them by $\mathscr{C}[v^\beta]$, $\mathscr{C}[v^\gamma]$, $\mathscr{C}[v^\delta]$, and $\mathscr{C}[v^\zeta]$, respectively.

The existence of α-, β-cores was shown in [3, 7], thus in view of Proposition 13.8, the next result directly follows.

Corollary 13.1 *Let δ-core be nonempty. Then $\mathscr{C}[v^\delta] \subseteq \mathscr{C}[v^\gamma] \subseteq \mathscr{C}[v^\alpha] = \mathscr{C}[v^\beta] \subseteq \mathscr{C}[v^\zeta]$.*

The above result notes a nested structure of the cores when the δ-core is nonempty. The existence of γ- and δ-cores can be guaranteed when the number of firms does not exceed 4 and/or when firms are symmetrical since the δ-game becomes convex in these cases (see [11]). As to the ζ-core, it always exists, and its nonemptiness follows from the existence of $\mathscr{C}[v^\alpha]$.

Table 13.1 The values of α-, β-, γ-, δ-, and ζ-characteristic functions

S	$v^{\alpha}(S)$	$v^{\beta}(S)$	$v^{\gamma}(S)$	$v^{\delta}(S)$	$v^{\zeta}(S)$
{1}	0	0	0.0676	0.0676	−0.25
{2}	0	0	0.0676	0.0676	−0.25
{3}	0	0	0.0256	0.0256	0
{4}	0	0	0.0036	0.0036	0
{1, 2}	0	0	0.1056	0.1521	−0.5
{1, 3}	0	0	0.09	0.1156	−0.25
{1, 4}	0	0	0.0756	0.0841	−0.25
{2, 3}	0	0	0.09	0.1156	−0.25
{2, 4}	0	0	0.0756	0.0841	−0.25
{3, 4}	0	0	0.0306	0.0361	0
{1, 2, 3}	0	0	0.16	0.2209	−0.5
{1, 2, 4}	0	0	0.1344	0.1764	−0.5
{1, 3, 4}	0	0	0.1111	0.1369	−0.25
{2, 3, 4}	0	0	0.1111	0.1369	−0.25
{1, 2, 3, 4}	0.25	0.25	0.25	0.25	0.25

Example 13.2 We consider an oligopoly with $N = \{1, 2, 3, 4\}$ where $a = 2$, $c_1 = c_2 = 1$, $c_3 = 1.1$, and $c_4 = 1.2$. Here the set $I_N = \{1, 2\}$. Table 13.1 summarizes the values of α-, β-, γ-, δ-, and ζ-characteristic functions. Figure 13.1 demonstrates ζ-, α-, γ-, and δ-cores where the largest set represents the ζ-core and the smallest one is the δ-core (recall that the α-core coincides with the β-core). On Fig. 13.2, we demonstrate the same cooperative set solutions in a more detailed view.[2]

Example 13.3 (Symmetric Firms) As one of the special cases of oligopoly often considered in the literature, we examine a symmetric game where unit costs of the firms are equal, i.e., $c_1 = \ldots = c_n = c$ with $c < a$. Under these assumptions, the condition (13.2) holds for any n. We also note that $I_S = S$ for any $S \subseteq N$. The equilibrium output of firm $i \in N$ determined by (13.1), takes the form $q_i^* = \frac{a-c}{n+1}$, whereas under the cooperative agreement, the output determined by (13.6) becomes $\bar{q}_i = \frac{a-c}{2n}$. The equilibrium profit of firm $i \in N$ equals $\pi_i(q^*) = \left(\frac{a-c}{n+1}\right)^2$ and the profit of i under the cooperative agreement becomes $\pi_i(\bar{q}) = \frac{(a-c)^2}{4n}$ exceeding $\pi_i(q^*)$. This fact means that all firms take advantage from cooperation even without reallocating the total profit of $\sum_{i \in N} \pi_i(\bar{q})$ according to a cooperative solution. We note that this result does not hold in a general case. However if firms come to a cooperative solution, the characteristic function should be determined. Then $v^{\alpha}(N) = v^{\beta}(N) = v^{\gamma}(N) = v^{\delta}(N) = v^{\zeta}(N) = (a-c)^2/4$. Further, for any $S \subset N$, we have $v^{\alpha}(S) = v^{\beta}(S) = 0$ while using (13.11), (13.15), and (13.18), it

[2]The figures were obtained with the use of TUGlab toolbox for Matlab http://mmiras.webs.uvigo.es/TUGlab/.

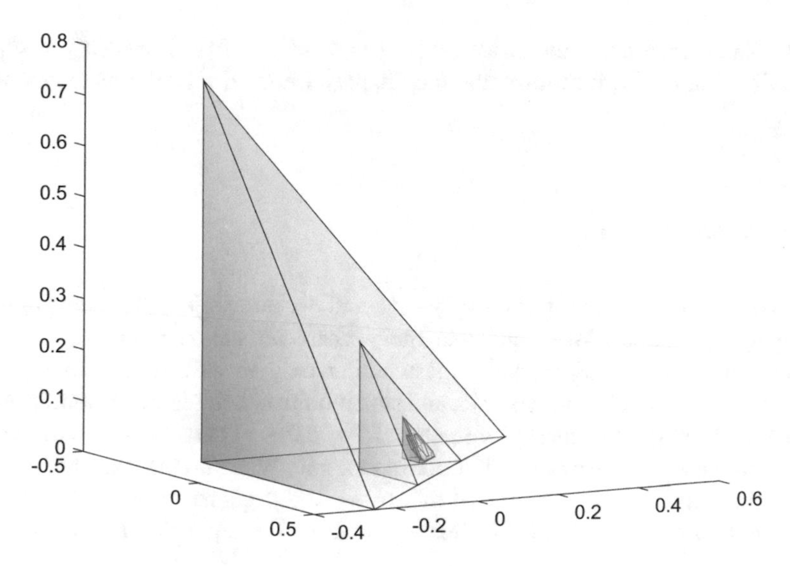

Fig. 13.1 ζ-, α-, γ-, and δ-cores (from largest to smallest)

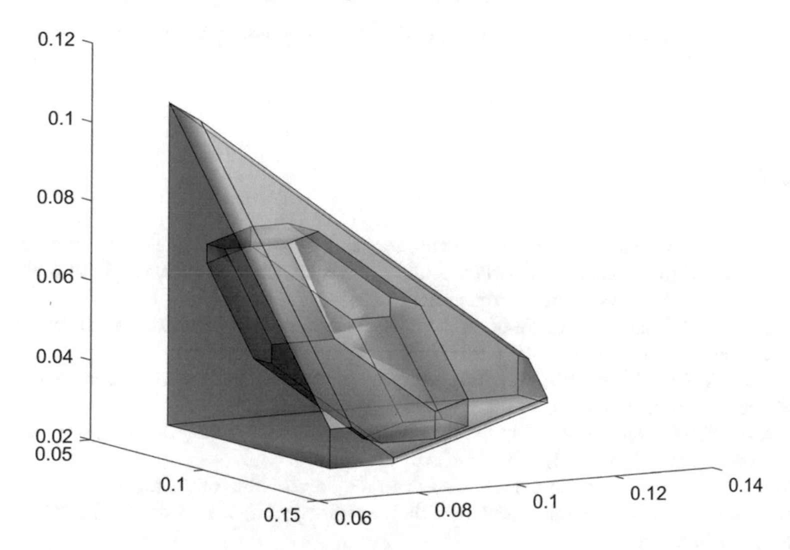

Fig. 13.2 γ-core (superset) and δ-core (subset)

follows that $v^{\gamma}(S) = \left(\frac{a-c}{n-s+2}\right)^2$, $v^{\delta}(S) = \left(\frac{(s+1)(a-c)}{2(n+1)}\right)^2$, and $v^{\zeta}(S) = -\frac{sc(a-c)}{n}$ where $s = |S|$.

By Proposition 13.7, where characteristic function v^{δ} is supermodular, the corresponding TU game (N, v^{δ}) is convex and therefore has a nonempty core $\mathscr{C}[v^{\delta}]$. From [11], the Shapley value $\Phi[v^{\delta}]$, whose components equal the cooperative profits $\pi_i(\bar{q})$, $i \in N$, belongs to $\mathscr{C}[v^{\delta}]$ being the center of gravity of its extreme points. Since firms are symmetrical, $\Phi[v^\alpha] = \Phi[v^\beta] = \Phi[v^\gamma] = \Phi[v^\delta] = \Phi[v^\zeta]$. From Corollary 13.1, it follows that the Shapley value $\Phi[v^\delta]$ belongs to any of the cores.

13.6 Conclusion

We have examined the properties of γ-, δ-, and ζ-characteristic functions in linear oligopoly TU games. We found that the γ-characteristic function is monotonic, however it is superadditive and supermodular only in case of duopoly. The δ-characteristic function is monotonic, and superadditive, but it is supermodular either when $n \leqslant 4$, or when firms are symmetrical. As to the ζ-characteristic function, it is superadditive and supermodular but not monotonic. When δ-characteristic function is supermodular, we also found that the γ-, δ-, and ζ-games have nonempty cores with a nested structure that is also expressed in their relationship to the α- and β-cores.

Acknowledgement This research was supported by the Russian Science Foundation (grant No. 17-11-01079).

References

1. Aumann, R.: Acceptable points in general cooperative n-person games. In: Tucker L. (ed.) Contributions to the Theory of Games. Annals of Mathematics Studies, vol. IV(40), pp. 287–324. Princeton University Press, Princeton (1959)
2. Chander, P., Tulkens, H.: A core of an economy with multilateral environmental externalities. Int. J. Game Theory **26**, 379–401 (1997)
3. Driessen, T.S.H., Meinhardt, H.I.: Convexity of oligopoly games without transferable technologies. Math. Soc. Sci. **50**, 102–126 (2005)
4. Driessen, T.S.H., Meinhardt, H.I.: On the supermodularity of homogeneous oligopoly games. Int. Game Theory Rev. **12**(4), 309–337 (2010)
5. Gromova, E., Petrosyan, L.: On a approach to the construction of characteristic function for cooperative differential games. Matematicheskaya Teoriya Igr I Ee Prilozheniya **7**(4), 19–39 (2015) [in Russian]
6. Lardon, A.: The γ-core in Cournot oligopoly TU-games with capacity constraints. Theor. Decis. **72**(3), 387–411 (2012)
7. Norde, H., Pham Do, K.H., Tijs, T.: Oligopoly games with and without transferable technologies. Math. Soc. Sci. **43**, 187–207 (2002)
8. Petrosjan, L., Zaccour, G.: Time-consistent Shapley value allocation of pollution cost reduction. J. Econ. Dyn. Control **27**(3), 381–398 (2003)
9. Rajan, R.: Endogenous coalition formation in cooperative oligopolies. Int. Econ. Rev. **30**(4), 863–876 (1989)
10. Reddy, P.V., Zaccour, G.: A friendly computable characteristic function. Math. Soc. Sci. **82**, 18–25 (2016)
11. Shapley, L.: Cores of convex games. Int. J. Game Theory **1**(1), 11–26 (1971)

12. Von Neumann, J., Morgenstern, O.: Theory of Games and Economic Behavior. Princeton University Press, Princeton (1944)
13. Zhao, J.: A necessary and sufficient condition for the convexity in oligopoly games. Math. Soc. Sci. **37**, 189–204 (1999)
14. Zhao, J.: A β-core existence result and its application to oligopoly markets. Games Econ. Behav. **27**, 153–168 (1999)

Chapter 14
The Position Value and the Myerson Value for Hypergraph Communication Situations

Erfang Shan and Guang Zhang

Abstract We characterize the position value and the Myerson value for uniform hypergraph communication situations by employing the "incidence graph game" and the "link-hypergraph game" which are induced by the original hypergraph communication situations. The incidence graph game and link-hypergraph game are defined on the "incidence graph" and the "link-hypergraph", respectively, obtained from the original hypergraph. Using the above tools, we represent the position value by the Shapley value of the incidence graph game and the Myerson value of the link-hypergraph game for uniform hypergraph communication situations, respectively. Also, we represent the Myerson value by the Owen value or the two-step Shapley value of the incidence graph game with a coalition structure for hypergraph communication situations.

14.1 Introduction

The study of TU-games with limited cooperation presented by means of a communication graph was initiated by Myerson [10], and an allocation rule for such games, now called the *Myerson value*, was also introduced simultaneously. Later on, various studies in this direction were done in the past nearly 40 years, such as [3, 6, 9, 13, 18] and [17]. Among them, the rule, named *position value* [9], is also widely used for communication situations. Both the Myerson value and the position value are extensively studied in the literature. On one hand, the two allocation rules are employed in many communication situations, such as for conference structures [11], hypergraph communication situations [19] and for union stable systems [1, 2]. On the other hand, both values have been characterized axiomatically in several ways (see [4, 15, 16]). Furthermore, the approaches of non-axiomatic characterization for both the Myerson value and the position value are investigated

E. Shan (✉) · G. Zhang
School of Management, Shanghai University, Shanghai, People's Republic of China
e-mail: efshan@i.shu.edu.cn

L. A. Petrosyan et al. (eds.), *Frontiers of Dynamic Games*,
Static & Dynamic Game Theory: Foundations & Applications,
https://doi.org/10.1007/978-3-319-92988-0_14

in [5] and [8], respectively. Casajus [5] gave a characterization of the position value by the Myerson value of a modification of communication situations, called the *link agent form* (LAF) on graph communication situations and the *hyperlink agent form* (HAF) on hypergraph communication situations. However, the structure of "HAF" is relatively complex. Kongo [8] provided unified and non-axiomatic characterizations of the position value and the Myerson value by using the *divided link game* and the *divided link game with a coalition structure*, respectively.

The aim of this paper is to provide non-axiomatic characterizations of the Myerson value and the position value for hypergraph communication situations. These characterizations extend results for graph communication situations, due to [8] and [5], to hypergraph communication situations. For this purpose, we first introduce a new tool—the *incidence graph* and the *incidence graph game* defined on the link set of the incidence graph, where the incidence graph is induced by the hypergraph and the incidence graph game is obtained from the original hypergraph communication situation. It turns out that the Shapley payoffs of the set of links incident to a player in the incidence graph sum up to the position value payoff of that player in an original uniform hypergraph communication situation. Similar in spirit to that of Kongo [8], we further define the *incidence graph game with a coalition structure* on the incidence graph and we characterize the Myerson value for a hypergraph communication situation in terms of the Owen value or the two-step Shapley value of the incidence graph game with a coalition structure. Based on the incidence graph, we define the *link-hypergraph* of the incidence graph. It turns out that the position value for a hypergraph communication situation is represented by the Myerson value of the *link-hypergraph game*, which is defined on the link-hypergraph and obtained from the original hypergraph communication situation.

As we will see in Sect. 14.3, for graph communication situations as a special case of hypergraph communication situations, the incidence graph game and the link-hypergraph game in this paper coincide with the divided link game [8] and the link agent form (LAF) [5], respectively. But, for a hypergraph communication situation, the link-hypergraph game differs from the hyperlink agent form (HAF) [5]. The link-hypergraph is more natural expression for hypergraph communication situations, since the HAF comprises much more agents than the link-hypergraph game.

This article is organized as follows. In Sect. 14.2, we introduce basic definitions and notation. In Sect. 14.3, we first introduce the incidence graph game and the link-hypergraph game, then we present characterizations of the position value and the Myerson value for hypergraph communication situations.

14.2 Basic Definitions and Notation

In this section, we recall some definitions and concepts related to TU-games, allocation rules for TU-games and, the Myerson value and position value for hypergraph communication situations.

A *cooperative game with transferable utility*, or simply a TU-game, is a pair (N, v) where $N = \{1, 2, \dots, n\}$ be a finite set of $n \geq 2$ players and $v : 2^N \to \mathbf{R}$ is a *characteristic function* defined on the power set of N such that $v(\emptyset) = 0$. For any $S \subseteq N$, S is called a *coalition* and the real number $v(S)$ represents the *worth* of coalition S. A *subgame* of v with a nonempty set $T \subseteq N$ is a game $v|_T(S) = v(S)$, for all $S \subseteq T$. The *unanimity game* (N, u_R) is the game defined by $u_R(S) = 1$ if $R \subseteq S$ and $u_R(S) = 0$ otherwise. We denote by $|S|$ the cardinality of S. A game (N, v) is *zero-normalized* if for any $i \in N$, $v(\{i\}) = 0$. Throughout this paper, we consider only zero-normalized games.

Let $\Sigma(N)$ be the set of all permutations on N. For some permutation σ on N, the corresponding marginal vector $m^{\sigma}(N, v) \in \mathbf{R}^n$ assigns to every player i a payoff $m_i^{\sigma}(N, v) = v(\sigma^i \cup \{i\}) - v(\sigma^i)$, where $\sigma^i = \{j \in N \mid \sigma(j) < \sigma(i)\}$, i.e., σ^i is the set of players preceding i in the permutation σ. The best-known single-valued solution is the *Shapley value* [14], which assigns to any game v the average of all marginal vectors. Formally, the Shapley value is defined as follows.

$$Sh_i(N, v) = \frac{1}{|\Sigma(N)|} \sum_{\sigma \in \Sigma(N)} m_i^{\sigma}(N, v), \quad \text{for each } i \in N.$$

Given a player set N, any partition $C = \{C_1, C_2, \dots, C_k\}$ of N into k sets is called a *coalition structure* of N. Let I_C be the set of all indices of C. A permutation $\sigma \in \Sigma(N)$ is *consistent* with respect to C if players in the same element of the coalition structure appear successively in σ. In other words, each C_i of C in σ is regarded as a "big player". Let $\Sigma(N, C)$ be a set of all permutations consistent with respect to C on N. A triple (N, v, C) is a *game with a coalition structure*. Owen [12] and Kamijo [7] generalized the Shapley value to games with coalition structures.

The *Owen value* ϕ is defined as follows.

$$\phi_i(N, v, C) = \frac{1}{|\Sigma(N, C)|} \sum_{\sigma \in \Sigma(N,C)} m_i^{\sigma}(N, v), \quad \text{for any } i \in N,$$

The *two-step Shapley value* χ is defined as follows.

$$\chi_i(N, v, C) = Sh_i(C_h, v|_{C_h}) + \frac{Sh_h(I_C, v_C) - v(C_h)}{|C_h|},$$
$$\text{for any } i \in N \text{ with } i \in C_h, h \in I_C,$$

where $v_C(S) = v\left(\bigcup_{h \in S} C_h\right)$ for any $S \subseteq I_C$. A pair (I_C, v_C) is called the *intermediate* or *quotient game*.

The communication possibilities for a TU-game (N, v) can be described by a (communication) *hypergraph* $\mathscr{H} = (N, H)$ where H is a family of non-singleton subsets of N, i.e., $H \subseteq H^N := \{e \subseteq N \mid |e| > 1\}$. The elements of N are called the *nodes* or *vertices* of the hypergraph, and the elements of H its *hyperlinks* or *hyperedges*. The hypergraph $\mathscr{H} = (N, H)$ is called a *conference structure* of a TU-game (N, v). Every hyperlink $e \in H$ represents a *conference*, the communication is only possible within a conference. The *rank* $r(\mathscr{H})$ of $\mathscr{H}$ is the maximum cardinality of a hyperlink in the hypergraph $\mathscr{H}$.

A hypergraph $\mathscr{H}$ is called an *r-uniform* or simply *uniform* if $|e| = r$ for all $e \in H$. Clearly, a graph $G = (N, L)$, $L \subseteq L^N := \{e \subseteq N \mid |e| = 2\} \subseteq H^N$, is a 2-uniform hypergraph where every hyperlink contains exactly two players. Hypergraphs are a natural generalization of graphs in which "edges" may consist of more than 2 nodes.

Let H_i be the set of hyperlinks containing player i in a hypergraph $\mathscr{H} = (N, H)$, i.e., $H_i := \{e \in H \mid i \in e\}$. The *degree* of i is defined as $|H_i|$, denoted by $deg(i)$. A node $i \in N$ is *incident* to a hyperlink $e \in H$ if $i \in e$. Two nodes i and j of N are *adjacent* in the hypergraph $\mathscr{H}$ if there is a hyperlink e in H such that $i, j \in e$. Two nodes i and j are *connected* if there exists a sequence $i = i_0, i_1, \ldots, i_k = j$ of nodes of $\mathscr{H} = (N, H)$ in which i_{l-1} is adjacent to i_l for $l = 1, 2, \ldots, k$. A *connected hypergraph* is a hypergraph in which every pair of nodes are connected. Given any hypergraph $\mathscr{H} = (N, H)$, a (connected) *component* of $\mathscr{H}$ is a maximal set of nodes of N in which every pair of nodes are connected. Let N/H denote the set of components in $\mathscr{H} = (N, H)$ and $(N/H)_i$ the component containing $i \in N$. For any $S \subseteq N$, let $H[S] = \{e \in H \mid e \subseteq S\}$, $(S, H[S])$ is called the *subhypergraph* induced by S. A hypergraph $\mathscr{H}' = (N, H')$ is called a *partial hypergraph* of $\mathscr{H} = (N, H)$ if $H' \subseteq H$. The notation $S/H[S]$ (or for short S/H) and N/H' are defined similarly.

A *hypergraph communication situation*, or simply a *hypergraph game*, is a triple (N, v, H) where (N, v) is a zero-normalized TU-game and H the set of hyperlinks (communication links) in the hypergraph $\mathscr{H} = (N, H)$.

The *Myerson value* μ [10, 11, 19] is defined by

$$\mu_i(N, v, H) = Sh_i(N, v^H), \text{ for any } i \in N,$$

where $v^H(S) = \sum_{T \in S/H} v(T)$ for any $S \subseteq N$. The game (N, v^H) is called the *Myerson restricted game*.

The *position value* [9, 19] is given by

$$\pi_i(N, v, H) = \sum_{e \in H_i} \frac{1}{|e|} Sh_e(H, v^N), \text{ for any } i \in N,$$

where $v^N(H') = \sum_{T \in N/H'} v(T)$ for any $H' \subseteq H$. The game (H, v^N) is called a *hyperlink game*.

14.3 Characterizations of the Position Value and Myerson Value

In this section we first introduce the definition of the incidence graph of an arbitrary hypergraph. We express the position value for a uniform hypergraph game in terms of the Shapley value of a game defined on the link set of the incidence graph (called the "incidence graph game"), which is obtained from the original hypergraph game. Based on the incidence graph of a hypergraph, we subsequently define the link-hypergraph of the incidence graph and express the position value for a uniform hypergraph game in terms of the Myerson value of a game restricted on the link-hypergraph (called the "link-hypergraph game"), which is induced by the original hypergraph game. Continuing to use the tool of the incidence graph, we further provide characterizations of the Myerson value in terms of the Owen value or two-step Shapley value of the incidence graph game with a coalition structure respectively, for an arbitrary hypergraph game.

14.3.1 The Incidence Graphs and Link-Hypergraphs of Hypergraphs

In this subsection we give the concept of the incidence graph and link-hypergraph of a hypergraph. By definition of the position value, the Shapley payoff that a hyperlink receives in the hyperlink game is equally divided among the players that lie in the hyperlink. This definition reflects the assumption that each hyperlink is composed of the players' cooperation in the hyperlink, and the contributions of the players in the hyperlink toward maintaining the hyperlink are considered to be the same. In light of the above observation, we introduce the concept of an incidence graph which is obtained from a given hypergraph $\mathscr{H} = (N, H)$ and then, we define a new game on the link set in the incidence graph.

A graph $G = (N, L)$ is *bipartite* if its node set N can be partitioned into two subsets X and Y so that every link has one end in X and one end in Y; such a partition (X, Y) is called a *bipartition* of the graph, and X and Y its *parts*.

Let $\mathscr{H} = (N, H)$ be a hypergraph, the *incidence graph* of $\mathscr{H}$ is a bipartite graph $I(\mathscr{H}) = (N \cup H, I(H))$ with node set $N \cup H$, and where $i \in N$ and $e \in H$ are adjacent, i.e., $\{i, e\} \in I(H)$ if and only if $i \in e$. The *link-hypergraph* $\mathscr{L}(I(\mathscr{H}))$ of the incidence graph $I(\mathscr{H})$ is the hypergraph with node set $I(H)$ and hyperlink set $L(I(\mathscr{H}))$, where $\{i, e\}, \{i', e'\} \in I(H)$ are joined as nodes by a link if and only if $i = i'$ and they are adjacent by the hyperlink $\big\{\{i, e\} \mid i \in e\big\}$ if and only if $e = e'$; that is,

$$L(I(\mathscr{H})) = \Big\{\{\{i, e\}, \{i, e'\}\} \mid e, e' \in H_i, i \in N\Big\} \bigcup \Big\{\{\{i, e\} \mid i \in e\} \mid e \in H\Big\}.$$

A hypergraph $\mathscr{F}_1 = (N, F_1)$ and its incidence graph $I(\mathscr{F}_1) = (N \cup F_1, I(F_1))$ are shown in Fig. 14.1, and the link-hypergraph $\mathscr{L}(I(\mathscr{F}_1)) = \big(I(F_1), L(I(\mathscr{F}_1))\big)$ of $I(\mathscr{F}_1)$ is shown in Fig. 14.2, where $N = \{1, 2, \ldots, 5\}$ and $F_1 = \big\{\{1, 3\}, \{2, 4\}, \{3, 4, 5\}\big\}$.

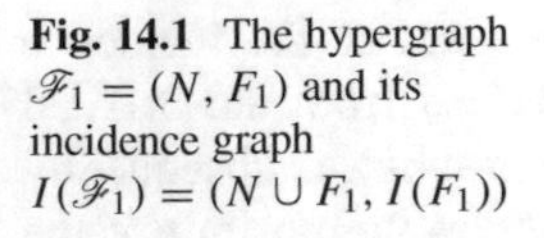

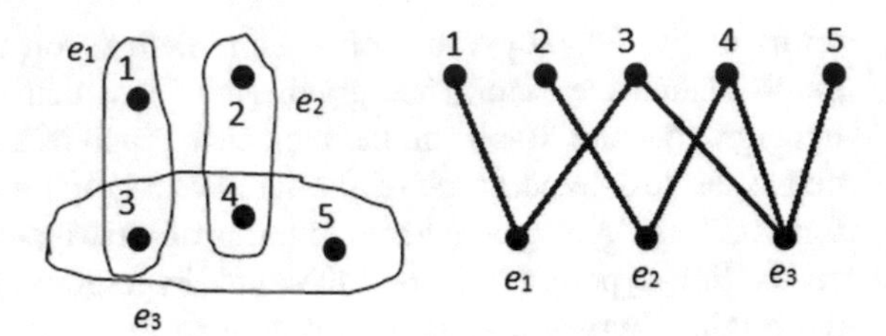

Fig. 14.1 The hypergraph $\mathscr{F}_1 = (N, F_1)$ and its incidence graph $I(\mathscr{F}_1) = (N \cup F_1, I(F_1))$

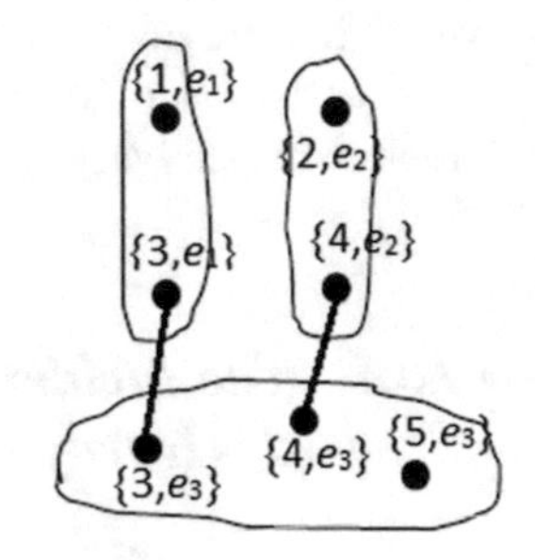

Fig. 14.2 The link-hypergraph $\mathscr{L}(I(\mathscr{F}_1))$ of the incidence graph $I(\mathscr{F}_1)$

In link-hypergraph $\mathscr{L}(I(\mathscr{H}))$ of the incidence graph $I(\mathscr{H})$, let

$$\widetilde{L} = \Big\{\{\{i, e\}, \{i, e'\}\} \mid e, e' \in H_i, i \in N\Big\}$$

and

$$\widetilde{H} = \Big\{\{\{i, e\} \mid i \in e\} \mid e \in H\Big\}.$$

Then $\widetilde{L}$ and $\widetilde{H}$ consist of links and hyperlinks, respectively, and $L(I(\mathscr{H})) = \widetilde{L} \cup \widetilde{H}$. Comparing with the original hypergraph $\mathscr{H} = (N, H)$, clearly $|\widetilde{H}| = |H|$, and each hyperlink $\widetilde{e} = \big\{\{i, e\} \mid i \in e\big\}$ in $\widetilde{H}$ corresponds to exactly the hyperlink e in H, the links of $\widetilde{L}$ are added in $\mathscr{L}(I(\mathscr{H}))$. Roughly speaking, the link-hypergraph $\mathscr{L}(I(\mathscr{H}))$ can be directly obtained from $\mathscr{H} = (N, H)$ by expanding each node i in N to $deg(i)$ new nodes and adding all possible links among the $deg(i)$ nodes. By definition, we have

$$|\widetilde{L}| = \sum_{i=1}^{n} \binom{deg(i)}{2} = \sum_{i=1}^{n} \binom{|H_i|}{2} = \frac{1}{2} \sum_{i=1}^{n} |H_i|(|H_i| - 1).$$

Given a hypergraph $\mathscr{H} = (N, H)$, we divide each hyperlink $e = \{i_1, i_2, \ldots, i_t\} \in H$ into t links, as $\{i_1, e\}, \{i_2, e\}, \ldots, \{i_t, e\}$. The resulting bipartite graph is the incidence graph $I(\mathscr{H}) = (N \cup H, I(H))$. Each divided link $\{i_j, e\}$ can be interpreted as a unilateral communication channel of player i_j through the hyperlink e. If player i_j is contained in a hyperlinks in $\mathscr{H} = (N, H)$, then it possesses a distinct unilateral communication channels. Clearly, each player i_j possesses exactly $deg(i_j)$ distinct unilateral communication channels. We introduce the incidence graph game defined on the incidence graph, it is employed to show that the position value for a uniform hypergraph game (N, v, H) is represented by the Shapley value of the incidence graph game. Furthermore, the game with a coalition structure serves to show that the Myerson value for hypergraph games (N, v, H) can be represented by either the Owen value or the two-step Shapley value of the incidence graph game with a coalition structure.

For the hypergraph $\mathscr{H} = (N, H)$, we split each player $i \in N$ in the original hypergraph (N, H) into $deg(i)$ separate "agents", each of which corresponds to exactly some link $\{i, e\}$ (a unilateral communication channel) in $I(H)$. All the agents generated by the same hyperlink e of H still form a conference (or hyperlink) and we link any two of agents generated by the same player. The resulting hypergraph is the link-hypergraph $\mathscr{L}(I(\mathscr{H})) = (I(H), L(I(\mathscr{H}))$. The link-hypergraph game defined on the link-hypergraph serves to show that the position value for a uniform hypergraph game (N, v, H) is represented by the Myerson value of the link-hypergraph game.

14.3.2 Characterizations of the Position Value

Let (N, v, H) be a hypergraph game. A pair $(I(H), u)$ is called an *incidence graph game* induced by the hyperlink game (H, v^N) where $u : 2^{I(H)} \to \mathbf{R}$ is defined as

$$u(L') = v^N\big(\{e \in H \mid \{i, e\} \in L', \text{ for all } i \in e\}\big),$$

for any $L' \subseteq I(H)$.

For a uniform hypergraph game (N, v, H), we present the following characterization of the position value in terms of the incidence graph game $(I(H), u)$.

Theorem 14.1 *For any uniform hypergraph game (N, v, H) and any $i \in N$,*

$$\pi_i(N, v, H) = \sum_{l \in I(H)_i} Sh_l\big(I(H), u\big),$$

where $I(H)_i$ is the set of all links incident to i in $I(H)$, i.e., $I(H)_i = \{\{i, e\} \in I(H) \mid e \in H_i\}$.

Proof Let f be a mapping from $\Sigma(I(H))$ to $\Sigma(H)$: For any hyperlinks $e, e' \in H$ and $\sigma \in \Sigma(I(H))$, $f(\sigma)(e) < f(\sigma)(e')$ if and only if

$$\max\{\sigma(\{i, e\}) \mid i \in e\} < \max\{\sigma(\{i', e'\}) \mid i' \in e'\}.$$

In other words, if all links adjacent to e in $I(H)$ precede any one of links adjacent to e' in $\sigma \in \Sigma(I(H))$, then the hyperlink e precedes the hyperlink e' in $f(\sigma)$.

For any $\sigma \in \Sigma(I(H))$ and any $\{k, e\} \in I(H)$, if $\sigma(\{i, e\})=\max\{\sigma(\{k, e\}) \mid k \in e\}$, then

$$m^{\sigma}_{\{i,e\}}(I(H), u) = m^{f(\sigma)}_{e}(H, v^N).$$

Otherwise, we have $m^{\sigma}_{\{i,e\}}(I(H), u) = 0$. Hence,

$$\sum_{\{k,e\}\in\{\{j,e\} \mid j\in e\}} m^{\sigma}_{\{k,e\}}(I(H), u) = m^{f(\sigma)}_{e}\big(H, v^N\big).$$

Note that $|\Sigma(I(H))| = \big(\sum_{i=1}^{N} |H_i|\big)!$ and $|\Sigma(H)| = |H|!$. For each $\delta \in \Sigma(H)$, since (N, H) is uniform, $\Sigma(I(H))$ has exactly $q = |\Sigma(I(H))|/|\Sigma(H)|$ permutations $\sigma_1, \sigma_2, \ldots, \sigma_q$ such that $f(\sigma_t) = \delta$ for $t = 1, 2, \ldots, q$. Therefore, we have

$$\begin{aligned}
&\frac{1}{|\Sigma(I(H))|} \sum_{\sigma\in\Sigma(I(H))} \Big(\sum_{\{i,e\}\in\{\{k,e\} \mid k\in e\}} m^{\sigma}_{\{i,e\}}(I(H), u)\Big) \\
&= \frac{1}{\Sigma(H)} \sum_{\delta\in\Sigma(H)} m^{\delta}_{e}\big(H, v^N\big),
\end{aligned}$$

or, equivalently,

$$\sum_{\{i,e\}\in\{\{k,e\} \mid k\in e\}} Sh_{\{i,e\}}(I(H), u) = Sh_e\big(H, v^N\big). \tag{14.1}$$

Note that $|e| > 1$ for each $e \in H$, so there exist at least two nodes $i, j \in e$ such that $\{i, e\}, \{j, e\} \in I(H)$. By the definition of u, for any $L' \subseteq I(H) \setminus \{\{k, e\} \mid k \in e\}$ and any two links $\{i, e\}, \{j, e\} \in \{\{k, e\} \mid k \in e\}$ in $I(H)$, we have $u(L' \cup \{i, e\}) = u(L')) = u(L' \cup \{j, e\})$. So $\{i, e\}$ and $\{j, e\}$ are symmetric in $(I(H), u)$. From the symmetry of the Shapley value, it follows that $Sh_{\{i,e\}}(I(H), u) = Sh_{\{j,e\}}(I(H), u)$ for any two links $\{i, e\}, \{j, e\} \in \{\{k, e\} \mid k \in e\}$. By Eq. (14.1), for any $\{i, e\} \in \{\{k, e\} \mid k \in e\}$, we have

$$Sh_{\{i,e\}}(I(H), u) = \frac{1}{|e|} Sh_e\big(H, v^N\big).$$

Consequently,

$$\begin{aligned}
\pi_i(N, v, H) &= \sum_{e\in H_i} \frac{1}{|e|} Sh_e\big(H, v^N\big) = \sum_{\{i,e\}\in I(H)_i} Sh_{\{i,e\}}(I(H), u) \\
&= \sum_{l\in I(H)_i} Sh_l(I(H), u).
\end{aligned}$$

This completes the proof of Theorem 14.1. □

Remark 14.1 For the communication situations, Kongo [8] introduced the divided link game u which is defined on the link set of a graph $D(L)$, where $D(L)$ is obtained from the original graph L by dividing each link $e = \{i, j\}$ into two directed links (i.e., unilateral communication channel in [8]), as ij and ji. By definition of the incidence graph, it is easy to see that the incidence graph $I(L)$ of L coincides with $D(L)$ by regarding $\{i, e\}$ and $\{j, e\}$ in $I(L)$ as the directed links ij and ji in $D(L)$ respectively. This shows that if we restrict to the graphs L, the incidence graph game coincides with the divided link game in [8]. Therefore, Theorem 14.1 implies the previous characterization of the position value for a graph game, due to [8]. Moreover, the following Example 14.1 shows that the scope of Theorem 14.1 cannot be extended to include non-uniform hypergraph communication situations.

By applying the incidence graph, we can define the duplicated hyperlink game $(I(H), \widetilde{u})$ and provide another description of the position value for a uniform hypergraph game. Specifically, for any hypergraph game (N, v, H), duplicated hyperlink game $\widetilde{u}$ is defined as follows.

$$\widetilde{u}(L') = v^N\big(\{e \in H \mid \text{there exists a node } i \in N \text{ such that } \{i, e\} \in L'\}\big),$$

for any $L' \subseteq I(H)$.

In order to realize this characterization, as in Theorem 14.1, we just define a mapping $g : \Sigma(I(H)) \rightarrow \Sigma(H)$ as follows: for any hyperlinks $e, e' \in H$ and $\sigma \in \Sigma(I(H))$, $g(\sigma)(e) < g(\sigma)(e')$ if and only if $\min\{\sigma(\{i, e\}) \mid i \in e\} < \min\{\sigma(\{i', e'\}) \mid i' \in e'\}$. The remainder of the proof for this result is almost the same as that of Theorem 14.1 and is omitted.

Corollary 14.1 *For any uniform hypergraph game (N, v, H) and any $i \in N$,*

$$\pi_i(N, v, H) = \sum_{l \in I(H)_i} Sh_l\big(I(H), \widetilde{u}\big).$$

Example 14.1 Let $N = \{1, 2, \ldots, 7\}$ and $\mathscr{F}_2 = (N, F_2)$ be a 3-uniform hypergraph (see Fig. 14.3), where $F_2 = \big\{\{1, 2, 5\}, \{5, 6, 7\}, \{3, 4, 6\}\big\}$. Consider the 3-uniform

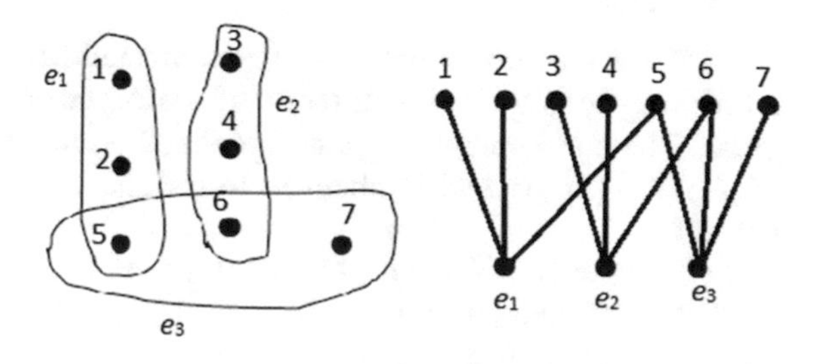

Fig. 14.3 The 3-uniform hypergraph $\mathscr{F}_2$ and its incidence graph $I(\mathscr{F}_2)$

hypergraph game (N, v, F_2) with $v = u_{\{1,3\}}$, where $u_{\{1,3\}}$ is the unanimity game of coalition $\{1, 3\}$.

The Shapley value associated to the hyperlink game v^N is determined by

$$Sh_{e_1}(F_2, v^N) = Sh_{e_2}(F_2, v^N) = Sh_{e_3}(F_2, v^N) = \frac{1}{3},$$

and it is easy to see that the Shapley values associated to the incidence graph game u and the duplicated hyperlink game $\widetilde{u}$ are $Sh_l(I(F_2), u) = Sh_l(I(F_2), \widetilde{u}) = \frac{1}{9}$ for all $l \in I(F_2)$. Therefore, by definition of the Myerson value and two different formula in Theorem 14.1 and Corollary 14.1, we obtain that the position value is

$$\pi(N, v, F_2) = \Big(\frac{1}{9}, \frac{1}{9}, \frac{1}{9}, \frac{1}{9}, \frac{2}{9}, \frac{2}{9}, \frac{1}{9}\Big).$$

However, Theorem 14.1 and Corollary 14.1 are not true for non-uniform hypergraphs. Consider the hypergraph game (N, v, F_1) (see Fig. 14.1) with $v = u_{\{1,2\}}$. Clearly, $Sh_{e_1}(F_1, v^N) = Sh_{e_2}(F_1, v^N) = Sh_{e_3}(F_1, v^N) = \frac{1}{3}$, so

$$\pi(N, v, F_1) = \Big(\frac{1}{6}, \frac{1}{6}, \frac{5}{18}, \frac{5}{18}, \frac{1}{9}\Big).$$

But, the Shapley value associated to the incidence graph game u is $Sh_l(I(F_1), u) = \frac{1}{7}$ for all $l \in I(F_1)$, and the Shapley value associated to the duplicated hyperlink game $\widetilde{u}$ is

$$\begin{aligned} Sh_{\{1,e_1\}}(I(F_1), \widetilde{u}) &= Sh_{\{3,e_1\}}(I(F_1), \widetilde{u}) \\ &= Sh_{\{2,e_2\}}(I(F_1), \widetilde{u}) = Sh_{\{4,e_2\}}(I(F_1), \widetilde{u}) = \frac{27}{140} \end{aligned}$$

and

$$Sh_{\{3,e_3\}}(I(F_1), \widetilde{u}) = Sh_{\{4,e_3\}}(I(F_1), \widetilde{u}) = Sh_{\{5,e_3\}}(I(F_1), \widetilde{u}) = \frac{8}{105}.$$

It is easy to see that formula in Theorem 14.1 and Corollary 14.1 do not hold for (N, v, F_1) and $\widetilde{u}$.

Next we turn our attention to the characterization of the position value for a uniform hypergraph game in terms of the Myerson value of the link-hypergraph game. The link-hypergraph game is defined on the link-hypergraph $\mathscr{L}\big(I(\mathscr{H})\big) = \big(I(H), L(I(\mathscr{H}))\big)$ and is obtained by original game v on a hypergraph $\mathscr{H} = (N, H)$.

In $\mathscr{L}\big(I(\mathscr{H})\big) = \big(I(H), L(I(\mathscr{H}))\big)$, let $N(V) = \{i \in N \mid I(H)_i \cap V \neq \emptyset\}$ for all $V \subseteq I(H)$ and let $H(V) = \{e \in H \mid \{i, e\} \in V,$ for all $i \in e\}$, where $I(H)_i = \{\{i, e\} \mid e \in H_i\}$. We define $\widehat{v}(V) = v(N(V))$, and we call $\big(I(H), \widehat{v}, L(I(\mathscr{H}))\big)$ the *link-hypergraph game*, where $I(H)$ is the node set of link-hypergraph $\mathscr{L}(I(\mathscr{H})) = \big(I(H), L(I(\mathscr{H}))\big)$.

Now we can express the position value for a uniform hypergraph game (N, v, H) in terms of the Myerson value of the link-hypergraph game $\big(I(H), \widehat{v}, L(I(\mathscr{H}))\big)$.

Theorem 14.2 *For any uniform hypergraph game (N, v, H) and any $i \in N$,*

$$\pi_i(N, v, H) = \sum_{\{i,e\} \in I(H)_i} \mu_{\{i,e\}}\big(I(H), \widehat{v}, L(I(\mathscr{H}))\big).$$

Proof For any $V \subseteq I(H)$ in $\mathscr{L}(I(\mathscr{H}))$, by definitions of the Myerson restricted game and the link-hypergraph game, we have

$$\widehat{v}^{L(I(\mathscr{H}))}(V) = \sum_{\widehat{S} \in V/\big(L(I(\mathscr{H}))[V]\big)} \widehat{v}(\widehat{S}) = \sum_{\widehat{S} \in V/\big(L(I(\mathscr{H}))[V]\big)} v\big(N(\widehat{S})\big)$$

$$= \sum_{S \in N(V)/H(V)} v(S) = \sum_{S \in N/H(V)} v(S) = v^N\big(H(V)\big),$$

where the forth equality holds since v is zero-normalized; the last equality holds follows the definition of the hyperlink game. On the other hand, by the definition of u, we have

$$u(V) = v^N(\{e \in H \mid \{i, e\} \in V, foralli \in e\}) = v^N(H(V)) = \widehat{v}^{L(I(\mathscr{H}))}(V).$$

Therefore, by Theorem 14.1, we have

$$\pi_i(N, v, H) = \sum_{l \in I(H)_i} Sh_l\big(I(H), u\big) = \sum_{\{i,e\} \in I(H)_i} \mu_{\{i,e\}}\big(I(H), \widehat{v}, L(I(H))\big).$$

This completes the proof of Theorem 14.2. □

Example 14.2 Consider the 3-uniform hypergraph game (N, v, F_2) with $v = u_{\{1,3\}}$ described in Example 14.1. We calculate the Myerson value associated to the link-hypergraph game $\widehat{v}$: $\mu_l\big(I(F_2), \widehat{v}, L(I(F_2))\big) = \frac{1}{9}$ for all $l \in I(F_2)$ and, therefore, the position value is $\pi(N, v, F_2) = \big(\frac{1}{9}, \frac{1}{9}, \frac{1}{9}, \frac{1}{9}, \frac{2}{9}, \frac{2}{9}, \frac{1}{9}\big)$ by Theorem 14.2.

Remark 14.2 When we restrict our attention to the uniform hypergraph (including the graph), it is easy to check that the link-hypergraph game is the same as the link agent form (LAF) or the hyperlink agent form (HAF) introduced by Casajus [5]. Here the definition of link-hypergraph game gives a clean and concise representation of the hyperlink agent form. However, by Example 14.1, we see that Theorem 14.2 does not hold for non-uniform hypergraph games.

14.3.3 Characterizations of the Myerson Value

In this subsection, by continuing to use "the incidence graph", we consider the relationships between the Myerson value of hypergraph communication situations and incidence graph games with coalition structures. By definition of the incidence graph $I(\mathscr{H}) = (N \cup H, I(H))$, clearly $\mathscr{I} = \{I(H)_1, I(H)_2, \ldots, I(H)_n\}$ forms a partition of the link set $I(H)$, where $n = |N|$. This means that the partition can be regarded as a coalition structure on $I(H)$. We define a triple

$$\big(I(H), u, \{I(H)_1, I(H)_2, \ldots, I(H)_n\}\big)$$

to be an *incidence graph game with a coalition structure*.

The following results give clear the relationships between the Myerson value of a hypergraph communication situation and the Owen value and the two-step Shapley value of the incidence graph game with a coalition structure corresponding to the original game. Since the proofs are similar to those of Theorems 3 and 4 in [8], we omit them.

Theorem 14.3 *For any hypergraph game (N, v, H) and any $i \in N$,*

$$\mu_i(N, v, H) = \sum_{l \in I(H)_i} \phi_l\big(I(H), u, \{I(H)_1, I(H)_2, \ldots, I(H)_n\}\big),$$

where $I(H)_i = \big\{\{i, e\} \in I(H) \mid e \in H_i\big\}$.

Theorem 14.4 *For any hypergraph game (N, v, H) and any $i \in N$,*

$$\mu_i(N, v, H) = \sum_{l \in I(H)_i} \chi_l(I(H), u, \{I(H)_1, I(H)_2, \ldots, I(H)_n\}),$$

where $I(H)_i = \big\{\{i, e\} \in I(H) \mid e \in H_i\big\}$.

Example 14.3 Consider the hypergraph game (N, v, F_1) with $v = u_{\{1,2\}}$ exhibited in Fig. 14.1. Then $u(L') = 1$ if $L' = I(F_1)$ and $u(L') = 0$ otherwise. The Owen value ϕ associated to the incidence graph game u is

$$\phi_{\{1,e_1\}} = \phi_{\{2,e_2\}} = \phi_{\{5,e_3\}} = \frac{1}{5}, \phi_{\{3,e_1\}} = \phi_{\{3,e_3\}} = \phi_{\{4,e_2\}} = \phi_{\{4,e_3\}} = \frac{1}{10},$$

where write $\phi_{\{i,e_j\}}$ for $\phi_{\{i,e_j\}}(I(F_1), u, \{I(F_1)_1, I(F_1)_2, \ldots, I(F_1)_5\})$. The two-step Shapley value χ to the incidence graph game u is $\chi_l = \frac{1}{5}$ for all $l \in I(F_1)$, where write χ_l for $\chi_l(I(F_1), u, \{I(F_1)_1, I(F_1)_2, \ldots, I(F_1)_5\})$. By the formula in Theorems 14.3 and 14.4, we obtain that the Myerson value is $\mu(N, v, F_1) = (\frac{1}{5}, \frac{1}{5}, \frac{1}{5}, \frac{1}{5}, \frac{1}{5})$.

For the duplicated hyperlink game, it fails to represent the Myerson value for hypergraph communication situations by the Owen value or the two-step Shapley value of the duplicated link game. In other words, Theorems 14.3 and 14.4 are not true for the duplicated link game $\widetilde{u}$. We consider the hypergraph game (N, v, F_1) with $v = u_{\{1,3\}}$ again. By definitions, it is easily calculated that the Owen value ϕ associated to the duplicated link game $\widetilde{u}$ is

$$\phi_{\{1,e_1\}} = \phi_{\{2,e_2\}} = \frac{1}{5}, \phi_{\{5,e_3\}} = \frac{1}{30}, \phi_{\{3,e_1\}} = \phi_{\{4,e_2\}} = \frac{27}{120}, \phi_{\{3,e_3\}} = \phi_{\{4,e_3\}} = \frac{7}{120},$$

and

$$\chi_{\{1,e_1\}} = \chi_{\{2,e_2\}} = \frac{1}{5}, \chi_{\{5,e_3\}} = \frac{1}{30}, \chi_{\{3,e_1\}} = \chi_{\{4,e_2\}} = \frac{27}{120}, \chi_{\{3,e_3\}} = \phi_{\{4,e_3\}} = \frac{7}{120}.$$

Obviously, the formula in Theorems 14.3 and 14.4 do not hold for (N, v, F_1) and $\widetilde{u}$.

Acknowledgements This research was supported in part by the National Nature Science Foundation of China (grant number 11571222).

References

1. Algaba, E., Bilbao, J.M., Borm, P., López, J.J.: The position value for union stable systems. Math. Meth. Oper. Res. **52**, 221–236 (2000)
2. Algaba, E., Bilbao, J.M., Borm, P., López, J.J.: The Myerson value for union stable systems. Math. Meth. Oper. Res. **54**(3), 359–371 (2001)
3. Béal, S., Rémila, E., Solal, P.: Fairness and fairness for neighbors: the difference between the Myerson value and component-wise egalitarian solutions. Econ. Lett. **117**(1), 263–267 (2012)
4. Borm, P., Owen, G., Tijs, S.: On the position value for communication situations. SIAM J. Discret. Math. **5**, 305–320 (1992)
5. Casajus, A.: The position value is the Myerson value, in a sense. Int. J. Game Theory **36**, 47–55 (2007)
6. Herings, P.J.J., van der Laan, G., Talman, A.J.J.: The average tree solution for cycle-free graph games. Games Econ. Behav. **62**(1), 77–92 (2008)
7. Kamijo, Y.: A two-step Shapley value for cooperative games with coalition structures. Int. Game Theory Rev. **11**, 207–214 (2007)
8. Kongo, T.: Difference between the position value and the Myerson value is due to the existence of coalition structures. Int. J. Game Theory **39**, 669–675 (2010)
9. Meessen, R.: Communication games. Master's thesis, Department of Mathematics, University of Nijmegen, The Netherlands (1988) (in Dutch)
10. Myerson, R.B.: Graphs and cooperation in games. Math. Oper. Res. **2**, 225–229 (1977)
11. Myerson, R.B.: Conference structures and fair allocation rules. Int. J. Game Theory **9**, 169–182 (1980)
12. Owen, G.: Value of games with a priori unions. In: Henn, R., Moeschlin, O. (eds.) Mathematical Economics and Game Theory, pp. 76–88. Springer, Berlin (1977)
13. Shan, E., Zhang, G., Dong, Y.: Component-wise proportional solutions for communication graph games. Math. Soc. Sci. **81**, 22–28 (2016)
14. Shapley, L.S.: A value for n-person games. In: Kuhn, H., Tucker, A.W. (eds.) Contributions to the Theory of Games II, pp. 307–317. Princeton University Press, Princeton (1953)
15. Slikker, M.: A characterization of the position value. Int. J. Game Theory **33**, 505–514 (2005)

16. Slikker, M., van den Nouweland, A.: Social and Economic Networks in Cooperative Game Theory. Kluwer, Norwell (2001)
17. van den Brink, R., van der Laan, G., Pruzhansky, V.: Harsanyi power solutions for graph-restricted games. Int. J. Game Theory **40**, 87–110 (2011)
18. van den Brink, R., Khmelnitskaya, A., van der Laan, G.: An efficient and fair solution for communication graph games. Econ. Lett. **117**(3), 786–789 (2012)
19. van den Nouweland, A., Borm, P., Tijs, S.: Allocation rules for hypergraph communication situations. Int. J. Game Theory **20**, 255–268 (1992)

Chapter 15
Bertrand Meets Ford: Benefits and Losses

Alexander Sidorov, Mathieu Parenti, and Jacques-Francois Thisse

Abstract The paper carries out the detailed comparison of two types of imperfect competition in a general equilibrium model. The price-taking Bertrand competition assumes the myopic income-taking behavior of firms, another type of behavior, price competition under a Ford effect, implies that the firms' strategic choice takes into account their impact to consumers' income. Our findings suggest that firms under the Ford effect gather more market power (measured by Lerner index), than "myopic" firms, which is agreed with the folk wisdom "Knowledge is power." Another folk wisdom implies that increasing of the firms' market power leads to diminishing in consumers' well-being (measured by indirect utility.) We show that in general this is not true. We also obtain the sufficient conditions on the representative consumer preference providing the "intuitive" behavior of the indirect utility and show that this condition satisfy the classes of utility functions, which are commonly used as examples (e.g., CES, CARA and HARA.)

A. Sidorov (✉)
Novosibirsk State University, Novosibirsk, Russia

Sobolev Institute for Mathematics, Novosibirsk, Russia
e-mail: a.sidorov@g.nsu.ru

M. Parenti
European Center for Advanced Research in Economics and Statistics (ECARES), Bruxelles, Belgium
e-mail: mathieu.parenti@ulb.ac.be

J.-F. Thisse
CORE-UCLouvain, Louvain-la-Neuve, Belgium

NRU-Higher School of Economics, Moscow, Russia
e-mail: jacques.thisse@uclouvain.be

L. A. Petrosyan et al. (eds.), *Frontiers of Dynamic Games*,
Static & Dynamic Game Theory: Foundations & Applications,
https://doi.org/10.1007/978-3-319-92988-0_15

15.1 Introduction

"The elegant fiction of competitive equilibrium" does not dominates now the frontier of theoretical microeconomics as stated by Marschak and Selten in [11] in early 1970s, being replaced by also elegant monopolistic competitive Dixit-Stiglitz "engine". The idea that firms are price-makers even if their number is "very large", e.g., continuum, is a common wisdom. But what if the monopolistic competitive equilibrium conception, where firms has zero impact to market statistics and, therefore, treat them as given, is just a brand new elegant fiction? When firms are sufficiently large, they face demands, which are influenced by the income level, depending in turn on their profits. As a result, firms must anticipate accurately what the total income will be. In addition, firms should be aware that they can manipulate the income level, whence their "true" demands, through their own strategies with the aim of maximizing profits [8]. This feedback effect is known as the *Ford effect*. In popular literature, this idea is usually attributed to Henry Ford, who raised wages at his auto plants to five dollars a day in January 1914. Ford wrote "our own sales depend on the wages we pay. If we can distribute high wages, then that money is going to be spent and it will serve to make... workers in other lines more prosperous and their prosperity is going to be reflected in our sales", see [7, p. 124–127]. To make things clear, we have to mention that the term "Ford effect" may be used in various specifications. As specified in [5], the Ford effect may have different scopes of consumers income, which is sum of wage and a share of the distributed profits. The first (extreme) specification is to take a whole income parametrically. This is one of solutions proposed by Marschak and Selten [11] and used, for instance, by Hart [9]. This case may be referred as "No Ford effect". Another specification (also proposed by Marschak and Selten [11] and used by d'Aspremont et al. [5]) is to suppose that firms take into account the effects of their decision on the total wage bill, but not on the distributed profits, which are still treated parametrically. This case may be referred as "Wage Ford effect" and it is exactly what Henry Ford meant in above citation. One more intermediate specification of The Ford effect is an opposite case to the previous one: firms take wage as given, but take into account the effects of their decisions on distributed profits. This case may be referred as "Profit Ford effect". Finally, the second extreme case, Full Ford effect, assumes that firms take into account total effect of their decisions, both on wages and on profits. These two cases are studied in newly published paper [4]. In what follows, we shall assume that wage is determined. This includes the way proposed by Hart [9], in which the worker fixed the nominal wage through their union. This assumption implies that only the Profit Ford effect is possible, moreover, firms maximize their profit anyway, thus being price-makers but not wage-makers, they have no additional powers at hand in comparison to No Ford case, with except the purely informational advantage—knowledge on consequences of their decisions. Nevertheless, as we show in this paper, this advantage allows firms to get more market power, which vindicate the wisdom "Knowledge is Power". As for welfare effect of this Knowledge, we show that it is ambiguous, but typically it is harmful for consumes. It should be mentioned also that being close in ideas with paper [4], we have no intersections in results, because the underlying economy model of this paper differers from our one, moreover, that research focuses on existence and uniqueness of equilibria with different specifications of Ford effect and does not concern the aspects of market power and welfare. We leave out of the scope of our research all consideration concerning Wage Ford effect, such as Big Push effect[1] and High Wage doctrine of stimulating consumer demand through wages. The idea that the firm could unilaterally use wages to increase demand for its own product enough to offset wage cost seems highly unlikely and was criticized by various reasons, including empirical evidences. For further discussions see [10, 15].

15.2 Model and Equilibrium in Closed Industry

15.2.1 Firms and Consumers

The economy involves one sector supplying a horizontally differentiated good and one production factor—labor. There is a continuum mass L of identical consumers endowed with one unit of labor. The labor market is perfectly competitive and labor is chosen as the numéraire. The differentiated good is made available under the form of a finite and discrete number $n \geq 2$ of varieties. Each variety is produced by a single firm and each firm produces a single variety. Thus, n is also the number of firms. To operate every firm needs a fixed requirement $f > 0$ and a marginal requirement $c > 0$ of labor. Without loss of generality we may normalize marginal requirement c to one. Since wage is also normalized to 1, the cost of producing q_i units of variety $i = 1, ..., n$ is equal to $f + 1 \cdot q_i$.

Consumers share the same additive preferences given by

$$U(\mathbf{x}) = \sum_{i=1}^{n} u(x_i), \tag{15.1}$$

where $u(x)$ is thrice continuously differentiable function, strictly increasing, strictly concave, and such that $u(0) = 0$. The strict concavity of u means that a consumer has a love for variety: when the consumer is allowed to consume X units of the differentiated good, she strictly prefers the consumption profile $x_i = X/n$ to any other profile $\mathbf{x} = (x_1, ..., x_n)$ such that $\sum_i x_i = X$. Because all consumers are identical, they consume the same quantity x_i of variety $i = 1, ..., n$.

[1]Suggesting that if firm profits are tied to local consumption, then firms create an externality by paying high wages: the size of the market for other firms increases with worker wages and wealth, see [12].

Following [17], we define the relative love for variety (RLV) as follows:

$$r_u(x) = -\frac{xu''(x)}{u'(x)}, \tag{15.2}$$

which is strictly positive for all $x > 0$. Technically RLV coincides with the Arrow-Pratt's relative risk-aversion concept, which we avoid to use due to possible misleading association in terms, because in our model there is no any uncertainty or risk considerations. Nevertheless, one can find some similarity in meaning of these concepts as the RLV measures the intensity of consumers' variety-seeking behavior. Under the CES, we have $u(x) = x^{\rho}$ where ρ is a constant such that $0 < \rho < 1$, thus implying a constant RLV given by $1 - \rho$. Another example of additive preferences is paper [2] where authors consider the CARA utility $u(x) = 1 - \exp(-\alpha x)$ with $\alpha > 0$ is the absolute love for variety (which is defined pretty much like the absolute risk aversion measure $-u''(x)/u'(x)$); the RLV is now given by αx.

A consumer's income is equal to her wage plus her share in total profits. Since we focus on symmetric equilibria, consumers must have the same income, which means that profits have to be uniformly distributed across consumers. In this case, a consumer's income y is given by

$$y = 1 + \frac{1}{L}\sum_{i=1}^{n} \Pi_i \geq 1,$$

where the profit made by the firm selling variety i is given by

$$\Pi_i = (p_i - 1)q_i - f, \tag{15.3}$$

p_i being the price of variety i. Evidently, the income level varies with firms' strategies.

A consumer's budget constraint is given by

$$\sum_{i=1}^{n} p_i x_i = y, \tag{15.4}$$

where x_i stands for the consumption of variety i.

The first-order condition for utility maximization yields

$$u'(x_i) = \lambda p_i, \tag{15.5}$$

where λ is the Lagrange multiplier of budget constraint. Conditions (15.4) and (15.5) imply that

$$\lambda = \frac{\sum_{j=1}^{n} u'(x_j)x_j}{y} > 0. \tag{15.6}$$

15.2.2 Market Equilibrium

The **market equilibrium** is defined by the following conditions:

1. each consumer maximizes her utility (15.1) subject to her budget constraint (15.4),
2. each firm i maximizes its profit (15.3) with respect to p_i ,
3. product market clearing: $Lx_i = q_i \quad \forall\, i = 1, ..., n$,
4. labor market clearing: $nf + \sum_{i=1}^{n} q_i = L$.

The last two equilibrium conditions imply that

$$\bar{x} \equiv \frac{1}{n} - \frac{f}{L} \tag{15.7}$$

is the only possible symmetric equilibrium demand, while the symmetric equilibrium output $\bar{q} = L\bar{x}$.

15.2.3 When Bertrand Meets Ford

As shown by (15.5) and (15.6), firms face demands, which are influenced by the income level, depending in turn on their profits. As a result, firms must anticipate accurately what the total income will be. In addition, firms should be aware that they can manipulate the income level, whence their "true" demands, through their own strategies with the aim of maximizing profits [8].

Let $\mathbf{p} = (p_1, ..., p_n)$ be a price profile. In this case, consumers' demand functions $x_i(\mathbf{p})$ are obtained by solving of consumer's problem—maximization of utility $U(\mathbf{x})$ subject to budget constraint (15.4)—with income y defined as

$$y(\mathbf{p}) = 1 + \sum_{j=1}^{n} (p_j - 1)x_j(\mathbf{p}).$$

It follows from (15.6) that the marginal utility of income λ is a market aggregate that depends on the price profile $\mathbf{p}$. Indeed, the budget constraint

$$\sum_{j=1}^{n} p_j x_j(\mathbf{p}) = y(\mathbf{p})$$

implies that

$$\lambda(\mathbf{p}) = \frac{1}{y(\mathbf{p})} \sum_{j=1}^{n} x_j(\mathbf{p})u'\left(x_j(\mathbf{p})\right),$$

while the first-order condition (15.5) may be represented as $\lambda(\mathbf{p})p_i = u'(x_i(\mathbf{p}))$. Since $u'(x)$ is strictly decreasing, the demand function for variety i is thus given by

$$x_i(\mathbf{p}) = \xi(\lambda(\mathbf{p})p_i), \tag{15.8}$$

where ξ is the inverse function to $u'(x)$. Thus, firm i's profits can be rewritten as

$$\Pi_i(\mathbf{p}) = (p_i - 1)x_i(\mathbf{p}) - f = (p_i - 1)\xi(\lambda(\mathbf{p})p_i) - f. \tag{15.9}$$

Remark 15.1 The definition of ξ implies that the Relative Love for Variety (15.2) may be equivalently represented as follows

$$r_u(x_i(\mathbf{p})) \equiv -\frac{\xi(\lambda(\mathbf{p})p_i)}{\xi'(\lambda(\mathbf{p})p_i)\lambda(\mathbf{p})p_i}. \tag{15.10}$$

Indeed, differentiating ξ as inverse to u' function, we obtain $\xi' = 1/u''$, while $x_i(\mathbf{p}) = \xi(\lambda(\mathbf{p})p_i)$, $u'(x_i(\mathbf{p})) = \lambda(\mathbf{p})p_i$.

Definition 15.1 For any given $n \geq 2$, a *Bertrand equilibrium* is a vector $\mathbf{p}^* = (p_1^*, ..., p_n^*)$ such that p_i^* maximizes $\Pi_i(p_i, \mathbf{p}_{-i}^*)$ for all $i = 1, ..., n$. This equilibrium is symmetric if $p_i^* = p_j^*$ for all i, j.

Applying the first-order condition to the profit (15.9) maximization problem, yields that the firm's i relative markup

$$m_i \equiv \frac{p_i - 1}{p_i} = -\frac{\xi(\lambda p_i)}{\xi'(\lambda p_i)\lambda p_i \cdot \left(1 + \frac{p_i}{\lambda}\frac{\partial\lambda}{\partial p_i}\right)}, \tag{15.11}$$

which involves $\partial\lambda/\partial p_i$ because λ depends on $\mathbf{p}$. Unlike what is assumed in partial equilibrium models of oligopoly, λ is here a function of $\mathbf{p}$, so that the markup depends on $\partial\lambda/\partial p_i \neq 0$. But how does firm i determine $\partial\lambda/\partial p_i$?

Since firm i is aware that λ is endogenous and depends on $\mathbf{p}$, it understands that the demand functions (15.8) must satisfy the budget constant as an identity. The consumer budget constraint can be rewritten as follows:

$$\sum_{j=1}^{n} p_j\xi(\lambda(\mathbf{p})p_j) = 1 + \sum_{j=1}^{n}(p_j - 1)\xi(\lambda(\mathbf{p})p_j),$$

which boils down to

$$\sum_{j=1}^{n} \xi(\lambda(\mathbf{p})p_j) = 1. \tag{15.12}$$

Differentiating (15.12) with respect to p_i yields

$$\xi'(\lambda p_i)\lambda + \frac{\partial\lambda}{\partial p_i}\sum_{j=1}^{n} p_j\xi'(\lambda p_j) = 0$$

or, equivalently,

$$\frac{\partial\lambda}{\partial p_i} = -\frac{\xi'(\lambda p_i)\lambda}{\sum_{j=1}^{n}\xi'(\lambda p_j)p_j}. \tag{15.13}$$

Substituting (15.13) into (15.11) and symmetrizing the resulting expression yields the candidate equilibrium markup:

$$\bar{m}^F = -\frac{\xi(\lambda p)}{\xi'(\lambda p)\cdot\lambda p\cdot\dfrac{n-1}{n}} = \frac{n}{n-1}r_u\left(\bar{x}\right), \tag{15.14}$$

where we use the identity (15.10) and $\bar{x} = \frac{1}{n} - \frac{f}{L}$ due to (15.7).

Proposition 15.1 *Assume that firms account for the Ford effect and that a symmetric equilibrium exists under Bertrand competition. Then, the equilibrium markup is given by*

$$\bar{m}^F = \frac{n}{n-1}r_u\left(\frac{1}{n} - \frac{f}{L}\right).$$

Note that $r_u\left(\frac{1}{n} - \frac{f}{L}\right)$ must be smaller than 1 for $\bar{m}^F < 1$ to hold. Since $\frac{1}{n} - \frac{f}{L}$ can take on any positive value in interval (0, 1), it must be

$$r_u(x) < 1\ \forall x \in (0, 1). \tag{15.15}$$

This condition means that the elasticity of a monopolist's inverse demand is smaller than 1 or, equivalently, the elasticity of the demand exceeds 1. In other words, the marginal revenue is positive. However, (15.15) is not sufficient for $\bar{m}^F$ to be smaller than 1. Here, a condition somewhat more demanding than (15.15) is required for the markup to be smaller than 1, that is, $r_u\left(\frac{1}{n} - \frac{f}{L}\right) < (n-1)/n$. Otherwise, there exists no symmetric price equilibrium. For example, in the CES case, $r_u(x) = 1 - \rho$ so that

$$\bar{m}^F = \frac{n}{n-1}(1-\rho) < 1,$$

which means that ρ must be larger than $1/n$. This condition is likely to hold because econometric estimations of the elasticity of substitution $\sigma = 1/(1-\rho)$ exceeds 3, see [1].

15.2.4 Income-Taking Firms

Now assume that, although firms are aware that consumers' income is endogenous, firms treat this income as a parameter. In other words, firms behave like *income-takers*. This approach is in the spirit of Hart (see [9]), for whom firms should take into account only some effects of their policy on the whole economy. Note that the income-taking assumption does not mean that profits have no impact on the market outcome. It means only that no firm seeks to manipulate its own demand through the income level. Formally, firms are income-takers when $\frac{\partial y}{\partial p_i} = 0$ for all i. Hence, the following result holds true. For the proof see Proposition 1 in [13].

Proposition 15.2 *Assume that firms are income-takers. If (15.15) holds and if a symmetric equilibrium exists under Bertrand competition, then the equilibrium markup is given by*

$$\bar{m}(n) = \frac{n}{n-1+r_u\left(\frac{1}{n}-\frac{f}{L}\right)} r_u \left(\frac{1}{n}-\frac{f}{L}\right). \tag{15.16}$$

Obvious inequality

$$\frac{n}{n-1+r_u\left(\frac{1}{n}-\frac{f}{L}\right)} < \frac{n}{n-1}$$

implies the following

Corollary 15.1 *Let number of firms n be given, then the income-taking firms charge the lesser price (or, equivalently, lesser markup) than the "Ford-effecting" firms.*

In other words, Ford effect provides to firms more marker power than in case of their income-taking behavior.

15.3 Free Entry Equilibrium

In equilibrium, profits must be non-negative for firms to operate. Moreover, if profit is strictly positive, this causes new firms to enter, while in the opposite case, i.e., when profit is negative, firms leave industry. The simple calculation shows that symmetric Zero-profit condition $\Pi = 0$ holds if and only if the number of firms satisfies

$$n^* = \frac{L}{f} m. \tag{15.17}$$

Indeed, let $L(p-1)\bar{x} - f = 0$ holds, where the symmetric equilibrium demand $\bar{x}$ is determined by (15.7). On the other hand, budget constraint (15.4) in symmetric case boils down to $n \cdot p\bar{x} = 1 + \sum_{i=1}^{n} \Pi_i = 1$ due to Zero-profit condition. Combining these identities, we obtain (15.17).

Assuming number of firms n is integer, we obtain generically that for two adjacent numbers, say n and $n+1$ the corresponding profits will have the opposite signs, e.g., $\Pi(n) > 0$, $\Pi(n+1) < 0$, and there is no integer number n^* providing the Zero-Profit condition $\Pi(n^*) = 0$. On the other hand, both markup expressions, (15.14) and (15.16), allow to use the arbitrary positive real values of n. The only problem is how to interpret the non-integer number of firms.[2] To simplify considerations, we assume that the fractional part $0 < \delta < 1$ of non-integer number of firms n^*, is a marginal firm, which entered to industry as the last, and its production is a linear extrapolation of typical firm, i.e., its fixed labor cost is equal to $\delta f < f$, while the production output is δq. In other words, marginal firm may be considered as "part-time-working firm".

Therefore, the equilibrium number of firms increases with the market size and the degree of firms' market power, which is measured by the Lerner index, and decreases with the level of fixed cost. Note also that

$$\bar{x} = \frac{f(1-m)}{Lm} > 0, \tag{15.18}$$

provided that m satisfies $0 < m < 1$. Substituting (15.17) and (15.18) into (15.14) and (15.16), we obtain that the equilibrium markups under free-entry must solve the following equations:

$$\bar{m}^F = \frac{f}{L} + r_u \left(\frac{f}{L} \frac{1 - \bar{m}^F}{\bar{m}^F} \right), \tag{15.19}$$

$$\bar{m} = \frac{f}{L} + \left(1 - \frac{f}{L}\right) r_u \left(\frac{f}{L} \frac{1 - \bar{m}}{\bar{m}} \right). \tag{15.20}$$

Under the CES, $\bar{m}^F = f/L + 1 - \rho$, while $\bar{m} = \rho f/L + 1 - \rho < \bar{m}^F$. It then follows from (15.17) and (15.18) that the equilibrium masses of firms satisfy $\bar{n}^F > \bar{n}$, while $\bar{q}^F < \bar{q}$. This result may be expanded to the general case. To prove

[2]Note that interpretation of non-integer *finite* number of oligopolies is totally different from the case of monopolistic competition, where mass of firms is *continuum* $[0, n]$, thus it does not matter whether n is integer or not. For further interpretational considerations see [13, subsection 4.3].

this, we assume additionally that

$$r_u(0) \equiv \lim_{x\to 0} r_u(x) < 1, \; r_{u'}(0) \equiv \lim_{x\to 0} r_{u'}(x) < 2 \tag{15.21}$$

Proposition 15.3 *Let conditions (15.21) hold and L be sufficiently large, then the equilibrium markups, outputs, and masses of firms are such that*

$$\bar{m}^F(L) > \bar{m}(L), \; \bar{q}^F(L) < \bar{q}(L), \; \bar{n}^F(L) > \bar{n}(L)$$

Furthermore, we have:

$$\lim_{L\to\infty} \bar{m}^F(L) = \lim_{L\to\infty} \bar{m}(L) = r_u(0).$$

Proof Considerations are essentially similar to the proof of Proposition 2 in [13]. Let's denote $\varphi = f/L$, then $L \to \infty$ implies $\varphi \to 0$ and condition "sufficiently large L" is equivalent to "sufficiently small φ."

It is sufficient to verify that function

$$G(m) \equiv \varphi + r_u\left(\varphi\frac{1-m}{m}\right) - m$$

is strictly decreasing at any solution of $\hat{m}$ of equation

$$m = \varphi + r_u\left(\varphi\frac{1-m}{m}\right) \tag{15.22}$$

Indeed, direct calculation show that

$$G'(m) = -\frac{1}{m}\left[\frac{1}{1-m}\frac{\varphi(1-m)}{m}r_u'\left(\varphi\frac{1-m}{m}\right) + m\right]. \tag{15.23}$$

Differentiating $r_u(x)$ and rearranging terms yields

$$r_u'(x)x = (1 + r_u(x) - r_{u'}(x))r_u(x)$$

for all $x > 0$. Applying this identity to $\hat{x} = \varphi\frac{1-\hat{m}}{\hat{m}}$ and substituting (15.22) into (15.23), we obtain

$$G'(\hat{m}) = -\frac{1}{\hat{m}}\left[\frac{r_u(\hat{x})\left(2 - \varphi - r_{u'}(\hat{x})\right)}{1-\hat{m}} + \varphi\right] < 0 \tag{15.24}$$

for all sufficiently small $\varphi = f/L$, or, equivalently, for all sufficiently large L. Moreover, inequality (15.24) implies, that there exists not more than one solution of Eq. (15.22), otherwise the sign of derivative $G'(m)$ must alternate for different roots.

An inequality $r_u(x) > 0$ for all x implies $G(0) \geq \varphi > 0$, while $G(1) = \varphi + r_u(0) - 1 < 0$, provided that $\varphi < 1 - r_u(0)$, therefore, for all sufficiently small φ there exists unique solution $\bar{m}^F(\varphi) \in (0, 1)$ of Eq. (15.22), which determines the symmetric Bertrand equilibrium under the Ford effect. In particular, inequality $m < \bar{m}^F(\varphi)$ holds if and only if $G(m) > 0$.

Existence an uniqueness of income-taking Bertrand equilibrium for all sufficiently small φ was proved in [13, Proposition 2]. By definition, the equilibrium markup $\bar{m}$ satisfies $F(\bar{m}(\varphi)) = 0$ for

$$F(m) \equiv \varphi + (1 - \varphi) r_u \left(\frac{\varphi(1 - m)}{m} \right) - m.$$

It is obvious that $G(m) > F(m)$ for all m and φ, therefore,

$$G(\bar{m}(\varphi)) > F(\bar{m}(\varphi)) = 0,$$

which implies $\bar{m}^F(\varphi) > \bar{m}(\varphi)$. The other inequalities follow from formulas (15.17) and (15.18).

The last statement of Proposition easily follows from the fact, that both equations $G(m) = 0$ and $F(m) = 0$ boil down to $m = r_u(0)$ when $\varphi \to 0$ (see proof of Proposition 2 in [13] for technical details.)

Whether the limit of competition is perfect competition (firms price at marginal cost) or monopolistic competition (firms price above marginal cost) when L is arbitrarily large depends on the value of $r_u(0)$. More precisely, when $r_u(0) > 0$, a very large number of firms whose size is small relative to the market size is consistent with a positive markup. This agrees with [3]. On the contrary, when $r_u(0) = 0$, a growing number of firms always leads to the perfectly competitive outcome, as maintained by Robinson [14]. To illustrate, consider the CARA utility given by $u(x) = 1 - \exp(-\alpha x)$. In this case, we have $r_u(0) = 0$, and thus the CARA model of monopolistic competition is not the limit of a large group of firms. By contrast, under CES preferences, $r_u(0) = 1 - \rho > 0$. Therefore, the CES model of monopolistic competition is the limit of a large group of firms.

15.4 Firms' Market Power vs. Consumers' Welfare

Proposition 15.3 also highlights the trade-off between *per variety* consumption and product diversity. To be precise, when free entry prevails, competition with Ford effect leads to a larger number of varieties, but to a lower consumption level per variety, than income-taking competition. Therefore, the relation between consumers' welfare values $\bar{V}^F = \bar{n}^F \cdot u(\bar{x}^F)$ and $\bar{V} = \bar{n} \cdot u(\bar{x})$ is a priori ambiguous.

In what follows we assume that the elemental utility satisfies $\lim_{x\to\infty} u'(x) = 0$, which is not too restrictive and typically holds for basic examples of utility functions. Consider the Social Planner's problem, who manipulates with masses of firms n trying to maximize consumers' utility $V(n) = n \cdot u(x)$ subject to the labor market clearing condition $(f + L \cdot x)n = L$, which is equivalent to maximization of

$$V(n) = n \cdot u\left(\frac{1}{n} - \varphi\right), \quad n \in (0, \varphi^{-1}),$$

where $\varphi = f/L$.

It is easy to see that

$$V(0) \equiv \lim_{n\to 0} n \cdot u\left(\frac{1}{n} - \varphi\right) = \lim_{x\to\infty} \frac{u(x)}{x + \varphi} = \lim_{x\to\infty} u'(x) = 0 = V(\varphi^{-1}),$$

where $x \equiv 1/n - \varphi$. Moreover,

$$V''(n) = \frac{1}{n^3} \cdot u''\left(\frac{1}{n} - \varphi\right) < 0,$$

which implies that graph of $V(n)$ is bell-shaped and there exists unique social optimum $n^* \in (0, \varphi^{-1})$, and $V'(n) \leq 0$ (resp. $V'(n) \geq 0$) for all $n \geq n^*$ (resp. $n \leq n^*$.)

This implies the following statement holds

Proposition 15.4

1. *If equilibrium number of the income-taking firms* $\bar{n} \geq n^*$, *then* $\bar{V}^F < \bar{V}$
2. *If equilibrium number of the Ford-effecting firms* $\bar{n}^F \leq n^*$, *then* $\bar{V}^F > \bar{V}$
3. *In the intermediate case* $\bar{n} < n^* < \bar{n}^F$ *the relation between* $\bar{V}^F$ *and* $\bar{V}$ *is ambiguous.*

In what follows, the first case will be referred as the "bad Ford" case, the second one—as the "good Ford" case.

Let's determine the nested elasticity of the elementary utility function

$$\Delta_u(x) \equiv \frac{x\varepsilon_u'(x)}{\varepsilon_u(x)},$$

where

$$\varepsilon_u(x) \equiv \frac{xu'(x)}{u(x)}.$$

The direct calculation shows that this function can be represented in different form

$$\Delta_u(x) = [1 - \varepsilon_u(x)] - r_u(x),$$

where $r_u(x)$ is Relative Love for Variety defined by (15.2), while $1 - \varepsilon_u(x)$ is so called *social markup*. Vives in [16] pointed out that social markup is the degree of preference for a single variety as it measures the proportion of the utility gain from adding a variety, holding quantity per firm fixed, and argued that 'natural' consumers' behavior implies increasing of social markup, or, equivalently, *decreasing of elasticity* $\varepsilon_u(x)$. In particular, the 'natural' behavior implies $\Delta_u(x) \leq 0$.

Lemma 15.1 *Let* $r_u(0) < 1$ *holds, then* $\Delta_u(0) \equiv \lim_{x\to 0} \Delta_u(x) = 0$.

Proof Assumptions on utility $u(x)$ imply that function $xu'(x)$ is strictly positive and

$$(xu'(x))' = 2u'(x) + xu''(x) = u'(x) \cdot (2 - r_{u'}(x)) > 0$$

for all $x > 0$, therefore there exists limit $\lambda = \lim_{x\to 0} x \cdot u'(x) \geq 0$. Assume that $\lambda > 0$, this is possible only if $u'(0) = +\infty$, therefore using the L'Hospital rule we obtain

$$\lambda = \lim_{x\to 0} x \cdot u'(x) = \lim_{x\to 0} \frac{x}{(u'(x))^{-1}} = \lim_{x\to 0} -\frac{(u'(x))^2}{u''(x)} = \lim_{x\to 0} \frac{xu'(x)}{-\frac{xu''(x)}{u'(x)}} = \frac{\lambda}{r_u(0)} > \lambda$$

because $r_u(0) < 1$ by (15.21). This contradiction implies that $\lambda = 0$. Therefore, using the L'Hospital rule, we obtain

$$\lim_{x\to 0} (1 - \varepsilon_u(x)) = 1 - \lim_{x\to 0} \frac{xu'(x)}{u(x)} = 1 - \lim_{x\to 0} \frac{u'(x) + xu''(x)}{u'(x)} = \lim_{x\to 0} r_u(x),$$

which implies $\Delta_u(0) = 0$.

The CES case is characterized by identity $\Delta_u(x) = 0$ for all $x > 0$, while for the other cases the sign and magnitude of $\Delta_u(x)$ may vary, as well as the directions of change for terms $1 - \varepsilon_u(x)$ and $r_u(x)$ may be arbitrary, see [6] for details.

Let $\delta_u \equiv \lim_{x\to 0} \Delta'_u(x)$, which may be finite or infinite. The following theorem provides the sufficient conditions for both "bad" and "good" Ford cases, while the obvious gap between (a) and (b) corresponds to the ambiguous third case of Proposition 15.4.

Theorem 15.1

(a) Let $\delta_u < r_u(0)$, *then for all sufficiently small* $\varphi = f/L$ *the 'bad Ford' inequality* $\bar{V} > \bar{V}^F$ *holds.*

(b) Let $\delta_u > \frac{r_u(0)}{1-r_u(0)}$, *then for all sufficiently small* $\varphi = f/L$ *the 'good Ford' inequality* $\bar{V}^F > \bar{V}$ *holds.*

Proof See Appendix.

It is obvious that in CES case $u(x) = x^\rho$ we obtain that $\delta_{CES} = 0 < r_{CES}(0) = 1-\rho$, thus CES is "bad For" function. Considering the CARA $u(x) = 1-e^{-\alpha x}, \alpha > 0$, HARA $u(x) = (x+\alpha)^\rho - \alpha^\rho, \alpha > 0$, and Quadratic $u(x) = \alpha x - x^2/2, \alpha > 0$, functions, we obtain $r_u(0) = 0$ for all these functions, while $\delta_{CARA} = -\alpha/2 < 0$, $\delta_{HARA} = -(1-\rho)/2\alpha < 0$ and $\delta_{Quad} = -1/2\alpha < 0$. This implies that these widely used classes of utility functions also belong to the "bad Ford" case.

To illustrate the opposite, "good Ford" case, consider the following function $u(x) = \alpha x^{\rho_1} + x^{\rho_2}$. Without loss of generality we may assume that $\rho_1 < \rho_2$, then

$$1 - \varepsilon_u(x) = \frac{\alpha(1-\rho_1) + (1-\rho_2)x^{\rho_2-\rho_1}}{\alpha + x^{\rho_2-\rho_1}},$$

$$r_u(x) = \frac{\alpha\rho_1(1-\rho_1) + \rho_2(1-\rho_2)x^{\rho_2-\rho_1}}{\alpha\rho_1 + \rho_2 x^{\rho_2-\rho_1}},$$

Using the L'Hospital rule we obtain

$$\lim_{x\to 0} \Delta'_u = \lim_{x\to 0} \frac{\alpha(\rho_2-\rho_1)^2 \cdot x^{-\rho_1-(1-\rho_2)}}{(\alpha + x^{\rho_2-\rho_1})(\alpha\rho_1 + \rho_2 x^{\rho_2-\rho_1})} = +\infty > \frac{r_u(0)}{1-r_u(0)} = \frac{1-\rho_1}{\rho_1}.$$

Corollary 15.2 *Let $\varepsilon'_u(0) < 0$, then $\bar{V} > \bar{V}^F$.*

Proof Using L'Hospital rule we obtain that

$$\delta_u = \lim_{x\to 0} \Delta'_u(x) = \lim_{x\to 0} \frac{\Delta_u(x)}{x} = \lim_{x\to 0} \frac{\varepsilon'_u(x)}{\varepsilon_u(x)} = \frac{1}{\varepsilon_u(0)} \lim_{x\to 0} \varepsilon'_u(x) < 0 \le r_u(0),$$

where $\varepsilon_u(0) = 1 - r_u(0) > 0$ due to assumption (15.21).

Remark 15.2 The paper [13] studied comparison of the Cournot and Bertrand oligopolistic equilibria under assumption of the income-taking behavior of firms. One of results obtained in this paper is that under Cournot competition firms charge the larger markup and produce lesser quantity, than under Bertrand competition, $\bar{m}^C > \bar{m}^B$, $\bar{q}^C < \bar{q}^B$, while equilibrium masses of firms $\bar{n}^C > \bar{n}^B$. This also implies ambiguity in comparison of the equilibrium indirect utilities $\bar{V}^C$ and $\bar{V}^B$. It is easily to see, that all considerations for $\bar{V}^F$ and $\bar{V}$ may be applied to this case and Theorem 15.1 (a) provides sufficient conditions for pro-Bertrand result $\delta_u < r_u(0) \Rightarrow \bar{V}^B > \bar{V}^C$. Moreover, considerations similar to proof of Theorem 15.1 (b) imply that inequality $\bar{V}^C > \bar{V}^B$ holds, provided that $\delta_u > 1$.

15.5 Concluding Remarks

Additive preferences are widely used in theoretical and empirical applications of monopolistic competition. This is why we have chosen to compare the market outcomes under two different competitive regimes when consumers are endowed with such preferences. It is important to stress, that unlike the widely used comparison of Cournot (quantity) and Bertrand (price) competitions, which are we compare two similar price competition regimes with "information" difference only: firms ignore or take into account strategically their impact to consumers' income. Moreover, unlike most models of industrial organization which assume the existence of an outside good, we have used a limited labor constraint. This has allowed us to highlight the role of the marginal utility of income in firms' behavior.

Acknowledgements This work was supported by the Russian Foundation for Basic Researches under grant No.18-010-00728 and by the program of fundamental scientific researches of the SB RAS No. I.5.1, Project No. 0314-2016-0018

Appendix

Proof of Theorem 15.1

Combining Zero-profit condition (15.17) $m = \frac{f}{L}n = \varphi n$ with formula for symmetric equilibrium demand $x = n^{-1} - \varphi \iff n = (x+\varphi)^{-1}$ we can rewrite the equilibrium mark-up equation for income-taking firms (15.20) as follows

$$\frac{\varphi}{x+\varphi} = \varphi + (1-\varphi)r_u(x).$$

Solving this equation with respect to x we obtain the symmetric equilibrium consumers' demand $x(\varphi)$, parametrized by $\varphi = f/L$, which cannot be represented in closed form for general utility $u(x)$, however, the inverse function $\varphi(x)$ has the closed-form solution

$$\varphi = \frac{1-x}{2} - \sqrt{\left(\frac{1-x}{2}\right)^2 - \frac{xr_u(x)}{1-r_u(x)}}. \tag{15.25}$$

It was mentioned above that graph of indirect utility $V(n)$ is bell-shaped and equilibrium masses of firms satisfy $n^* \leq \bar{n} \leq \bar{n}^F$ if and only if $V'(\bar{n}) \leq 0$. Calculating the first derivative $V'(n) = u(n^{-1} - \varphi) - n^{-1} \cdot u'(n^{-1} - \varphi)$ and substituting both $n = (x+\varphi)^{-1}$ and (15.25) we obtain that

$$n^* \leq \bar{n} \leq \bar{n}^F \iff u(x) \leq \left(\frac{1+x}{2} - \sqrt{\left(\frac{1-x}{2}\right)^2 - \frac{xr_u(x)}{1-r_u(x)}}\right) u'(x),$$

at $x = \bar{x}$—the equilibrium consumers demand in case of income-taking firms. The direct calculation shows that this inequality is equivalent to

$$\Delta_u(x) \leq (1 - r_u(x))\frac{1-x}{2}\left[1 - \sqrt{1 - \frac{4xr_u(x)}{(1-r_u(x))(1-x)^2}}\right]. \tag{15.26}$$

We shall prove that this inequality holds for all sufficiently small $x > 0$, provided that $\Delta'(0) < r_u(0)$. To do this, consider the following function

$$A_u(x) = \frac{x \cdot r_u(x)}{1-x},$$

which satisfies $A_u(0) = 0 = \Delta_u(0)$, $\Delta_u'(0) < A_u'(0) = r_u(0)$. This implies that inequality $\Delta_u(x) \leq A_u(x)$ holds for all sufficiently small $x > 0$.

Applying the obvious inequality $\sqrt{1-z} \leq 1 - z/2$ to

$$z = \frac{4xr_u(x)}{(1-r_u(x))(1-x)^2},$$

we obtain that the right-hand side of inequality (15.26)

$$(1 - r_u(x))\frac{1-x}{2}\left[1 - \sqrt{1 - \frac{4xr_u(x)}{(1-r_u(x))(1-x)^2}}\right] \geq A_u(x) \geq \Delta_u(x)$$

for all sufficiently small $x > 0$, which completes the proof of statement (a).

Applying the similar considerations to Eq. (15.19), which determines the equilibrium markup under a Ford effect, we obtain the following formula for inverse function $\varphi(x)$

$$\varphi = \frac{1 - r_u(x) - x}{2} - \sqrt{\left(\frac{1 - r_u(x) - x}{2}\right)^2 - xr_u(x)}$$

Using the similar considerations, we obtain that

$$\bar{n}^F \leq n^* \iff u(x) \geq \left(\frac{1 - r_u(x) + x}{2} - \sqrt{\left(\frac{1 - r_u(x) - x}{2}\right)^2 - xr_u(x)}\right) u'(x)$$

at $x = \bar{x}^F$—the equilibrium demand under Bertrand competition with Ford effect. The direct calculation shows that the last inequality is equivalent to

$$\Delta_u(x) \geq \frac{1 - r_u(x) - x}{2}\left[1 - \sqrt{1 - \frac{4xr_u(x)}{(1 - r_u(x) - x)^2}}\right]. \tag{15.27}$$

Now assume

$$\delta_u > \frac{r_u(0)}{1 - r_u(0)},$$

which implies that

$$\alpha \equiv \frac{r_u(0) + (1 - r_u(0))\delta_u}{2r_u(0)} > 1.$$

Let

$$B_u(x) \equiv \frac{\alpha x r_u(x)}{1 - r_u(x) - x},$$

it is obvious that $\Delta_u(0) = B_u(0) = 0$, and

$$B_u'(0) = \frac{\alpha r_u(0)}{1 - r_u(0)} = \frac{r_u(0) + (1 - r_u(0))\delta_u}{2(1 - r_u(0))} < \delta_u = \Delta_u'(0),$$

which implies that inequality $\Delta_u(x) \geq B_u(x)$ holds for all sufficiently small x.

On the other hand, the inequality $\sqrt{1 - z} \geq 1 - \alpha z/2$ obviously holds for any given $\alpha > 1$ and $z \in \left[0, \frac{4(\alpha-1)}{\alpha^2}\right]$. Applying this inequality to

$$z = \frac{4x r_u(x)}{(1 - r_u(x) - x)^2}, \alpha = \frac{r_u(0) + (1 - r_u(0))\delta_u}{2r_u(0)},$$

we obtain that the right-hand side of (15.27) satisfies

$$\frac{1 - r_u(x) - x}{2}\left[1 - \sqrt{1 - \frac{4x r_u(x)}{(1 - r_u(x) - x)^2}}\right] \leq B_u(x) \tag{15.28}$$

for all sufficiently small $x > 0$, because $x \to 0$ implies $z \to 0$. This completes the proof of Theorem 15.1.

References

1. Anderson, J.E., van Wincoop, E.: Trade costs. J. Econ. Lit. **42**, 691–751 (2004)
2. Behrens, K., Murata, Y.: General equilibrium models of monopolistic competition: a new approach. J. Econ. Theory **136**, 776–787 (2007)
3. Chamberlin, E.: The Theory of Monopolistic Competition. Harvard University Press, Cambridge (1933)
4. d'Aspremont, C., Dos Santos Ferreira, R.: The Dixit-Stiglitz economy with a 'small group' of firms: a simple and robust equilibrium markup formula. Res. Econ. **71**(4), 729–739 (2017)

5. d'Aspremont, C., Dos Santos Ferreira, R., Gerard-Varet, L.: On monopolistic competition and involuntary unemployment. Q. J. Econ. **105**(4), 895–919 (1990)
6. Dhingra, S., Morrow, J.: Monopolistic competition and optimum product diversity under firm heterogeneity. Res. Econ. **71**(4), 718–728 (2017)
7. Ford, H.: My Life and Work. Doubleday, Page, Garden City (1922)
8. Gabszewicz, J., Vial, J.: Oligopoly à la Cournot in general equilibrium analysis. J. Econ. Theory **4**, 381–400 (1972)
9. Hart, O.: Imperfect competition in general equilibrium: an overview of recent work. In: Arrow, K.J., Honkapohja, S. (eds.) Frontiers in Economics. Basil Blackwell, Oxford (1985)
10. Magruder, J.R.: Can minimum wage cause a big push? Evidence from Indonesia. J. Dev. Econ. **4**(3), 138–166 (2013)
11. Marschak T., Selten R.: General Equilibrium with Price-Making Firms. Lecture Notes in Economics and Mathematical Systems. Springer, Berlin (1972)
12. Murphy, K.M., Shleifer, A., Vishny, R.W.: Industrialization and the big push. J. Polit. Econ. **97**(5), 1003–1026 (1989)
13. Parenti, M., Sidorov, A.V., Thisse, J.-F., Zhelobodko, E.V.: Cournot, Bertrand or Chamberlin: toward a reconciliation. Int. J. Econ. Theory **13**(1), 29–45 (2017)
14. Robinson, J.: What is perfect competition? Q. J. Econ. **49**, 104–120 (1934)
15. Taylor, J.E.: Did Henry Ford mean to pay efficiency wages? J. Lab. Res. **24**(4), 683–694 (2003)
16. Vives, X.: Oligopoly Pricing. Old Ideas and New Tools. The MIT Press, Cambridge (2001)
17. Zhelobodko, E., Kokovin, S., Parenti M., Thisse, J.-F.: Monopolistic competition in general equilibrium: beyond the constant elasticity of substitution. Econometrica **80**, 2765–2784 (2012)

Chapter 16
On Multilateral Hierarchical Dynamic Decisions

Krzysztof Szajowski

Abstract Many decision problems in economics, information technology and industry can be transformed to an optimal stopping of adapted random vectors with some utility function over the set of Markov times with respect to filtration build by the decision maker's knowledge. The optimal stopping problem formulation is to find a stopping time which maximizes the expected value of the accepted (stopped) random vector's utility.

There are natural extensions of optimal stopping problem to stopping games-the problem of stopping random vectors by two or more decision makers. Various approaches dependent on the information scheme and the aims of the agents in a considered model. This report unifies a group of non-cooperative stopping game models with forced cooperation by the role of the agents, their aims and aspirations (v. Assaf and Samuel-Cahn (1998), Szajowski and Yasuda (1995)) or extensions of the strategy sets (v. Ramsey and Szajowski (2008)).

16.1 Introduction

The subject of the analysis is the problem of making collective decisions by the team of agents in which the position (significance) of the members is not equal. An object that is subject to management generates a signal that changes over time. Agents deal with capturing signals. Everyone can capture and save one of them, and its value is relative, determined by the function that takes into account the results of all decisions. Both the ability to observe signals and their capture determines the rank of agents who compete in this process. It is also possible that unequalizes of the decision makers is a consequence of social agreement or policy (v. [10]).

K. Szajowski (✉)
Wrocław University of Science and Technology, Wrocław, Poland
e-mail: Krzysztof.Szajowski@pwr.edu.pl

L. A. Petrosyan et al. (eds.), *Frontiers of Dynamic Games*,
Static & Dynamic Game Theory: Foundations & Applications,
https://doi.org/10.1007/978-3-319-92988-0_16

Earlier studies of such issues (cf. [6, 11, 13, 20, 29, 36]) showed their complexity, and detailed models of the analyzed cases a way to overcome difficulties in modeling and setting goals with the help of created models. The basic difficulty, except for cases when the decision is made by one agent, consists in determining the goals of the team, which can not always be determined so that the task can be reduced to the optimization of the objective function as the result of scalarisation. Most often, individual agents are to achieve an individual goal, but without the destabilization of the team. When modeling such a case, one should remember about establishing the rational goal of the agents in connection with the existence of the team (v. [5]). In the considerations of this study, we use methods of game theory with a finite number of players. However, the classic model of the antagonistic game is not the best example of progress. The team has interactions of agents resulting even from the hierarchy of access to information and the order in which decisions are made. The proposed overcoming of this difficulty consists in the appropriate construction of strategy sets and the payment function of players so that, taking into account the interactions, construct a multi-player game in which players have sets of acceptable strategies chosen regardless of the decisions of other players. Due to the sequential nature of the decision-making process, this player's decision-making independence is at the time of making it, but it is conditioned by the team's existing decision-making process.

Due to the fact that the goal of each agent, aspiration assessment by defining a withdrawal function, is to accept the most important signal from its point of view, the result of modeling is the task of repeatedly stopping the sequence of random vectors. In fact rating aspirations by defining the functions of payment is the one of the preliminary work on the mathematical modeling of management problem. Taking this into account, it should be mentioned here that this task was first put forward by Haggstrom [16], although Dynkin's [7] considerations can also be included in this category. Despite the undoubtedly interesting implications of such a model in applications, the subject has not been explored too much in its most general formulation, at least it has not been referred to. We will try to point out considerations that support such implicit modeling.[1]

In the game models applied to business decisions there are important models formulated and investigated by economist von Stackelberg [41].[2] Formulation of the game related to the secretary problem by Fushimi [14] with restricted set of strategies, namely threshold stopping times, opened research on the stopping games with leader by Szajowski (see papers [31, 35]). Similar games are subject of the research by Enns and Ferenstein [8], Radzik and Szajowski [28]. The extension of idea of Stackelberg was assumption that the lider is not fixed but the priority to the player is assigned randomly. Such version of the stopping game is investigated in

[1]The stopping games as the special case of the stochastic game has been presented by Jaśkiewicz and Nowak [17].

[2]This is his habilitation (see also the dissertation [42]), translated recently to English and published by Springer [43] (v. [9] for the review of the edition).

[36–38]. The Nash equilibria are obtained in the set of randomized strategies (cf. [25, 26]).

Two or multi-person process stopping, originally formulated by Dynkin [7],[3] met with more interest and research on multi-player games with stopping moments as players' strategies are quite well described in the literature. Both for random sequences and for certain classes of processes with continuous time. We will use this achievement in our deliberations.

In the following Sects. 16.2–16.4, we will discuss hierarchical diagrams in multi-person decision problems and their reduction to an antagonistic game. We will use the lattice properties of the stopping moments and we will obtain an equilibrium point in the problems under consideration based on the fixed point theorem for the game on the complete lattice.

16.2 Decision Makers' Hierarchy in Multi-Choice Problem

Let us consider N agents multiple-choice decision model on observation of stochastic sequence. The decision makers (DMs) are trying to choose the most profitable state based on sequential observation. In the case when more than one player would like to accept the state there are priority system which choose the beneficiary and the other players have right to observe further states of the process trying to get their winning state.

Agents' goals are defined by their payout functions. The rationality is subject of arbitrary decision when the mathematical model is formulated and should emphasize the requirement of the agents. One of the popular way is transformation of such multilateral problem to a non-zero-sum game. When there are two DMs it could be also zero-sum stopping game.

16.2.1 Zero-Sum Dynkin's Game

The originally Dynkin [7] has formulated the following optimization problem. Two players observe a realization of two real-valued processes (X_n) and (R_n). Player 1 can stop whenever $X_n \geq 0$, and player 2 can stop whenever $X_n < 0$. At the first stage τ in which one of the players stops, player 2 pays player 1 the amount R_τ and the process terminates. If no player ever stops, player 2 does not pay anything.

[3]See also models created by McKean [24] and Kifer [18].

A *strategy* of player 1 is a stopping time τ that satisfies $\{\tau = n\} \subset \{X_n \geq 0\}$ for every $n \geq 0$. A strategy σ of player 2 is defined analogously. The termination stage is simply $\nu = \min\{\tau, \sigma\}$. For a given pair (τ, σ) of strategies, denote by $K(\tau, \sigma) = \mathbf{E}\mathbb{I}_{\{\nu<\infty\}} R_\nu$ the expected payoff to player 1.

Dynkin [7] proved that if $\sup_{n\geq 0} |R_n| \in L_1$ then this problem has a value v i.e.

$$v = \sup_\tau \inf_\sigma K(\tau, \sigma) = \inf_\sigma \sup_\tau K(\tau, \sigma)$$

16.2.2 Non-zero Sum Stopping Game

Basement process under which the game is formulated can be defined as follows. Let $(X_n, \mathscr{F}_n, \mathbf{P}_x)_{n=0}^T$ be a homogeneous Markov process defined on a probability space $(\Omega, \mathscr{F}, \mathbf{P})$ with state space $(\mathbb{E}, \mathscr{B})$. At each moment $n = 1, 2, ..., T$, $T \in \tilde{\mathbb{N}} = \mathbb{N} \cup \{\infty\}$, the decision makers (henceforth called players) are able to observe the consecutive states of Markov process sequentially. There are N players. Each player has his own utility function $g_i : \mathbb{E}^N \to \Re$, $i = 1, 2, \ldots, N$, dependent on his own and others choices of state the Markov process. At moment n each decides separately whether to accept or reject the realization x_n of X_n. We assume the functions g_i are measurable and bounded.

- Let $\mathscr{T}^i$ be the set of pure strategies for ith player, the stopping times with respect to the filtration $(\mathscr{F}_n^i)_{n=1}^T$, $i = 1, 2, \ldots, N$. Each player has his own sequence of σ-fields $(\mathscr{F}_n^i)_{n=1}^T$ (the available information).
- The randomize extension of $\mathscr{T}_i$ can be constructed as follows (see [34, 44]). Let $(A_n^i)_{n=1}^T$, $i = 1, 2, \ldots, N$, be i.i.d.r.v. from the uniform distribution on $[0, 1]$ and independent of the Markov process $(X_n, \mathscr{F}_n, \mathbf{P}_x)_{n=0}^T$. Let $\mathscr{H}_n^i$ be the σ-field generated by $\mathscr{F}_n^i$ and $\{(A_s^i)_{s=1}^n\}$. A randomized Markov time $\tau(p^i)$ for strategy $p^i = (p_n^i) \in \mathscr{P}^{T,i} \subset \mathfrak{M}_i^T$ of the ith player is $\tau(p^i) = \inf\{T \geq n \geq 1 : A_n^i \leq p_n^i\}$.

Clearly, if each p_n^i is either zero or one, then the strategy is pure and $\tau(p^i)$ is in fact an $\{\mathscr{F}_n^i\}$-Markov time. In particular an $\{\mathscr{F}_n^i\}$-Markov time τ_i corresponds to the strategy $p^i = (p_n^i)$ with $p_n^i = \mathbb{I}_{\{\tau_i=n\}}$, where $\mathbb{I}_A$ is the indicator function for the set A.

Two concepts are take into account in this investigation. It can be compared with real investments and investment on the financial market. In real investment the choice of state is not reversible and sharable. In the financial market the choice of state by many players can be split of profit to all of them according some rules. Here, it is separately considered models of payoffs definition.

The payoff functions should be adequate to the information which players have and their decision. The player who do not use his information should be penalize.

- Let the players choose the strategies $\tau_i \in \mathscr{T}^i, i = 1, 2, \ldots, N$. The payoff of the ith player is $G_i(\tau_1, \tau_2, \ldots, \tau_N) = g_i(X_{\tau_1}, X_{\tau_2}, \ldots, X_{\tau_N})$.
- If the ith player control the ith component of the process, than the function $G_i(i_1, i_2, \ldots, i_n) = h_i(X_{i_1}, X_{i_2}, \ldots, X_{i_n})$ forms the random field. Such structure of payoffs has been considered by Mamer [22]. Under additional assumptions concerning monotonicity of incremental benefits of players Mamer has proved the existence of Nash equilibrium for two player non-zero sum game.
- Let $\mathscr{G}_n = \sigma(\mathscr{F}_n^1 \cup \mathscr{F}_n^2 \cup \ldots \cup \mathscr{F}_n^N)$ and $\mathscr{T}$ be the set of stopping with respect to $(\mathscr{G}_n)_{n=1}^T$. For a given choice of strategies by players the effective stopping time $\nu = \psi(\tau_1, \tau_2, \ldots, \tau_N)$ and $G_i(\tau_1, \tau_2, \ldots, \tau_N) = g_i(X_\nu)$. In some models the process X_n can be multidimensional and the payoff of ith player is the ith component of the vector X_n.

Definition 16.1 (Nash Equilibrium) The strategies $\tau_1^\star, \tau_2^\star, \ldots, \tau_N^\star$ are equilibrium in stopping game if for every player i

$$\mathbf{E}_x G_i(\tau_1^\star, \tau_2^\star, \ldots, \tau_N^\star) \geq \mathbf{E}_x G_i(\tau_1^\star, \tau_2^\star, \ldots, \tau_i, \ldots, \tau_N^\star). \tag{16.1}$$

16.2.3 Rights Assignment Models

However, there are different systems of rights to collect information about underlined process and priority in acceptance the states of the process. The various structures of decision process can have influence the knowledge of the players about the process which determine the pay-offs of the players. It is assumed that the priority decide about the investigation of the process and decision of the state acceptance. The details of the model, which should be precise are listed here.

1. The priority of the players can be defined before the game (in deterministic or random way) or it is dynamically managed in the play.
2. The priority of the players is decided after the collection of knowledge about the item by all players.
3. The random assignment of the rights can run before observation of each item and the accepted observation is not known to players with lowest priority. It makes that after the first acceptance some players are better informed than the others.
   (a) The information about accepted state is known to all players.
   (b) The information is hidden to the players who do not accepted the item.

The topics which are analyzed could be pointed out as follows:

1. Dynkin's game;
2. The fix and dynamic priority of the players: deterministic and random;

16.3 The Fix and Dynamic Priority of the Players

16.3.1 Deterministic Priority

16.3.1.1 Static

Among various methods of privileges for the players one of the simples is permutation of players' indices (rang). Let us propose a model of assignments the priority (rang) to the players as follows. In non-zero two person Dynkin's game the role of an *arbiter* was given to a random process. The simplest model can assume that the players are ordered before the play to avoid the conflict in assignment of presented sequentially states. At each moment the successive state of the process is presented to the players, they decide to stop and accept the state or continue observation. The state is given to the players with highest rang (we adopt here the convention that the player with rang 1 has the highest priority). In this case each stopping decision reduce the number of players in a game. It leads to recursive algorithm of construction the game value and in a consequence to determining the equilibrium (see [27, 33] for review of such models investigation).

The players decision and their priorities define an effective stopping time for player i in the following way.

- Let $P = \{1, 2, \ldots, N\}$ be the set of players and π a permutation of P. It determines the priority $\pi(i)$ of player i.

The considered model can be extended to fix deterministic priority. The effective stopping time for player i in this case one can get as follows.

- Let $(p^i_n)_{n=1}^T$ be the pure stopping strategy. If it is randomized stopping time we can find pure stopping time with respect to an extended filtration. The effective stopping strategy of the player i is following:

$$\tau_i((p)) = \inf\{k \geq 1 : p^i_k \prod_{j=1}^{N}(1 - p^j_k)\mathbb{I}_{\{j:\pi(j)<\pi(i)\}} = 1\}, \tag{16.2}$$

  where $\mathbf{p} = (p^1, p^2, \ldots, p^N)$ and each $p^i = (p^i_n)_{n=1}^T$ is adapted to the filtration $(\mathscr{F}^i_n)_{n=1}^T$. The effective stopping time of the player i is the stopping time with respect to the filtration $\tilde{\mathscr{F}}^i_n = \sigma\{\mathscr{F}^i_n, \{(p^j_k)^n_{k=1,\{j:\pi(j)<\pi(i)\}}\}\}$.
- The above construction of effective stopping time assures that each player will stop at different moment. It translates the problem of fixed priority optimization problem to the ordinary stopping game with payoffs $G_i(\tau_1, \tau_2, \ldots, \tau_N) = g_i(X_{\tau_1}, X_{\tau_2}, \ldots, X_{\tau_N})$.

16.3.1.2 Dynamic

In this case the effective stopping time for player i is obtained from parameters of the model similarly.

- Let $(p_n^i)_{n=1}^T$ be the pure stopping strategy. If it is randomized stopping time we can find pure stopping time with respect to an extended filtration. The effective stopping strategy of the player i is following:

$$\tau_i(\mathbf{p}) = \inf\{k \geq 1 : p_k^i \prod_{j=1}^{N} (1 - p_k^j)\mathbb{I}_{\{j:\pi_k(j)<\pi_k(i)\}} = 1\}, \tag{16.3}$$

  where $\mathbf{p} = (p^1, p^2, \ldots, p^N)$ and each $p^i = (p_n^i)_{n=1}^T$ is adapted to the filtration $(\mathscr{F}_n^i)_{n=1}^T$. The effective stopping time of the player i is the stopping time with respect to the filtration $\tilde{\mathscr{F}}_n^i = \sigma\{\mathscr{F}_n^i, \{(p_k^j)_{k=1,\{j:\pi_k(j)<\pi_k(i)\}}^n\}\}$.
- The above construction of effective stopping time assures that each player will stop at different moment. It translates the problem of fixed priority optimization problem to the ordinary stopping game with payoffs $G_i(\tau_1, \tau_2, \ldots, \tau_N) = g_i(X_{\tau_1}, X_{\tau_2}, \ldots, X_{\tau_N})$.

16.3.2 *The Random Priority of the Players*

16.3.2.1 Static (Fixed) and Dynamic

The random permutation of the players' can be model of the random fix priority when before the play the assignment of priority is based on the random permutation. The fixed permutation is valid for one turn of the game. The effective stopping time for player i has the following construction in this case.

- It is still fixed permutation of the player but its choice is random. The drawing of the permutation Π is done once for each play. Let $(p_n^i)_{n=1}^T$ be the pure stopping strategy. If it is randomized stopping time we can find pure stopping time with respect to an extended filtration. The effective stopping strategy of the player i is following:

$$\tau_i(\mathbf{p}) = \inf\{k \geq 1 : p_k^i \prod_{j=1}^{N} (1 - p_k^j)\mathbb{I}_{\{j:\Pi(j)<\Pi(i)\}} = 1\}, \tag{16.4}$$

  with rest of denotations the same as in the previous section, i.e. where $\mathbf{p} = (p^1, p^2, \ldots, p^N)$ and each $p^i = (p_n^i)_{n=1}^T$ is adapted to the filtration $(\mathscr{F}_n^i)_{n=1}^T$. The effective stopping time of the player i is the stopping time with respect to the filtration $\tilde{\mathscr{F}}_n^i = \sigma\{\mathscr{F}_n^i, \Pi, \{(p_k^j)_{k=1,\{j:\Pi(j)<\Pi(i)\}}^n\}\}$.

- The above construction of effective stopping time assures that each player will stop at different moment. It translates the problem of fixed priority optimization problem to the ordinary stopping game with payoffs $G_i(\tau_1, \tau_2, \ldots, \tau_N) = g_i(X_{\tau_1}, X_{\tau_2}, \ldots, X_{\tau_N})$.

When the priority is changing at each step of the game we have the dynamic random priority. The question is if the moment of the assignments, before the arrival of the observation and its presentation to the players or after, has a role. The effective stopping time for player i proposed here assume that the priority is determined before arrival of the observation, and the observation is presented according this order.

- If the priority is dynamic and random it is defined by the sequence $\Pi = (\Pi_k)_{k=1}^T$. The effective stopping strategy of the player i is following in this case:

$$\tau_i(\mathbf{p}, \Pi) = \inf\{k \geq 1 : p_k^i \prod_{j=1}^{N} (1 - p_k^j)\mathbb{I}_{\{j: \Pi_k(j) < \Pi_k(i)\}} = 1\}, \tag{16.5}$$

It is the stopping time w.r.t. $\tilde{\mathscr{F}}_n^i = \sigma\{\mathscr{F}_n^i, \Pi_k, \{(p_k^j)_{k=1,\{j:\Pi_k(j)<\Pi_k(i)\}}^n\}\}$.
- Each player stops at different moment. It translates the problem of fixed priority optimization problem to the ordinary stopping game with payoffs

$$G_i(\tau_1(\mathbf{p}, \Pi), \tau_2(\mathbf{p}, \Pi), \ldots, \tau_N(\mathbf{p}, \Pi)) = g_i(X_{\tau_1(\mathbf{p},\Pi)}, X_{\tau_2(\mathbf{p},\Pi)}, \ldots, X_{\tau_N(\mathbf{p},\Pi)}).$$

16.3.3 Restricted Observations of Lower Priority Players

16.3.3.1 Who Has Accepted the Observation?

In a sequential decision process taken by the players the consecutive acceptance decision are effectively done by some players. For every stopping time $\tau^i(\mathbf{p}, \Pi)$ the representation by the adapted random sequence $(\delta_k^i)_{k=1}^T$, $i = 1, 2, \ldots, N$, is given. Let us denote $\gamma_k = \inf\{1 \leq i \leq N : \delta_k^i = 1\}$, the player who accepted the observation at moment k, if any. Similar index can be defined for the fix deterministic and random priority as for dynamic, deterministic priority as well.

16.3.3.2 Restricted Knowledge

In the class of such games the natural question which appears is the accessibility of the information. It could be that the accepted observation by the high rang players are hidden for the lower rang players when has been accepted. However, some information are acquired taking into account the players' behavior.

- As the result of the decision process players collect information about the states of the process and some of them accept some states. In considered models it was assumed that players are equally informed about the process. Further it will be admitted that the player has access to information according the priority assigned to him. States accepted by others are not fully accessible to the players which have not seen it before. However, some conjectures are still available assuming rational behavior of the players and it is given by the function $\varphi^i_{\gamma_k}(X_k)$ for the player i when its priority is lower than player's γ_kth.
- Effective information available for the player i at moment k can be presented as follows.

$$\tilde{X}^i_k = X_k \mathbb{I}_{\{i:i\leq\gamma_k\}} + \varphi^i_{\gamma_k}(X_k)\mathbb{I}_{\{i:i>\gamma_k\}}.$$

- The player investigation and interaction with other players gives him filtration $\bar{\mathscr{F}}^i_n = \sigma\{\tilde{\mathscr{F}}^i_n, \varphi^i_{\gamma_k}(X_k)\}$.
- Each player stops at different moment. It translates the problem of random priority optimization problem, with restricted access to observation, to the ordinary stopping game with payoffs

$$G_i(\tau_1(\mathbf{p},\Pi),\tau_2(\mathbf{p},\Pi),\ldots,\tau_N(\mathbf{p},\Pi)) = g_i(\tilde{X}^i_{\tau_1(\mathbf{p},\Pi)},\tilde{X}^i_{\tau_2(\mathbf{p},\Pi)},\ldots,\tilde{X}^i_{\tau_N(\mathbf{p},\Pi)}).$$

Let us analyze who has accepted the observation? In a sequential decision process taken by the players the consecutive acceptance decision are effectively done by some players. For every stopping time $\tau^i(\mathbf{p},\Pi)$ the representation by the adapted random sequence $(\delta^i_k)^T_{k=1}$, $i = 1, 2, \ldots, N$, is given. Let us denote $\gamma_k = \inf\{1 \leq i \leq N : \delta^i_k = 1\}$ the player who accepted the observation at moment k if any. Similar index can be defined for the fix deterministic and random priority as for dynamic, deterministic priority as well.

16.4 Monotone Stopping Games with Priority

16.4.1 General Assumption

Boundedness assumptions–maximal payoff.

$$\mathbf{E}(\sup_{\substack{1\leq j_i\leq T\\ i=1,\ldots,N}} G_k(j_1,j_2,\ldots,j_N)) < \infty \tag{16.6}$$

$$\forall_{\substack{1\leq j_i\leq T\\ j\neq i}} \mathbf{E}(\inf_{1\leq n\leq T} G_k(j_1,\ldots,j_{i-1},n,j_{i+1},\ldots,j_N)) > -\infty \tag{16.7}$$

where $k = 1, 2, \ldots, N$.

In order to assure that each player has a best response to any strategy chosen by other players it is required:

$$\forall_{\substack{\tau_j\in\mathscr{T}^j\\ j\neq i}} \quad \mathbf{E}[\sup_n G_i(\tau_1,\ldots,\tau_{i-1},n,\tau_{i+1},\ldots,N)|\mathscr{F}_n^i]^+ \leq \infty \tag{16.8}$$

$$\forall_{\substack{\tau_j\in\mathscr{T}^j\\ j\neq i}} \quad \limsup_{n\to T} \mathbf{E}\left[G_i(\tau_1,\ldots,\tau_{i-1},n,\tau_{i+1},\ldots,N)|\mathscr{F}_n^i\right] \tag{16.9}$$

$$\leq \mathbf{E}(G_i(\tau_1,\ldots,\tau_{i-1},T,\tau_{i+1},\ldots,N)|\mathscr{F}_T^i) \text{ a.e..}$$

For further analyses the conditional expectation of the payoffs for player i should be determined. The sequence $\eta_n = \mathbf{E}(G_i(\tau_1,\ldots,\tau_{i-1},n,\tau_{i+1},\ldots,\tau_N)|\mathscr{F}_n)$ is the conditional expected return to player i if he decide to stop after his nth observation and other players uses the stopping rules of their choice. The sequence η_n is $\mathscr{F}_n^i$ adapted and, under the boundedness assumption presented above, the exists an optimal stopping rule for this sequences for $i = 1, 2, \ldots, N$.

Definition 16.2 (Regular Stopping Time) The stopping time $\tau_i \in \mathscr{T}^i$ is regular with respect to $\tau_1 \in \mathscr{T}^1, \ldots, \tau_{i-1} \in \mathscr{T}^{i-1}, \tau_{i+1} \in \mathscr{T}^{i+1}, \ldots, \tau_N \in \mathscr{T}^N$ if

$$\mathbf{E}(\eta_{\tau_i}|\mathscr{F}_n^i) \geq \mathbf{E}(\eta_n|\mathscr{F}_n^i) \text{ on } \{\omega : \tau_i > n\} \text{ for all } n. \tag{16.10}$$

Let $\overrightarrow{\tau_{-i}} = (\tau_1,\ldots,\tau_{i-1},\tau_{i+1},\ldots,\tau_n)$.

Maximal regular best response will be considered. By the results of [22] and [19] it can be established:

Lemma 16.1 *Under (16.6)–(16.9) each player has a unique, maximal regular best response* $\hat{\tau}_i(\overrightarrow{\tau_{-i}})$ *to any vector of stopping times* $\overrightarrow{\tau_{-i}}$ *chosen by his opponents.*

This does not immediately imply the Nash equilibrium existence.

16.4.2 Monotone Structure of Best Responses

The incremental benefit to player should be analyzed. It is assumed that increments of payoffs have the following properties. Let us consider the following increments of payoffs.

$$\forall_{\substack{m<T\\ 1\leq k\leq T}} \Delta_m^i(k,\overrightarrow{j_{-i}}) = G_i(j_1,\ldots,j_{i-1},m+k,j_{i+1},\ldots,j_N)$$
$$-G_i(j_1,\ldots,j_{i-1},m,j_{i+1},\ldots,j_N)$$

ND Let us assume that $\Delta_m^i(k,\mathbf{j}_{-i})$ is nondecreasing in $\overrightarrow{j_{-i}}$;

NI Let us assume that $\Delta_m^i(k,\mathbf{j}_{-i})$ is nonincreasing in $\overrightarrow{j_{-i}}$;

Lemma 16.2 *If (16.6)–(16.9) and condition ND are fulfilled and* $\sigma \in \mathscr{T}^i$ *is regular with respect of* $\overrightarrow{\tau_{-i}}_1 \in \mathscr{T}^{-i}$ *then it is also regular with respect to any* $\overrightarrow{\tau_{-i}}_2 \in \mathscr{T}^{-i}$ *such that* $\overrightarrow{\tau_{-i}}_1 \preceq \overrightarrow{\tau_{-i}}_2$ *a.e. (Under NI* $\overrightarrow{\tau_{-i}}_2 \preceq \overrightarrow{\tau_{-i}}_1$ *a.e.)*

Lemma 16.3 *Let* $\overrightarrow{\tau_{-i}}_k \in \mathscr{T}^{-i}$, $k = 1, 2$, *such that* $\overrightarrow{\tau_{-i}}_1 \preceq \overrightarrow{\tau_{-i}}_2$ *a.e. and ND is fulfilled then the best response* $\hat{\sigma}(\overrightarrow{\tau_{-i}}_1) \preceq \hat{(\sigma)}(\overrightarrow{\tau_{-i}}_2)$ *a.e. (Under NI* $\hat{\sigma}(\overrightarrow{\tau_{-i}}_1) \succeq \hat{\sigma}(\overrightarrow{\tau_{-i}}_2)$ *a.e.)*

16.4.2.1 Tarski's Fixed Point Theorem

The fixed point theorem which will be helpful for proving the existence of the equilibrium is obtained for the complete lattices and an isotone functions. We consider the partial order of random variables: $\tau \preceq \sigma$ iff $\tau \leq \sigma$ *a.e.* The operations of supremum of random variables and infimum of random variables are inner operations in $\mathscr{T}^i$. If the essential supremum is considered we have also for every subset $\mathscr{A} \subset \mathscr{S}$ that $\vee\mathscr{A} \in \mathscr{S}$ and $\wedge\mathscr{A} \in \mathscr{S}$.

Lemma 16.4 (Stopping Set Is a Complete Lattice) *The partially ordered sets* $\mathscr{T}^i$ *with order* $\preceq$ *and operations essential supremum* $\vee$ *and essential infimum* $\wedge$ *defined in it are complete lattices.*

Definition 16.3 (Isotone Function) Let $\mathscr{S}$ be lattice. f is isotone function from $\mathscr{S}$ into $\mathscr{S}$ if for $\tau, \sigma \in \mathscr{S}$ such that $\tau \preceq \sigma$ implies $f(\tau) \leq f(\sigma)$.

Theorem 16.1 ([40]) *If* $\mathscr{S}$ *is a complete lattice and if* f *is an isotone function from* $\mathscr{S}$ *into* $\mathscr{S}$, *then* f *has a fixed point.*

16.4.3 Main Result

Monotonicity of increments with integrability of payoff functions guarantee existence of Nash equilibrium in stopping game with various models of priority (rule of assignments) based on the theorem.

Theorem 16.2 ([22]) *Suppose that assumptions (16.6)–(16.9) with ND or NI holds. Then there is a Nash equilibrium pair of stopping times. There is an vector of stopping times* $\boldsymbol{\tau}^\star$ *which forms an equilibrium such that* $\tau_i^\star = \hat{\sigma}(\overrightarrow{\tau^\star_{-i}})$, $i = 1, 2, \ldots, N$.

16.5 Conclusion

Based on the consideration of the paper we know that the various priority approach model in the multiple choice problem can be transformed to the multiperson antagonistic game with the equilibrium point as the rational treatment. The equilibrium point in all these problems exist. The construction of them need individual treatment and it is not solved in general yet.

The presented decision model can be found with a slightly different interpretation, namely games with the arbitration procedure. Details can be found e.g. in the works of Sakaguchi [32] and Mazalov et al. [23].[4]

The close to the models are some multivariate stopping problem with cooperation [1, 30, 39]. In cooperative stopping games the players have to use the decision suggested by coordinator of the decision process (cf. [2, 15]).

In [21] the idea of voting stopping rules has been proposed. The game defined on the sequence of iid random vectors has been defined with the concept of the Nash equilibrium as the solution. There are generalization of the results obtained by Szajowski and Yasuda [39]. Conditions for a unique equilibrium among stationary threshold strategies in such games are given by Ferguson [12].

Acknowledgements The authors' thanks go to many colleagues taking part in discussion of the topics presented in the paper.

References

1. Assaf, D., Samuel-Cahn, E.: Optimal cooperative stopping rules for maximization of the product of the expected stopped values. Stat. Probab. Lett. **38**(1), 89–99 (1998). https://doi.org/10.1016/S0167-7152(97)00158-2. Zbl 0912.60061
2. Assaf, D., Samuel-Cahn, E.: Optimal multivariate stopping rules. J. Appl. Probab. **35**(3), 693–706 (1998). ISSN 0021-9002; 1475-6072/e. https://doi.org/10.1239/jap/1032265217. Zbl 0937.60040
3. Brams, S.J., Merrill, S., III: Arbitration procedures with the possibility of compromise. In: Stefański, J. (ed.) Bargaining and Arbitration in Conflicts. Control and Cybernetics, vol. 21(1), pp. 131–149. Systems Research Institute of the Polish Academy of Sciences, Warszawa (1992). http://control.ibspan.waw.pl:3000/contents/export?filename=1992-1-09_brams_merrill.pdf. ISSN 0324-8569. MR 1218889
4. Chatterjee, K.: Comparison of arbitration procedures: models with complete and incomplete information. IEEE Trans. Syst. Man Cybern. **11**(2), 101–109 (1981). ISSN 0018-9472. https://doi.org/10.1109/TSMC.1981.4308635. MR 611435
5. Diecidue, E., van de Ven, J.: Aspiration level, probability of success and failure, and expected utility. Int. Econ. Rev. **49**(2), 683–700 (2008). ISSN 0020-6598. https://doi.org/10.1111/j.1468-2354.2008.00494.x. MR 2404450
6. Dorobantu, D., Mancino, M.E., Pontier, M.: Optimal strategies in a risky debt context. Stochastics **81**(3–4), 269–277 (2009). ISSN 1744-2508. https://doi.org/10.1080/17442500902917433. MR 2549487
7. Dynkin, E.: Game variant of a problem on optimal stopping. Sov. Math. Dokl. **10**, 270–274, (1969). ISSN 0197-6788. Translation from Dokl. Akad. Nauk SSSR **185**, 16–19 (1969). Zbl 0186.25304

[4] See also [3] and [4] for details concerning arbitration procedure.

8. Enns, E.G., Ferenstein, E.Z.: On a multiperson time-sequential game with priorities. Seq. Anal. **6**(3), 239–256 (1987). ISSN 0747-4946. https://doi.org/10.1080/07474948708836129. MR 918908
9. Etro, F.: Book review of: Heinrich von Stackelberg, Market structure and equilibrium. J. Econ. **109**(1), 89–92 (2013). ISSN 0931-8658; 1617-7134/e. https://doi.org/10.1007/s00712-013-0341-9
10. Feng, Y., Xiao, B.: Revenue management with two market segments and reserved capacity for priority customers. Adv. Appl. Probab. **32**(3), 800–823 (2000). ISSN 0001-8678. https://doi.org/10.1239/aap/1013540245. MR 1788096
11. Ferenstein, E.Z.: Two-person non-zero-sum sequential games with priorities. In: Strategies for Sequential Search and Selection in Real Time (Amherst, MA, 1990). Contemporary Mathematics, vol. 125, pp. 119–133. American Mathematical Society, Providence (1992). https://doi.org/10.1090/conm/125/1160615. MR 1160615
12. Ferguson, T.S.: Selection by committee. In: Szajowski, K., Nowak, A.S. (eds.) Advances in Dynamic Games: Applications to Economics, Finance, Optimization, and Stochastic Control. Annals of the International Society of Dynamic Games, vol. 7, pp. 203–209. Birkhäser, Boston (2005). Zbl 1123.91003
13. Ferguson, T.S.: The sum-the-odds theorem with application to a stopping game of Sakaguchi. Math. Appl. (Warsaw) **44**(1), 45–61 (2016). ISSN 1730-2668. https://doi.org/10.14708/ma.v44i1.1192. MR 3557090
14. Fushimi, M.: The secretary problem in a competitive situation. J. Oper. Res. Soc. Jpn. **24**, 350–358 (1981). Zbl 0482.90090
15. Glickman, H.: Cooperative stopping rules in multivariate problems. Seq. Anal. **23**(3), 427–449 (2004)
16. Haggstrom, G.: Optimal sequential procedures when more then one stop is required. Ann. Math. Stat. **38**, 1618–1626 (1967)
17. Jaśkiewicz, A., Nowak, A.: Non-zero-sum stochastic games. In: Başsar, T., Zaccour, G. (eds.) Handbook of Dynamic Game Theory, 64pp. Birkhäuser and Springer International Publishing AG of Springer Nature, Bassel (2016). https://doi.org/10.1007/978-3-319-27335-8_33-1
18. Kifer, J.I.: Optimal behavior in games with an infinite sequence of moves. Teor. Verojatnost. i Primenen. **14**, 284–291 (1969). ISSN 0040-361x
19. Klass, M.J.: Properties of optimal extended-valued stopping rules for S_n/n. Ann. Probab. **1**, 719–757 (1973). ISSN 0091-1798; 2168-894X/e. https://doi.org/10.1214/aop/1176996843. Zbl 0281.62085
20. Krasnosielska-Kobos, A., Ferenstein, E.Z.: Construction of Nash equilibrium in a game version of Elfving's multiple stopping problem. Dyn. Games Appl. **3**(2), 220–235 (2013). ISSN 2153-0785. https://doi.org/10.1007/s13235-012-0070-7
21. Kurano, M., Yasuda, M., Nakagami, J.: Multi-variate stopping problem with a majority rule. J. Oper. Res. Soc. Jpn. **23**, 205–223 (1980). Zbl 0439.90105
22. Mamer, J.W.: Monotone stopping games. J. Appl. Probab. **24**, 386–401 (1987). ISSN 0021-9002; 1475-6072/e. https://doi.org/10.2307/3214263. Zbl 0617.90092
23. Mazalov, V.V., Sakaguchi, M., Zabelin, A.A.: Multistage arbitration game with random offers. Int. J. Math. Game Theory Algebra **12**(5), 409–417 (2002). ISSN 1060-9881. MR 1951832. Zbl 1076.91502
24. McKean, H.P., Jr.: Appendix: a free boundary problem for the heat equation arising from a problem of mathematical economics. Ind. Manag. Rev. **6**, 32–39 (1965)
25. Neumann, P., Porosiński, Z., Szajowski, K.: A note on two person full-information best choice problems with imperfect observation. In: Operations Research. Extended Abstracts of the 18th Symposium on Operations Research (GMOOR), Cologne, 1–3 September, 1993, pp. 355–358. Phisica, Heidelberg (1994)
26. Neumann, P., Ramsey, D., Szajowski, K.: Randomized stopping times in Dynkin games. Z. Angew. Math. Mech. **82**(11–12), 811–819 (2002). ISSN 0044-2267. https://doi.org/10.1002/1521-4001(200211)82:11/12<811::AID-ZAMM811>3.0.CO;2-P; 4th GAMM-Workshop "Stochastic Models and Control Theory" (Lutherstadt Wittenberg, 2001)

27. Nowak, A., Szajowski, K.: Nonzero-sum stochastic games. In: Bardi, T.P.M., Raghavan, T.E.S. (eds.) Stochastic and Differential Games. Theory and Numerical Methods. Annals of the International Society of Dynamic Games, pp. 297–342. Birkhäser, Boston (1998). MR 200d:91021. Zbl 0940.91014
28. Radzik, T., Szajowski, K.: On some sequential game. Pure Appl. Math. Sci. **28**(1–2), 51–63 (1988). ISSN 0379-3168
29. Ramsey, D., Cierpiał, D.: Cooperative strategies in stopping games. In: Advances in Dynamic Games and Their Applications. Annals of the International Society of Dynamic Games, vol. 10, pp. 415–430. Birkhäuser Boston, Boston (2009)
30. Ramsey, D.M., Szajowski, K.: Selection of a correlated equilibrium in Markov stopping games. Eur. J. Oper. Res. **184**(1), 185–206 (2008). https://doi.org/10.1016/j.ejor.2006.10.050
31. Ravindran, G., Szajowski, K.: Nonzero sum game with priority as Dynkin's game. Math. Japonica **37**(3), 401–413 (1992). ISSN 0025-5513
32. Sakaguchi, M.: A time-sequential game related to an arbitration procedure. Math. Japonica **29**(3), 491–502 (1984). ISSN 0025-5513
33. Sakaguchi, M.: Optimal stopping games – a review. Math. Japonica **42**(2), 343–351 (1995). ISSN 0025-5513. Correction to "Optimal stopping games – a review". Zbl 0879.60044. Zbl 0865.60035
34. Shmaya, E., Solan, E.: Two-player nonzero-sum stopping games in discrete time. Ann. Probab. **32**(3B), 2733–2764 (2004). ISSN 0091-1798; 2168-894X/e. https://doi.org/10.1214/009117904000000162. Zbl 1079.60045
35. Szajowski, K.: On non-zero sum game with priority in the secretary problem. Math. Japonica **37**(3), 415–426 (1992)
36. Szajowski, K.: Double stopping by two decision-makers. Adv. Appl. Probab. **25**(2), 438–452 (1993). ISSN 0001-8678. https://doi.org/10.2307/1427661
37. Szajowski, K.: Markov stopping games with random priority. Z. Oper. Res. **37**(3), 69–84 (1993)
38. Szajowski, K.: Optimal stopping of a discrete Markov process by two decision makers. SIAM J. Control Optim. **33**(5), 1392–1410 (1995). ISSN 0363-0129. https://doi.org/10.1137/S0363012993246877
39. Szajowski, K., Yasuda, M.: Voting procedure on stopping games of Markov chain. In: Anthony, S.O., Christer, H., Thomas, L.C. (eds.) UK-Japanese Research Workshop on Stochastic Modelling in Innovative Manufacturing, 21–22 July 1995, Moller Centre, Churchill College, University of Cambridge, UK. Lecture Notes in Economics and Mathematical Systems, vol. 445, pp. 68–80. Springer, Berlin (1997). ISBN 3-540-61768-X/pbk. MR 98a:90159. Zbl 0878.90112
40. Tarski, A.: A lattice-theoretical fixpoint theorem and its applications. Pac. J. Math. **5**, 285–309 (1955). ISSN 0030-8730. https://doi.org/10.2140/pjm.1955.5.285. Zbl 0064.26004
41. von Stackelberg, H.F.: Marktform und Gleichgewicht. Springer, Wien (1934)
42. von Stackelberg, H.F.: Grundlagen einer reinen Kostentheorie. Meilensteine Nationalokonomie. Springer, Berlin (2009); Originally published monograph. Reprint of the 1st Ed. Wien, Verlag von Julius Springer (1932). http://www.springerlink.com/content/978-3-540-85271-1
43. von Stackelberg, H.: Market Structure and Equilibrium (Translated from the German by Damian Bazin, Lynn Urch and Rowland Hill). Springer, Berlin (2011). ISBN 978-3-642-12585-0/hbk; 978-3-642-12586-7/ebook. https://doi.org/10.1007/978-3-642-12586-7
44. Yasuda, M.: On a randomized strategy in Neveu's stopping problem. Stoch. Process. Appl. **21**, 159–166 (1985). ISSN 0304-4149. https://doi.org/10.1016/0304-4149(85)90384-9. Zbl 0601.60039

Printed in the United States
By Bookmasters